D1541452

Telecommunications
and
Economic Development
Second Edition

A WORLD BANK PUBLICATION

Telecommunications
and
Economic
Development
Second Edition

Robert J. Saunders
Jeremy J. Warford
Björn Wellenius

Published for The World Bank
The Johns Hopkins University Press
Baltimore and London

Copyright © 1994 by the International Bank
for Reconstruction and Development/THE WORLD BANK
1818 H Street, N.W., Washington, D.C. 20433, U.S.A.
All rights reserved
Manufactured in the United States of America

The Johns Hopkins University Press
Baltimore, Maryland 21218, U.S.A.

The views and interpretations in this book are the authors'
and should not be attributed to the World Bank, to its affili-
ated organizations, or to any individual acting on their
behalf.

First printing April 1994

WISSER MEMORIAL LIBRARY
HE8635
.S28
1994
copy 1

EDITOR Elizabeth Forsyth
FIGURES Kathy Rosen
COVER DESIGN Joyce Petruzzelli

The authors wish to thank the International Telecommunication Union for its kind permission to use the cover drawing by Martin Widiarsono of Indonesia. The drawing was
included in the ITU's 1991 "Youth in the Electronic Age"
international art exhibition.

Library of Congress Cataloging-in-Publication Data

Saunders, Robert J.
  Telecommunications and economic development / Robert J. Saunders,
Jeremy J. Warford, Björn Wellenius. — 2d ed.
    p. cm.
  Includes bibliographical references and index.
  ISBN 0-8018-4665-X
  1. Telecommunication—Developing countries. 2. Telecommunication—
Mathematical models. I. Warford, Jeremy J. II. Wellenius, Björn.
III. Title.
HE8635.S28 1994
384'.09172'4—dc20   93-40704
                                                           CIP

ISBN 0-8018-4665-X

Contents

Part V. Telecommunications Tariff Policy

Part VI. Mobilizing Resources and Promoting Efficiency: Alternative Approaches

Appendixes

Tables

Figures

Examples

Boxes

Preface to the Second Edition

WHEN THIS BOOK FIRST APPEARED IN 1983, it quickly became the standard reference on the analysis of telecommunications in developing countries. Translations into Japanese and Spanish were published in 1985 and 1987, respectively. By 1988, although the book continued to elicit much interest, the English edition was out of print.

Since 1983, the approach to telecommunications taken by policymakers, operating enterprises, and international development organizations has changed dramatically. Telecommunications, earlier regarded as a minor component of infrastructure, became in the 1980s a strategic factor of development at all levels, from individual firms to regions and countries. Traditional telecommunications sector structures, based largely on state monopolies, began giving way to more complex and flexible arrangements featuring reduced monopoly privileges, diversified and increasingly competitive supply of services and networks, increased participation of private capital and private enterprise, and a shift of government responsibilities from ownership and operation to policy and regulation. In developing countries, efforts to build up the telecommunications sector had, for decades, focused on securing a larger share of funds from government and aid agencies and on improving the internal organization and management of state enterprises. Now this strategy is turning increasingly toward improving the policy and regulatory framework so that operating enterprises can enjoy the freedoms and incentives, as well as be subject to the market disciplines, of a modern business.

Given these major changes, we debated the extent to which this book should be revised for a second edition. Surely a book on telecommunications and economic development written in the early 1990s would bear little resemblance to one written in the late 1970s and early 1980s. We decided, however, that a limited revision, rather than a major overhaul, would serve readers best. An updated version would meet the needs of readers interested in, for example, the role that telecommunications play in economic development, especially regarding benefits, costs, and pricing, because much of the original material is still relevant. Moreover, the issues and quantitative techniques discussed in the first edition mainly from the viewpoint of developing countries are now receiving attention in industrial countries as well, where initiatives are being undertaken to redevelop rural areas and inner cities and to create new competitive advantages in states or provinces where traditional industries are declining. As for the rather different issues of policy, structure, and regulation of the sector, a whole new body of literature is quickly developing.

The second edition that we now offer has, therefore, the same structure of, and contains much of the material included in, the original English edition. All statistical tables, however, have been updated, and new ones added. The case material was partly replaced and considerably expanded to reflect more recent experiences. Several new examples were added. The three chapters on pricing (chapters 13, 14, and 15) were extensively rewritten to emphasize practical approaches to setting tariffs and to reflect the shift of regulatory attention from costs to prices. The chapter on technology and costs (chapter 2) was rewritten to reflect the rapid innovation that continues to occur. The new version of the last chapter (chapter 16) now provides a link to the rapidly growing literature on telecommunications restructuring. The notes and references were extensively revised and updated. All remaining original material was edited, some of it substantially.

We would like to thank the many colleagues and friends who provided assistance, advice, and material for this volume. Greta Nettleton, as consultant to the World Bank, was responsible for researching the large volume of literature, collating and updating statistical data, writing new material, and substantively editing the text cover to cover. At the World Bank, Björn Wellenius was responsible for preparing the revised manuscript; he also wrote new versions of chapters 2 and 16 and various smaller parts of the whole. Peter Smith rewrote chapters 13 through 15 on pricing and tariffs. Dinshaw Joshi and Hugh Lantzke gave generously of their experience in engineering and management and contributed substantially to the revisions, especially in chapter 2. Additional assistance in engineering, management, finance, and economics was provided by other colleagues at the Bank, especially Gerald Buttex, James Cowie, Alberto Cruzat, Rogati Kayani, Philippe Lecharny, David Lomax, Cecile Ramsay, Syed Sathar, A. Shanmugarajah, Mark Tomlinson, and Eloy Vidal. Many individuals outside the Bank also contributed to the project, including James

Alleman, Jerry Cohen, Christina Hennig, Bridger Mitchell, Mitchell Moss, Lionel Nicol, Edwin Parker, Kirsten Pehrsson, Joseph Pelton, Michael Tyler, and Jeffrey Rohlfs. Heather Hudson, of the University of San Francisco, rewrote and expanded appendix A for this edition. Douglas Conn, of the Center for Telecommunications and Information Study (part of the Columbia University School of Graduate Studies) provided access to extensive research material. We thank them all.

Acronyms and Abbreviations

AID	Agency for International Development (United States)
AT&T	American Telephone and Telegraph Company (United States)
CCITT	International Consultative Committee on Telephone and Telegraph
CEE	Central and Eastern Europe
CTSC	Community teleservice centres, or telecottages (Scandinavia)
EC	European Communities
GDP	Gross domestic product
GNP	Gross national product
IRR	Internal rate of return
ISDN	Integrated Services Digital Network
ITU	International Telecommunication Union
OECD	Organization for Economic Co-operation and Development
OFTEL	Office of Telecommunications (United Kingdom)
OPEC	Organization of Petroleum Exporting Countries
OPT	Office of Posts and Telecommunications
P&T	Posts and Telecommunications department, board, or corporation
PCO	Public call office
PTT	Posts, telegraph, and telephones (administration)
TLRIC	Total long-run incremental cost
VSAT	Very small aperture terminal

Part I
An Introductory Perspective

Chapter 1

The Role of Telecommunications in Economic Development

TELECOMMUNICATIONS ARE INCREASINGLY RECOGNIZED as a key component in the infrastructure of economic development, yet telecommunications services in most developing countries continue to fall far short of needs. The result is that development is constrained significantly throughout these economies. The imbalance between telecommunications development in industrial and in developing countries was the focus of global attention in the 1985 Maitland Commission report, which noted that "in most developing countries, the telecommunications system is not adequate even to sustain essential services."[1] However, despite increasing awareness of the connection between telecommunications and economic development, governments in developing countries and development agencies rarely pay attention to these shortfalls and the measures required to redress them.

Can the economic value of the benefits derived from investments in telecommunications be demonstrated and quantified? Which segments of the population derive these benefits? What is the impact on the development of other economic and social sectors? How do the benefits compare with the economic costs of expanding and improving the telecommunications system? What investment and pricing policies would increase benefits and contain costs? What level of investment can be justified by economic analysis? These central issues of telecommunications economics in developing countries are the main themes of this book.

3

Signals for Expanding Investment

The developing countries have about 75 percent of the world's population and 16 percent of its product, but only 12 percent of the total number of telephone main lines.[2] Of the more than 435 million lines in existence in 1988, about 50 million were in the developing countries of Africa, Asia, and Latin America, while 386 million were in the industrial and newly industrialized countries of North America, Europe, Asia, and Oceania (see table 1-1).

Table 1-1. *Distribution of the World's Telephone Main Lines by Region, Population, and Income, 1988*

| | Main lines | | Population | | GNP | |
| | Millions | Percent | Millions | Percent | Billions of dollars | Percent |
Region						
Industrial countries						
Africa						
South Africa	2.5	0.6	30	0.6	63	0.5
Asia						
Japan	49.9	11.5	123	2.6	1,925	14.1
Others[a]	17.9	4.1	74	1.6	209	1.5
Europe						
Former U.S.S.R.	27.7	6.4	285	5.9	—	—
Others	149.7	34.4	421	8.8	4,146	30.4
North America						
Canada	11.4	2.6	26	0.5	390	2.9
United States	118.4	27.2	246	5.1	4,486	32.9
Oceania						
Australia	6.8	1.6	16	0.3	176	1.3
New Zealand	1.4	0.3	3	0.1	27	0.2
Total	385.7	88.6	1,223	25.6	11,422	83.9
Developing countries						
Africa	3.9	0.9	568	11.9	287	2.1
Asia	22.6	5.2	2,597	54.3	1,188	8.7
Latin America	23.0	5.3	393	8.2	725	5.3
Total	49.5	11.4	3,558	74.4	2,199	16.1
World total[b]	435.2	100.0	4,782	100.0	13,621	100.0

— Not available.

a. Israel, Republic of Korea, Hong Kong, Singapore, and Taiwan.

b. A small percentage of nonreporting countries are not included.

Source: American Telephone and Telegraph Corporation, *The World's Telephones* (various years); World Bank, *World Bank Atlas 1988*; ITU, *Yearbook of Public Telecommunication Statistics* (1990); and World Bank data.

In early 1988, the industrial countries had on average about 32 main lines per 100 inhabitants.[3] Canada, France, Scandinavia, Switzerland, and the United States all had a density of about 50 main lines and between 78 and 89 telephone sets (equipment on the customer's premises) per 100 inhabitants. Most other industrial countries had over 30 lines per 100 inhabitants, while the developing countries, in contrast, averaged only about 1.5 lines per 100 inhabitants (see table 1-2).

Telephone density also varies widely among and within developing countries. A ratio of more than 50:1 exists between countries such as Argentina and Uruguay, which have more than 10 main lines per 100 persons, and countries such as the Central African Republic, Ethiopia, Madagascar, Nigeria, and Tanzania, which have just 0.1 to 0.2 lines per 100 persons (table 1-2). In 1988, the density of telephone main lines in Africa (excluding South Africa) was 0.7 per 100 persons; it was 0.9 in Asia (excluding Japan, and the newly industrialized countries of Hong Kong, Singapore, South Korea, and Taiwan) and 5.9 in Latin America. Also, in developing countries, telephones are concentrated in a few large cities, and much of the population lives in areas with little or no service. In industrial countries, the telephone density is fairly uniform across each country; in the developing world, in contrast, telephone density is several times greater in the main cities than in provincial towns and rural areas (see table 1-3).

The gap that exists between industrial and developing countries in basic telephone service, although still large, is narrowing. Between 1969 and 1988, developing countries almost doubled their share of the world's telephone lines, from about 7 percent to 12 percent, while their share of world population and product remained roughly unchanged. Despite such improvements, the disparities remain extreme.

Level of Investment

Although a decade ago developing countries invested only about 0.3 percent of gross domestic product (GDP) in telecommunications, by the end of the 1980s this level had doubled to about 0.6 percent, or some $12 billion a year.[4] This proportion is similar to the level of investment found in industrial countries, which remained relatively steady during the same period, at about 0.6 percent of GDP. In 1989, industrial countries invested about $100 billion in their public telecommunications sectors.[5]

Table 1-2. *Proportion of Telephone Main Lines and* GNP *per Capita in Selected Countries, 1987–88*

Region and country	Number of main lines per 100 persons (January 1, 1988)	GNP per capita (1987 U.S. dollars)
World	9.2	2,848
Industrial countries	31.5	9,339
North America	47.7	17,927
Canada	51.2	15,080
United States	48.1	18,430
Asia and Oceania	35.1	10,819
Australia	42.8	10,900
Hong Kong	35.4	8,260
Japan[a]	40.8	15,770
Korea	20.7	2,690
New Zealand	41.6	8,230
Singapore	33.3	7,940
Taiwan	25.0	—
Europe	25.1	9,848
Austria	38.4	11,970
Belgium	34.5	11,360
Czechoslovakia	13.0	—
Denmark	55.1	15,010
Finland	47.9	14,370
France	44.7	12,860
Germany[b]	34.2	14,460
Greece	34.7	4,350
Hungary	7.7	2,240
Ireland	22.5	6,030
Italy	33.3	10,420
Netherlands[a]	42.4	11,860
Norway	46.4	17,110
Portugal	16.1	2,890
Spain	26.2	6,010
Sweden	65.1	15,690
Switzerland	52.9	21,250
United Kingdom	42.4	10,430
Yugoslavia	12.9	2,480
U.S.S.R.[c]	9.7	—
Developing countries	1.5	618
Africa	0.7	505
Algeria[a]	2.7	2,760
Benin	0.3	300
Botswana[a]	1.2	1,030
Cameroon[e]	0.3	960
Central African Republic	0.2	330
Côte d'Ivoire[f]	0.6	750
Egypt	2.2	710
Ethiopia[a].	0.2	120
Gabon	1.8	2,750
Kenya	0.7	340
Liberia	0.4	440
Madagascar	0.2	200
Malawi	0.3	160
Mauritius[b]	4.5	1,470
Morocco	1.1	620
Namibia	3.7	—
Nigeria[a,f]	0.2	370
Rwanda	0.1	310
Senegal	0.4	510
Seychelles	8.2	3,180

6

Table 1-2 (continued)

Region and country	Number of main lines per 100 persons (January 1, 1988)	GNP per capita (1987 U.S. dollars)
Developing countries (continued)		
Africa (continued)		
Sierra Leone[b]	0.4	300
Sudan	0.2	330
Tanzania	0.2	220
Togo	0.3	300
Tunisia	3.0	1,210
Zambia	0.7	240
Zimbabwe[a]	1.4	590
Asia	0.9	458
China	0.5	300
Fiji	4.7	1,510
India	0.4	300
Indonesia	0.4	450
Iran	2.9	—
Iraq	4.8	—[d]
Jordan	6.7	1,540
Malaysia	7.2	1,800
Oman[a]	5.1	5,780
Pakistan	0.6	350
Papua New Guinea	0.9	730
Philippines[a]	0.8	590
Saudi Arabia	12.4	6,930[f]
Sri Lanka	0.5	400
Syria	4.3	1,820
Thailand	1.7	840
Turkey[a]	7.0	1,200
Latin America	5.9	1,845
Argentina[a]	9.6	2,370
Bolivia	2.2	570
Brazil	5.6	2,020
Chile	4.9	1,310
Colombia	7.2	1,220
Costa Rica	8.6	1,590
Cuba	3.0	—
Ecuador	4.4	1,040
El Salvador	2.1	850
Guyana	2.5	380
Haiti	0.5	360
Honduras	1.1	780
Mexico	4.9	1,820
Nicaragua	1.3	830
Panama	8.6	2,240
Paraguay	2.4	1,000
Peru	2.3	1,430
Suriname	7.8	2,360
Uruguay[a]	10.6	2,180
Venezuela	9.2	3,230

— Not available.

a. Population estimated. b. Estimated for Germany after unification. c. Includes Asian U.S.S.R. d. As of January 1, 1985. e. According to Siemens data. f. As of January 1, 1986.

Source: American Telephone and Telegraph Corporation, The World's Telephones (various years); World Bank, World Bank Atlas 1988; ITU, Yearbook of Public Telecommunication Statistics (1990).

Table 1-3. *Access to Telephone Services in Main Cities and Other Areas in Selected Countries, as of January 1, 1988*
(number of main lines per 100 persons)

Region and country	National	Main cities[a]	Other areas
Industrial countries			
Austria	38.38	54.20	31.32
Canada	44.49	59.20	43.54
Denmark	55.13	59.58	52.36
France	44.68	47.98	29.27
Germany[b]	39.27	50.20	35.98
Italy	33.28	41.48	30.65
Japan	40.81	56.13	37.48
Norway	46.41	55.81	41.89
Spain	26.18	31.84	21.02
Switzerland	52.87	65.45	46.73
Developing countries			
Africa			
Algeria	2.70	7.13	1.58
Ethiopia	0.24	3.39	0.04
Kenya	0.66	4.95	0.19
Malawi	0.28	2.20	0.07
Morocco	1.14	3.17	0.42
Sudan	0.24	1.32	0.04
Togo	0.28	1.27	0.00
Tunisia	3.01	7.00	0.79
Zambia	0.73	1.36	0.17
Zimbabwe	1.45	6.39	0.41
Asia			
Iran[a]	3.15	6.31	1.10
Malaysia	7.21	22.65	5.17
Pakistan	0.61	2.69	0.19
Papua New Guinea	0.91	5.91	0.22
Sri Lanka	0.54	1.12	0.29
Thailand	1.67	6.94	0.45
Turkey	7.01	7.46	6.56
Latin America			
Brazil	5.59	10.17	4.14
Colombia	7.20	13.26	1.83
Costa Rica	8.62	15.28	2.57
Ecuador	4.41	8.27	1.91
Peru	2.30	4.90	0.52
Uruguay	10.61	16.05	5.24
Venezuela	9.19	16.20	5.08

a. Defined by the national administration; population thresholds, and consequently the number of cities included, vary widely among countries.

b. Estimated from combined Federal Republic of Germany (January 1987) and German Democratic Republic (January 1988) data.

Source: American Telephone and Telegraph Corporation, *The World's Telephones* (various years); ITU, *Yearbook of Telecommunication Statistics* (1990); and World Bank data.

Given the low base from which they begin, basic telephone systems in developing countries can grow at relatively high rates, averaging more than 9 percent; from 1979 to 1987, regional growth rates ranged between 7 and 11 percent for Asia, Latin America, and Africa.[6] Some countries have sustained rates of 15 percent or more for years. These growth rates are well above the 3 to 6 percent typically found among industrial countries, which already have relatively large telecommunications infrastructures in place (see table 1-4).

Telex has grown even faster than voice telephone service (see table 1-5). More advanced services introduced in recent years, such as facsimile and data transmission, have also expanded at record rates, as illustrated in tables 1-6 and 1-7. Growth in the use of facsimile machines has been explosive in countries such as Brazil, Colombia, Malaysia, and Singapore, and strong in many countries throughout Asia, Latin America, and Sub-Saharan Africa.[7] Growth in the number of leased circuits for data communications has been more uneven, reflecting differences in government policies. Between 1987 and 1988, the number of circuits increased more than 1,000 percent in Thailand but remained close to or at 0 percent in most Sub-Saharan African countries.

Lack of demand does not explain the historically low level of investment in the telecommunications sector in most developing countries. Throughout the developing world the demand for telephone and more advanced services typically far exceeds the supply, and the number of unmet applications for telephone connections often exceeds the number of existing lines (see table 1-8).

New applicants frequently wait two to five years—and sometimes more than ten—to obtain service. In these situations, a large proportion of the potential demand for telecommunications services remains unrecorded and emerges only when the system is perceived to be expanding rapidly.[8] Also, applications are rarely accepted in areas where service is neither available nor planned in the near future. Besides, in towns or villages where most of the population is not familiar with telephone service, significant demand arises only after service is introduced at the initiative of the government, an operating company, or a special interest group and the public has gradually gained experience using telephones and begun to appreciate the cost savings and other benefits that can be obtained from their use. The introduction of more advanced services also tends to expand demand as the modern business sector finds that it needs these services to compete in the global marketplace.

When the shortage of telephone lines is acute, the proportion of

Table 1-4. *Growth of Telephone Main Lines in Service in Selected Developing Countries, 1979–88*
(thousands of main lines unless otherwise indicated)

Region and country	As of January 1, 1979	As of December 1, 1988	Annual growth rate (percent)
Africa			
Algeria	260.4	697.2	11.6
Benin	7.8	13.6	6.4
Botswana	6.4	18.2	12.3
Egypt[a]	392.0	1,118.0	14.0
Ghana	36.4	40.4	1.2
Kenya	69.7	157.4	9.5
Malawi[a]	10.6	22.5	9.9
Mozambique	30.6	40.8	3.2
Senegal	17.3	28.9	5.9
Tanzania	38.3	66.1	6.2
Zimbabwe	92.9	118.4	2.7
Asia			
China	2,460.0[c]	5,550.0	9.5
Fiji	21.3	35.7	5.9
India[a]	1,868.0	3,487.9	8.1
Indonesia	317.9	828.8	11.2
Iran	1,233.9	1,803.6	4.3
Malaysia	325.0	1,247.7	16.1
Myanmar (Burma)	28.2	74.7	11.4
Pakistan	286.0	636.6	9.3
Papua New Guinea	22.8	31.0	3.5
Thailand	332.0	1,005.9	13.1
Asia (newly industrialized countries)			
Hong Kong	1,173.0	2,153.8	7.0
Korea	2,341.0	10,486.2	18.1
Singapore	464.8	924.0	7.9
Taiwan[b]	2,630.0	4,909.0	11.0
Latin America			
Argentina	1,797.0	2,747.5	4.8
Brazil	3,829.0	9,081.6	10.1
Chile	351.0	625.5	6.6
Colombia	1,022.3	2,070.4	8.2
Costa Rica	133.6	256.5	7.5
El Salvador[a]	66.1	104.5	5.9
Mexico[a]	2,494.0	3,774.0	5.3
Peru[a]	287.0	461.9	6.1
Venezuela	701.0	1,457.8	8.5

a. From January 1, 1979, to December 31, 1987.
b. From January 1, 1981, to December 31, 1987.
c. Estimated from World Bank data.
Source: ITU, *Yearbook of Public Telecommunication Statistics* (1990), and World Bank data.

Table 1-5. *Growth of Telex Lines in Service in Selected Developing Countries, 1979–88*

	Number of telex lines		Annual growth rate (percent)
Region and country	January 1, 1979	December 31, 1988	
Africa			
Algeria	2,400	8,244	14.7
Botswana[a]	140	741	23.2
Burundi[c]	85	191	17.6
Cameroon[d]	563	1,940	16.7
Ethiopia	372	864	9.8
Ghana	172	563	14.1
Mauritius	215	784	15.5
Niger[b]	179	297	8.8
Rwanda	60	138	9.7
Swaziland	193	353	6.9
Tanzania	426	1,384	14.0
Asia			
Bangladesh[b]	256	1,090	27.3
Fiji	235	655	12.1
India[d]	16,500	34,044	9.5
Indonesia	3,612	15,441	17.5
Malaysia	2,908	9,930	14.6
Myanmar (Burma)	47	138	12.7
Nepal[e]	161	405	20.3
Oman	460	1,319	12.4
Papua New Guinea	999	1,129	1.4
Philippines	5,733	12,199	8.8
Sri Lanka	350	1,535	17.9
Syria	699	2,695	16.2
Latin America			
Brazil	27,362	121,200	18.0
Chile	2,157	11,648	20.6
Colombia	3,800	6,452	6.1
Costa Rica	900	1,500	5.8
Ecuador[d]	1,292	3,152	11.8
El Salvador[d]	610	906	5.1
Uruguay	897	2,236	10.7
Venezuela[f]	7,848	18,800	13.3

a. January 1, 1980, to December 31, 1988.
b. January 1, 1979, to December 31, 1985.
c. January 1, 1980, to December 31, 1986.
d. January 1, 1979, to December 31, 1987.
e. January 1, 1982, to December 31, 1987.
f. January 1, 1981, to December 31, 1988.
Source: ITU, *Yearbook of Public Telecommunication Statistics* (1990).

Table 1-6. *Number of Facsimile Stations in Selected Industrial and Developing Countries, 1986–88*

Country	1986	1987	1988
Industrial countries			
Australia	60,000	100,723	—
Canada	55,000	84,000	—
Finland	6,339	9,866	16,759
France	60,000	96,000	185,000
Germany, Fed. Rep. of	44,453	85,295	291,213
Hong Kong	12,673	32,076	59,876
Hungary	—	22	1,096
Italy	25,203	48,269	92,813
Singapore	5,349	12,069	18,986
Spain	6,515	18,549	53,678
Switzerland	3,858	19,246	40,285
Developing countries			
Botswana	122	—	812
Brazil	—	1,736	20,090
Burkina Faso	—	4	15
Cyprus	356	872	1,858
Colombia	—	—	17,090[a]
Egypt	3	767	—
Ethiopia	—	34	128
Fiji	1	—	612
Ghana	1	1	49
Guatemala	—	—	2,503
Indonesia	—	2,453	4,255
Iran	150	131	620
Malawi	13	89	—
Malaysia	1,415	4,707	13,702
Mauritius	—	220	268
Oman	450	900	1,500
Paraguay	—	—	158
Rwanda	—	—	158
Tanzania	—	89	172
Thailand	1,521	—	5,453
United Arab Emirates	1,101	2,254	6,604
Zambia	—	111	198
Zimbabwe	—	232	387

— Not available.
Note: Figures are as of December 31.
a. Estimated.
Source: ITU, *Yearbook of Public Telecommunication Statistics* (1990).

Table 1-7. *Growth in the Number of Privately Leased Circuits for Data Communications in Selected Industrial and Developing Countries, 1981 and 1986–88*

Country	1981	1986	1987	1988
Industrial countries				
Australia	40,252	86,076	96,000	—
Finland	9,800	29,900	37,500	95,800
Germany, Fed. Rep. of	3,203	4,593	4,229	4,807
Hong Kong	—	—	68,940	67,642
Hungary	342	616	698	772
Ireland	1,098	4,822	5,450	7,827
New Zealand	833	2,919	18,887	25,233
Poland	400	570	570	888
Portugal	1,315	5,600	6,700	8,390
Singapore	417	24,184	30,380	36,000
Spain	28,700	44,600	53,389	58,261
Switzerland	11,879	18,383	—	—
Developing countries				
Botswana	—	11	—	351
Brazil	58	12,000[a]	16,177	20,821
Burkina Faso	—	5	5	—
Cyprus	18	235	292	724
Colombia	150	14,427	—	16,314
Ethiopia	—	1	13	13
Fiji	13	210	230	233
Ghana	—	17	17	15
India	—	907	1,268	—
Indonesia	—	934	1,132	1,628
Malawi	291	362	245	—
Malaysia	—	4,544	6,875	8,206
Mauritius	11	—	75	112
Oman	12	156	480	559
Paraguay	—	4	—	53
Rwanda	3	3	3	3
Senegal	—	65	79	512
Tanzania	11	716	715	715
Thailand	38	—	951	11,913
United Arab Emirates	271	364	393	385
Zambia	—	277	296	—
Zimbabwe	—	644	589	816

— Not available.

Note: Figures are as of December 31.

a. Estimated.

Source: ITU, *Yearbook of Public Telecommunication Statistics* (1990).

Table 1-8. *Supply of and Expressed Demand for Telephone Main Lines in Selected Developing Countries, Various Years, 1985–89*
(thousands unless otherwise indicated)

Region, country, and date	Main lines in service	Unmet applications	Total expressed demand[a]	Percentage of expressed demand met
Africa				
Algeria, December 1988	697	524	1,221	57
Botswana, March 1989	18	3	21	87
Egypt, December 1987	1,118	1,137	2,255	50
Ethiopia, December 1988	113	88	200	56
Ghana, December 1988	40	30	70	58
Kenya, December 1988	157	64	221	71
Lesotho, December 1988	11	5	17	68
Malawi, December 1987	22	6	28	80
Mali, December 1988	9	4	13	68
Mauritius, December 1988	50	42	91	54
Morocco, December 1988	286	205	491	58
Senegal, December 1988	29	8	37	79
Sudan, December 1985	58	22	79	73
Swaziland, March 1987	10	3	12	79
Tanzania, December 1988	66	76	142	47
Asia				
Fiji, December 1988	36	12	47	75
India, March 1987	3,488	1,125	4,612	76
Iran, March 1989	1,804	314	2,118	85
Malaysia, December 1988	1,248	69	1,317	95
Myanmar (formerly Burma), December 1988	75	300	375	20
Nepal, July 1987	30	41	72	42
Oman, December 1988	88	9	96	91
Pakistan, December 1988	637	531	1,168	54
Papua New Guinea, December 1988	31	2	33	94
Philippines, December 1988	591	582	1,173	50
Sri Lanka, December 1988	104	51	155	67
Syria, December 1988	478	1,339	1,817	26
Thailand, December 1988	1,006	377	1,383	73
Latin America				
Argentina, December 1988	2,747	789	3,537	78
Chile, December 1988	625	236	862	73
Colombia, December 1988	2,070	660	2,730	76
Costa Rica, December 1988	257	14	270	95
El Salvador, December 1986	96	11	107	90
Mexico, December 1986	3,821	757	4,577	83
Peru, December 1986	445	273	718	62
Uruguay, December 1988	345	93	439	79
Venezuela, December 1988	1,458	453	1,911	76

a. The number of lines in service plus the number of outstanding applications for service. In developing countries, actual demand is generally much higher than expressed demand.

Source: ITU, *Yearbook of Public Telecommunication Statistics* (1990).

subscribers who are very intensive users (mainly business and government) tends to be high, as does the number of users for each telephone line.[9] Hence, the average number of calls per line is high, and
the traffic of local and long-distance calls is frequently congested.
This congestion results partly from the inability of major components
of the system, especially telephone exchanges and trunks, to handle
call traffic and a tendency in some developing countries to provide insufficient private branch exchange facilities. To a large extent, however, it also results from the high proportion of time that the
telephone called is engaged; thus subscribers repeatedly attempt to
place calls, which, when added to the repeat calls resulting from
equipment congestion, further strains the network's capacity. Congestion at major centers or long-distance routes tends to propagate
across the national system.[10] When call traffic is congested, the cost
per completed call is high in terms of both the system's use and the
users' time, many of the potential benefits of investing in telecommunications do not materialize, and maintenance problems increase
markedly.

Returns on Investment

In developing countries, inadequate investment in the telecommunications sector is not caused by telecommunications entities that
lose money or require government subsidies, any more than it is
caused by lack of demand (some of the more binding constraints on
expansion are discussed further in chapter 3). In general, reasonably
well-managed telecommunications entities can generate large financial surpluses in local currency. In the usual situation, where demand
greatly exceeds supply, it is easy, with proper pricing policies, to recover the full cost of providing telecommunications services (including the cost of capital) from tariffs, even if internal inefficiencies
often result in high costs. It is likewise feasible to generate internally a
large proportion of the funds required for improving and expanding
the system in the future.

On average, in thirteen programs recently funded in part by the
World Bank, telecommunications investments are expected to yield
internal financial rates of return between 13 and 25 percent, averaging about 20 percent. Approximately 60 percent of the new construction funds required will be generated internally. The operating entities
are conservatively estimated to average more than a 15 percent annual
rate of return on overall net plant in service, revalued to current
prices, over the five years after the project is appraised by the Bank.[11]

Furthermore, many of the telecommunications entities will make substantial net contributions to the government, which will tend to be used in sectors less able to mobilize domestic resources.[12]

Telecommunications investments also yield high returns to the economy as a whole. After correcting for general distortions of the price system and pure transfer payments, the economic rates of return attributable to the thirteen projects approved for World Bank support are expected to range between 15 and 30 percent. Where it has been possible to quantify meaningfully some of the benefits accruing to the country apart from the revenues realized by the telecommunications entity, the economic rates of return are expected to range between 20 and 40 percent (examples of such exercises are presented in chapters 8 and 9). These figures are higher than the usual 10 to 14 percent rate of return used as the threshold for acceptable investments in developing countries. Very large economic returns also result from the telecommunications components of investment programs in other sectors (for example, railways, power, tourism, banking, and rural development).

Hence, in developing countries a large excess demand exists for telecommunications services, and the measurable private and social returns on the investment required to provide those services are relatively high. Were economic efficiency the sole goal of development, such evidence of market forces would be sufficient to justify expanding the sector rapidly.

Benefits from Investment

Other arguments are sometimes used to support increasing telecommunications investment in developing countries. Proponents of such investment contend that as economic development takes place, some form of telecommunications gradually becomes the most cost-effective means of communicating for increasing proportions of the population. Telecommunications services can substitute for other forms of communication (mainly postal service and personal travel) and are often more effective and more efficient than those forms in their use of time, energy, and materials and in their effect on the quality of the environment. Further, there is some evidence that a reliable telecommunications system generates new communication and builds stronger, more complex, and more productive patterns of communication, partly because it interacts directly and indirectly with numerous production and distribution functions.

It is also argued that having accessible and reliable telephone service removes some of the physical constraints on organizational communication in various sectors of the economy, permitting increased productivity through better management in both the public and private sectors, making it possible to adopt different structures and locations, and aiding the evolution of increasingly complex organizations. Markets become more effective as communication improves, more rapid responses to market signals become possible, and access to market information is extended at village, town, city, regional, national, and global levels. Also, household operations become more efficient as telecommunications improve access to goods and services and support forms of work that require some integration of workplace and residence.[13]

In the late 1970s, as fuel prices escalated, a particularly timely argument in favor of increasing investment was that telecommunications could partly substitute for transport and also bring about the more efficient use of transport facilities, thus reducing the amount of energy required to sustain a given level of communication (see chapter 7 for a more detailed discussion). On the one hand, transport costs, led by increased petroleum prices and petroleum-related steel and materials costs, had risen significantly. On the other hand, dramatic changes in telecommunications technology had lowered the real cost of providing telecommunications services. Hence, increased investment in telecommunications was seen to create opportunities for reducing the costs of transport and transport-related materials and supplies and thus deserved serious attention. Although the energy crisis has since abated, technological innovation has further reduced the cost of telecommunications, and energy conservation, transportation efficiency, and related environmental matters continue to be of major concern to national policymakers (some of the recent changes in technology and their effect on costs are discussed in chapter 2).

More general discussions sometimes emphasize that virtually every aspect of economic and social development in any one sector of an economy interacts with development in other sectors.[14] For example, most development activities require services from the economy's infrastructure networks: roads, water, power, and—important, but not well researched or understood—telecommunications. In particular, as agricultural development increases the marketable agricultural surplus, it gives rise to widespread trade in agricultural commodities, seeds, fertilizers, fuels, and other goods and services, which, to be efficient, inevitably requires reliable means of communicating rapidly and over long distances. A need also exists for reliable and rapid information about weather conditions, disease outbreaks, and new agricultural techniques.

The utility of telecommunications services is also apparent for commerce and industry. Industrial development requires coordination of numerous activities: acquisition of supplies, recruitment and coordination of labor, control of stocks, processing of materials, billing, record-keeping, delivery of goods to buyers, and general market search activities. Commerce, however, is inherently an information-processing activity. Effective buying, selling, brokerage, and transport require a continuous supply of up-to-date information on the availability and price of numerous goods and services. In the absence of accessible and reliable telecommunications services such activities suffer a variety of inefficiencies, including the creation of markets in which a few information-rich individuals are able to gain significant advantage over the majority of individuals who are information poor. This is particularly important as new telecommunications technologies emerge; global business patterns increasingly require access to new services, and regions where such services are not available risk being shut out of effective competitive trade.

Finally, as is examined further in chapters 10, 11, and 12, advocates of telecommunications investment stress that the direct contribution that telecommunications make to the welfare of individuals and families cannot be dismissed by derogatory references to "merely social" telephone calls. The well-being of the family is assisted by telecommunications, which provide rapid access to services often needed to preserve life, health, and property and enhance contact with kin, friends, and special interest groups. Telecommunications contribute to the development of a shared environment that reaches a country's most remote areas and can facilitate political, cultural, economic, and social integration.

Although telecommunications still appear to command a lower order of priority in many developing countries than is justified in economic terms, the high economic and financial returns from telecommunications investment and the benefits they imply are being increasingly recognized. In particular, countries often assign high priority to investment in the sector when, for one reason or another, they overcome some of the perceived constraints hindering the development of their telecommunications networks. For example, in the 1970s and 1980s, oil-surplus developing countries (such as Indonesia, Iran, Kuwait, Nigeria, Saudi Arabia, and Venezuela) undertook large investment programs in telecommunications once foreign exchange became readily available. Once-developing countries such as Hong Kong, Korea, Singapore, and Taiwan used telecommunications as a key part of their overall economic strategy to build up what is now a highly competitive position in the world market for high-technology industries and services. Likewise, a number of countries that did not recognize the importance of telecommunications, such as Egypt and India, have embarked on large investment programs to overcome what they now regard as a significant constraint on their economic growth and the decentralization of economic activity outside major urban areas. The most striking example of this approach among industrial countries is the telecommunications investment program of France since 1974.[15]

Moreover, driven largely by technological innovations, related changes in cost structures, and growing and diversified demands of business, many industrial countries, including some of those with the most advanced telecommunications services, undertook major changes in the 1980s in how the sector was structured and regulated. Liberalization and the breakup of the American Telephone and Telegraph Company (AT&T) in the United States were followed by privatization and the introduction of competition to the telecommunications sector in the United Kingdom and Japan, and this trend has gradually extended to most of the countries in the Organisation for Economic Co-operation and Development (OECD). These events, coupled with the widespread globalization and increased information content of economic activity, led developing countries from the mid-1980s on to consider a range of options for reforming the telecommunications sector with the aim of helping overcome long-standing constraints on development. By 1990, some forty developing countries had completed, were embarked on, or were actively preparing major reforms of the telecommunications sector.[16] More than half of the world's main lines were operated by privately held companies by the beginning of 1989.[17] These alternatives to having state monopolies operate telecommunications systems will be examined again in chapter 16.

Views about Telecommunications Development

Views concerning the most desirable rate of development for the telecommunications sector can roughly be grouped into three general categories. First, one group feels that telecommunications investment should be held well below what would be indicated by the market forces outlined above. Second, another group contends that telecommunications should grow mostly as indicated by the market, with operating entities behaving in most respects like commercial enterprises with relatively unhindered access to capital markets for investment funds, subject to some government controls to ensure wide access to basic services and to protect the public's interest. More and more, this point of view is coming to predominate in policymaking circles and in the literature. And finally, a more activist technology-oriented group in part promotes rapidly advancing telecommunications technology as a prime means to achieve a wide range of social and economic goals in numerous socially oriented sectors, including the delivery of education and health services. More generally, this last approach would not only implement the growth of a telecommunications system that is called for by market forces but would also in some instances take that growth in a different direction with respect to the types of services provided, the types of users served, pricing and control of access, network planning, technology, and other matters. In some cases representatives of this group also advocate government subsidies for selected telecommunications services.

The first, restrictive, view, in one of its forms, argues that expansion of telecommunications services does not deserve priority in resource allocation, partly because it has too little measurable economic effect and partly because in some countries rapid two-way communication among the population could facilitate political instability. The contention is that this service is used for the most part for economically and socially worthless purposes. Surprisingly, although this view is often superficially expressed, it rarely appears in print.[18] A variant of the restrictive philosophy contends that telecommunications serve economic and social needs, but in an undesirable way: "Telephony development ... is generally by and for the elite groups ... primarily confined to the more modern and urban areas of society. ... By creating an urban-based communications infrastructure, which is only accessible to a limited segment of society, economic opportunity becomes further concentrated in urban settings, and hence urban migration is encouraged."[19] Some analysts caution that introducing telecommunications services, particularly advanced technologies, in developing regions will not necessarily bring social benefits and economic development. Reducing the costs for high-volume users may increase the costs for small-volume users, reducing the affordability of access and possibly creating structural dependencies between more-developed and less-developed regions.[20] One question arising from such views is that even if they were true of some forms or mixes of telecommunications investment (and that has not been demonstrated), would they be true for all? And if not, how can preferred strategies and policies be identified that would preclude these adverse effects?

The second, market-response, view suggests that the demonstrated market demand for telecommunications should be met and that new technical applications should be provided when they are the most cost-effective way to meet market demand and to provide minimum telephone access nationally as needed for broader development purposes. The position taken by the International Telecommunication Union (ITU), an entity that will be quoted on several occasions in this book, has, for many years, been generally along these lines.[21] In contradiction to those who argue that telecommunications focus economic opportunity on only a few urban areas, the ITU contended that telecommunications form a vital part of the national economic infrastructure and that, when provided, they produce widespread benefits. The ITU argued through the years that failure to recognize the full range of benefits from telecommunications leads to chronic underinvestment: "a permanent shortage of investment funds . . . leads to capital rationing which can be quite arbitrary in effect. This often results in the treasury of the country allocating a given sum to telecommunications, even though a much larger amount would be warranted." More recently, the ITU's Advisory Group on Telecommunications Policy took a vigorous pro-market approach and recommended adjusting current policies and structures to increase private participation in the sector, develop competition, and shift the government's role from ownership and operation to policy and regulation. The ITU's plenipotentiary conference of 1989 endorsed these recommendations and brought them to the attention of all developing countries.[22]

The third, more activist, point of view is represented in early writings such as those of Philip Okundi, in Kenya, and the late Ithiel de Sola Pool, Manfred Kochen, Edwin Parker, and Melvin M. Webber, in the United States. Okundi was one of the first advocates of an African domestic satellite system and vigorously supported the expansion of the Pan-African Telecommunications Network.[23] Pool argued, "It would seem clear that international telecommunication is of great importance to the developing world. It can bring deficient information facilities (e.g., for access to scientific data bases) at a leap up to the best, or its lack can lead to slipping further and further behind."[24] Kochen contended that computer conferencing facilities should be made available throughout the developing world and that they would facilitate knowledge transfer and more efficient resource allocation.[25] Parker argued for accelerated development of rural satellite ground station installations with applications in health, education, and general rural development, as did Hudson and others.[26] Finally, Webber suggested that modern telecommunications has the potential for permitting developing countries to organize their urban areas on a more spatially dispersed basis, which he hoped would allow people to live closer to their work and would reduce the need for large investment in mass transit and in high-density and high-cost city centers.[27] Videotex in France is a prime example of activism by an industrial country's government. By 1990, several million Minitels (basic data communication terminals) were in place in French homes and businesses, providing a market base for the development of electronic mail, home banking, and a host of new telecommunications and information services. This was the result of a policy decision made in the mid-1970s, as part of a drive to modernize and expand the telecommunications system and to replace the printed telephone directory with an electronic version and provide the necessary terminals free of charge.[28]

Who is right? The importance of answering this question correctly can scarcely be exaggerated. If a strong telecommunications infrastructure is indeed essential for rapid and efficient development to take place, its neglect may severely hinder the success of development efforts in both directly productive and social sectors, impose inefficient spatial settlement patterns on the rapidly growing urban areas in the developing world, and reduce the ability of these countries to participate in the world economy. If, however, the present level of telecommunications service in developing countries is sufficient (although in many towns, villages, and informal urban settlements it is virtually nonexistent), then massive investment in the premature expansion of a major infrastructure system would not only misdirect resources but also create a serious burden of unnecessary administration, training, and maintenance.[29]

Benefits from Improved Telecommunications

Many telecommunications professionals or development experts have observed individual cases in which specific benefits of improved communication have been associated with changes in telecommunication infrastructure. The following examples of such instances are grouped somewhat arbitrarily under four convenient (but not inclusive or mutually exclusive) headings.[30]

Market Information for Buying and Selling

• A distributor of industrial spare parts and machinery in Nairobi, Kenya, found that after additional lines were installed to his office from the local exchange, his business expanded 35 percent. This permitted him to hire six new employees and add three light trucks to his company's fleet.

• In Greece, the packet-switching network only serves central, urban regions, leaving businesses located in peripheral regions (termed less favored regions by the European Communities) isolated from business networks that are important for their operations. Without packet switching, banks in less favored regions must pay much higher costs to install electronic banking systems. This weakens the national banking system overall, since many banks cannot afford to participate in the national banking network. Similarly, travel agents and insurance companies in these regions have trouble competing because they do not have access to network-based quotation and booking systems that depend on packet switching.[31]

• A bank manager in a small town in India was able to reach a prospective customer in the regional market city 60 kilometers away rapidly after a rural automatic digital exchange was installed in his town. The manager maintained that if direct dialing had not been available, the resulting delay in communications would have caused the depositor to take his substantial business to another bank.[32]

• The introduction of telephone service to several rural towns and villages in Sri Lanka allowed small farmers to obtain, among other things, current and direct information on wholesale and retail prices of fruits, coconuts, and other produce in Colombo, the capital city. As a result, the farmers began to demand and receive higher prices for their products. Before they had telephone service, they sold at prices averaging 50 to 60 percent of the Colombo price. Following telephone access they regularly sold at 80 to 90 percent of the Colombo price.

• A small grocer in Rosario, Uruguay, who sold and delivered groceries to homes, was able to serve a large clientele beyond his immediate neighborhood primarily because residential telephones were available locally and customers could order his goods by telephone for home delivery.

• When the Paraguayan National Development Bank began to upgrade and extend its activities in rural areas, it found that telecommunications links enhanced the effectiveness of its rural branches. In addition, without access to a telephone, local farmers were unable to relate effectively with markets, were not knowledgeable about current market prices (many farmers sold their production to truck owners at prices substantially below prevailing market field prices), and had difficulty procuring fertilizers and other supplies in a timely manner.

• A Ministry of Communications survey showed that public call offices in rural areas of Korea (Republic of) averaged 85 local and 160 long-distance calls per month and that the calls helped remove feelings of isolation. They improved villagers' income by providing access to timely market information and by enabling them to bypass middlemen. The local government administration was also able to disseminate information to the villagers more quickly, as well as to save time and funds by reducing the personal visits that government staff made to the villages.

Transport Efficiency and Regional Development

• In peripheral neighborhoods of metropolitan Lima, Peru, more than 100 Community Telephone Centers have been installed to serve low-income users; each center serves, on average, more than 640 customers, who receive messages left at the center's number, use the center to make outgoing calls, and are listed in the Lima telephone directory under the center's number. The typical subscriber is underemployed and travels extensively by bus to search for temporary work in centrally located districts. Subscribers can save up to ten times the cost of their subscription in a year by using the telephone to look for work instead of making daily bus trips. Subscribers also use the centers to report fires, crimes, and health emergencies and thus gain a heightened sense of security.[33]

• Cameroon is experiencing a severe shortage of properly trained managers who serve the needs of the telecommunications authority and carry out strategic development. This shortage hampers their efficiency. Sending managers abroad to study incurs significant costs and exacerbates personnel shortages. Recently, seven managers took part in a training course in project management. Offered at a distance, this course used audio conferences and facsimile communications via telephone lines, along with traditional course materials, to link students with their instructors overseas. The local course administrator found that using the distance option was much cheaper than sending the same seven managers away from their jobs to study overseas. Participants learned as much as they would have abroad, without having to leave their jobs and families.[34]

• In Sri Lanka, a survey of 176 new telephone subscribers (77 business and 99 residential) in the country's two major cities, Colombo and Kandy, found that almost 70 percent increased the number of contacts they had with other people after getting their phones and nearly half increased the volume of their business. Moreover, 40 percent increased their profits or incomes, about 33 percent increased the regional extent of their business, and a little over 10 percent increased the variety of their business.[35]

• Brazil has a national fleet of over 180,000 trucks that must cover extremely long distances, often through rural, remote areas with no telephone service. A major problem for drivers is that their vehicles sometimes break down on the road. When this happens, several days may pass before they can contact their company headquarters for help. This poses a major security risk in addition to slowing deliveries, since cargoes are frequently stolen off the truck while the driver is seeking help. In 1986, a very high-frequency, manually switched radio-telephone system was introduced to serve truck drivers and their company headquarters. It allows a driver to access the national phone network from anywhere on the road and summon assistance without leaving the truck, and it allows dispatchers to locate any truck and relay instructions or other information.

• The Road Transport Corporation in Myanmar (formerly Burma) operates more than 2,000 freight trucks out of more than twenty branch stations. In the past, telecommunications facilities were not reliable, and trucks often returned to base empty even though nearby cargoes were waiting to be picked up. Breakdowns could not be quickly reported, and there were delays in analyzing the mechanical problem and sending a repair team with suitable parts and tools. Finally, transport station managers relied heavily on messengers to carry simple communications; at the Yangon (formerly Rangoon) station, from thirty to forty messages a day, which could have been conveyed by telephone, had to be sent by messenger.

• According to state government officials in India, the availability of telecommunications services is the primary reason that industrial projects progressed rapidly in some locales in Maharashtra state. They observed that industries were reluctant to move to places inadequately served by telecommunications and that a lack of reliable telephone service in some areas was curtailing the state-supported program to decentralize industry. Furthermore, when telephone service was interrupted for technical reasons, businessmen frequently traveled by automobile to nearby locales to communicate with suppliers and customers, thus using expensive fuel and taking time away from their businesses.

• After disturbances in Ogaden, which blocked the road between the port of Assab and Addis Ababa, the Ethiopian government established a series of checkpoints linked by radio along the road to speed the flow of supplies and to improve security. The progress of individual trucks was monitored, and, in case of a breakdown, spare parts were ordered and brought on the next truck. The radio-linked system cut the average journey time in half, and in a matter of weeks the port of Assab was cleared of the goods that had accumulated there.

• A Ministry of Agriculture official in Tanzania, who is responsible for project implementation, requires daily information from parastatals and ministry departments. Inadequate telecommunications links necessitate frequent visits by car; the official averages three hours of travel each day. Without adequate long-distance telephone connections several other officials must make between eight and twelve costly and time-consuming safaris by automobile each year.

• In Myanmar, the Inland Water Transport Corporation operates on the country's principal navigable waterways, carrying all of the oil produced in the country and much of the cement and rice. The corporation controls the movements of its boats from its headquarters in Yangon through key stations. In the past, the lack of telecommunications links meant that it was often not possible to divert a boat for an unscheduled stop to pick up extra cargo. Unexpected changes or emergencies were frequent, however, and the delay in providing information made it difficult for management to make alternative arrangements, notify shippers and customers, arrange for spare parts, and take other steps to minimize damage and losses.

• In Uganda, an estimated 2,000 local government officials and senior staff make about 40,000 extra trips a year from district (provincial) headquarters to Kampala (an average distance of about 250 kilometers), from county to district towns (50 kilometers), and between county and subcounty towns (20 kilometers). These trips are made to handle administrative matters that could be dealt with effectively by mail or telephone if these services were available and reliable in the approximately 700 mostly rural centers. An estimated 250 man-years of qualified government manpower is wasted every year, at a direct cost (salaries, allowances, and public transport) of about $0.6 million.

Isolation and Emergency Security

• Newly installed direct dialing service in a small town in India makes it possible to summon medical care rapidly. Before this service was available, snakebite victims often died before help could arrive.

Recently, a man bitten by a cobra was saved because the doctor was quickly contacted by phone.[36]

• In the remote province of San Martín in Peru, health workers in small villages work in primitive medical facilities and have inadequate training; they must try to help their patients in isolation, without consulting other doctors and medical workers. As part of a pilot program in rural telephony, the Peruvian state telecommunications enterprise, ENTEL, installed radio and satellite telephone links between the larger town of Juanjui, where there was a small hospital, and four small villages. As a result, rural health workers and their patients were able to consult directly with doctors at the hospital via audio teleconferences. Doctors gave direct advice not only to the workers in charge of a specific case but also to health workers in nearby towns who listened in and thus enlarged their own experience. Although the program boosted the morale and confidence of isolated health workers, technical problems ended the service, which is sorely missed.[37] Other well-known telemedicine programs exist in Alaska and Guyana, where doctors and health workers use conference calls to work together.[38] Appendix A discusses telecommunications and the health sector in more detail.

• In Ghana, a forest-resources management program operates in rural areas, far from conventional telephone service. A private network of radios held by hand and located in vehicles and base stations was set up to tie workers in the field with field stations and park headquarters. Users can coordinate their activities much more effectively than before, and the mobile Wildlife Protection Squad, which enforces game and wildlife regulations, can respond much more quickly to reported incidents of poaching and other violations.

• In Fiji, during a hurricane, one of the smaller islands suffered severe damage with some loss of life and many injuries. The only telecommunications link with the main island was an old radio, which was destroyed during the storm. One week passed before assistance was sent to the distressed island because the authorities on the main island were unaware of the severity of the problem. A properly engineered long-distance telecommunications system would have withstood the hurricane and could have been used to summon immediate assistance.

• In Rwanda, the political head of a region has no access to a telephone. He relies on passing travelers to hand carry messages to the capital. There is no assurance that his messages will get through or that he will receive a reply. Often he must wait up to two weeks before finding someone who will carry his messages.

• When a small Alaskan village received a telephone, the number of applications for government grants-in-aid increased dramatically; previously with the once-a-month mail barge that served the village, the turnaround time between legal authorities in Anchorage (the state's largest city) and the village was usually too long for grant applications to arrive on time. Also, the relative slowness of the mail barge and the lack of other access to the outside world were more widely perceived only after a telephone link was established. Hence, the villagers began actively to lobby for building an airplane landing strip near the village.[39]

• Local leaders of the fairly isolated native population of northern Canada used their newly installed village telephones to plan, discuss priorities for, and coordinate political and economic strategies. Previously, the native leaders had no means by which to coordinate their negotiating positions before meeting with representatives of government or commercial agencies. After the introduction of telephones, however, the native leaders were able to present their positions at meetings in a much more coordinated manner.[40]

• Leticia, in the Amazon area of Colombia near the borders of Peru and Brazil, has good timber and fishing resources. Supplies arrive only by river and air, and most workers can only visit their families at most once every two years. The installation several years ago of telephone facilities made regular personal and business communication possible. Morale improved, and the supplies received corresponded more closely to short-term variations in needs.

Coordination of International Activity

• A copper mining company in Papua New Guinea is able to administer its operations directly from a location close to the mine on the island of Bougainville primarily because access to telecommunications is good. Among other activities, the company manages its investment portfolio by international telex on a continuous, twenty-four-hour basis.

• In Nepal, the tourist industry has the potential to be a major source of foreign exchange. Unfortunately, however, the operations of airlines, hotels, and travel services have been significantly constrained by inadequate and unreliable international communication. There have been frequent problems with room and airplane reservations and with overbooking. By improving international communication links, as well as the domestic telecommunications network, Nepal is now substantially increasing the capacity of its communication-dependent tourist industry to generate foreign exchange for the country.

- Political problems are sometimes exaggerated by poor communications among countries. The Organization of African Unity stated in 1980 that "The political backlash that plagues inter-country disputes in Africa can be attributed to the lack of adequate communication facilities between the capitals of neighbouring countries."[41] According to the OAU, relying on parties outside the continent, which often have preconceived notions of African conflicts, threatens the peace and stability essential for development. Vested interests often distort minor incidents, which could be solved reasonably and speedily if clearly defined positions could be communicated. Having the means to do so would alleviate the fears of neighboring countries in times of crisis.

- In Jamaica in 1966 a tourist hotel opened before telecommunications facilities could be provided. The hotel was about 20 miles from the nearest sizable town, but it was attractively located with excellent amenities and a first-class beach. Until telephone and telex facilities were provided, however, this hotel faced very low occupancies because visitors with business, political, and social interests or with health problems were not prepared to be cut off from communicating rapidly with the rest of the world. New arrivals who had not checked on the availability of communication facilities in advance often left after a day or two for this reason. The hotel itself was also unable to handle bookings and cancellations directly. During the peak tourist season when similar hotels with communication facilities were operating at 95 to 100 percent capacity, the communication-deficient hotel had difficulty exceeding 65 percent.

Scope and Organization of the Book

The review of selected literature on telecommunications investment and economic development, which is summarized in the main body of this book, suggests a paucity of sound analytical material and relevant empirical data on which to base policy decisions about investment in telecommunications.[42] Many of the questions that are the most important for analyzing and justifying investment have received little attention from economists, statisticians, and engineers. Furthermore, telecommunications are not an intrinsic part of the development debate. The study of telecommunications and development has been undertaken mostly by telecommunications specialists, and dissemination of the findings has been largely confined to their peers. This is in stark contrast with, say, transportation and energy, which are regularly covered in the literature on economic development.[43]

Scope of the Book

Given this shortage of analytic material, the scope of this book is limited to two basic objectives: a review of the available evidence on the role of telecommunications in economic development and an outline of the principles and techniques of economic analysis that yield information on the telecommunications sector useful for governments seeking to improve their allocation of resources and formulation of policy. The primary questions posed are the following:

a. What are the principal constraints on the development and performance of the telecommunications sector? What measures can be taken to help overcome these constraints?

b. What evidence is currently available to help policymakers assess the effects that investments in telecommunications have on economic development?

c. To what extent can the established tools of economic analysis be used to quantify the benefits of telecommunications and thus determine more accurately the appropriate composition and level of investment in telecommunications needed to allocate national resources efficiently?

d. What is known about how telecommunications services are used by different categories of users and about how the benefits are shared by them? How can available evidence be used to assess the distributional effects of expanded or improved services?

e. What pricing policies are appropriate for telecommunications services in developing countries? To what extent can pricing enable the market mechanism to yield correct signals for the use and expansion of telecommunications networks? How effective are current pricing practices in this respect?

f. How much could traditional state telecommunications monopolies improve? What changes in policy and structure could further promote efficiency and responsiveness and mobilize capital for investment?

Organization of the Book

Organized to address these questions, this book proceeds in pragmatic steps by examining a succession of approaches ranging from the study of sets of countries, single countries, and regions to the examination of specific sectors and the detailed analysis of specific projects.

A traditional cost-benefit framework has not been followed throughout. This is partly because the effects of a developing country's telecommunications investment program are likely to be diverse and widespread. Hence, only an extensive and assumption-dependent general equilibrium analysis could begin to identify all of them.[44] The primary reason is, however, that existing empirical evidence on the effects of telecommunications investment does not fit neatly into a classic cost-benefit framework. Traditionally, benefits have been viewed in the cost-benefit literature as belonging either to a group variously labeled as primary, direct, or internal (benefits accruing to the users of the service) or to a group variously labeled as secondary, indirect, or external (benefits accruing to others).[45] However, a review of the literature on the effects of telecommunications indicates that an attempt to categorize existing analytical work according to a primary/secondary or direct/indirect dichotomy, although appealing in a partial equilibrium project-related sense, has several problems. Such a review would indicate a degree of empirical precision that does not exist and would not provide a suitable framework for much of the material reviewed (for example, cross-country correlations, input-output analysis, and some literature on the energy sector).

The book has six parts. Part I gives a background on investment and benefits (presented earlier in this chapter) and explains why telecommunications is, in general, an industry with declining costs (chapter 2). It also identifies the main constraints on the expansion and performance of the telecommunications sector in developing countries and discusses measures that help overcome these constraints within the limits of a state monopoly of supply (chapter 3).

The core of the book deals with the benefits to be gained from telecommunications. Part II examines the relationship between telecommunications and the economy from several macroeconomic perspectives: statistical analysis of a series of country data (chapter 4), structural analysis of the economy of individual countries (chapter 5), and studies of the spatial organization of economic activity (chapter 6) and of the transport and energy sectors (chapter 7). Part III takes a less aggregate view, presenting microeconomic approaches to measuring benefits and using them to analyze individual telecommunications investment projects (chapters 8 and 9). Part IV addresses the question of distribution of benefits by characterizing telecommunications users (chapter 10) and usage (chapters 11 and 12) as revealed by surveys around the world.

Part V looks at the decision to invest in telecommunications from the standpoint of pricing policy in general (chapter 13) and examines how tariffs can contribute to economic efficiency (chapter 14). In addition, it covers the practical application of marginal cost pricing, tar-

NEW YORK INSTITUTE
OF TECHNOLOGY LIBRARY

iff analysis in developing countries, and regulatory issues such as price-cap mechanisms and rate-of-return mechanisms in the context of the rapidly changing regulatory and policy environment (chapter 15).

Lastly, Part VI discusses the limits to overcoming constraints on telecommunications development when traditional state monopolies supply services and outlines alternative policies and structures that may result in considerable gains in performance, responsiveness, and capital mobilization (chapter 16). This theme is further developed in two separate books.[46]

Limitations of Scope

To keep this book a manageable size, three further limitations have been adopted. First, in discussions of the economics of telecommunications infrastructure, the somewhat contentious subject of radio and television broadcasting policy has been avoided, except incidentally as part of comments on the role of general telecommunications infrastructure, which permits national reception and networking of broadcast programs.[47]

Second, the book focuses primarily on telephone service. Despite the dramatic growth of facsimile, data transmission, and other advanced services, telecommunications in both the developing and the industrial worlds are dominated by telephone service and use of the telephone network for transmitting text, image, and data; this use typically accounts for more than 90 percent of telecommunications investment, traffic, and revenues. Although some advanced services are now well established in a number of countries, they still operate on a very small scale compared with telephone service, even in industrial countries. More important, however, most of the essential issues and concepts surrounding the voice, text, image, and data transmission aspects of telecommunications policy can be understood through a discussion and analysis of telephone service.[48]

A final limitation of scope concerns the economics of telecommunications technology. The increasingly complex technical decisions facing developing countries are susceptible to economic analysis. The analyst will find relevant material in this book, but not an analysis of technological choices themselves. Many of these are country-specific choices, and the correct solution will also change as technology advances and relative costs vary.

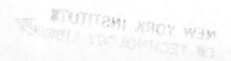

Notes

1. International Telecommunication Union (1984). The Maitland Commission, formally known as the Independent Commission for World Wide Telecommunications Development, was set up by the ITU in 1982 to examine global development of telecommunications. The commission's final report established the general goal for world policymakers as being to bring "the whole of mankind within easy reach of a telephone . . . by the early part of the 21st Century." The 1989 report of the Advisory Group on Telecommunications Policy to the ITU reiterated some of the points made by the Maitland Commission, noting that although "there is a widespread belief that a relationship exists between investing in telecommunications and boosting the overall economic health of a country," there is also "considerable skepticism about the benefits to the economy . . . in comparison with other urgent needs." ITU (1989).

2. In this book, the word "telephone" is used to designate telephone sets or instruments connected to the public telephone network. The expression "telephone line" (or "main line" or "line" for short) is used to mean a connection between a local telephone exchange and (a) a subscriber telephone or other equipment with a distinct calling number, (b) such a telephone with one or more in-house extensions sharing the same number, or (c) a larger subscriber's private branch exchange or key system, in which several lines are shared by many otherwise independent telephones. This corresponds to what is also called "exchange access line" or "main station" by the International Telecommunication Union, "line" in the United States, "direct exchange line" in Great Britain and Commonwealth countries, and "raccordement principal" in France and some African countries.

3. Until the mid-1980s, the statistics describing telecommunications infrastructure commonly measured the number of telephone sets. Current practice is to count the number of main lines instead, because the number of nonvoice terminal equipment such as facsimile machines and computer modems has proliferated rapidly, as has the number of extensions and private branch exchanges. Each telephone main line may be associated with more than one telephone. National averages normally fall in the range of 1.2 to 1.5 telephones per main line and occasionally reach 2.0 or more. Thus, for any particular country at a given time, the density of main lines per 100 persons will be less than the density of telephone sets per 100 persons. This edition of the book gives main line densities, which are lower than the telephone densities given for the same countries in the first edition appearing ten years ago. The two densities are not directly comparable.

4. World Bank (1989). All dollar amounts are U.S. dollars. A billion is 1,000 million.

5. Wilson and Inan (1989). Investment in developing countries generally expands the system, while investment in industrial countries mainly replaces outdated equipment and provides new services.

6. Luhan (1989).

7. Statistics for facsimile machines often understate the actual number of pieces of equipment in use, since in many cases, sets do not have to be registered officially with the telephone authority. Where licenses are required, machines are often installed without registration to avoid fees and import restrictions.

8. For example, in 1967, the unrecorded demand for new residential telephone main lines in greater Santiago (Chile) was approximately three times higher than the number of outstanding applications. See Wellenius (1969a). In numerous instances, the waiting list in different countries actually grew after a large project to expand the system made new lines available.

9. For example, when a single telephone must be shared by many employees or when insufficient lines link a private automatic branch exchange to the public network.

10. For instance, repeated attempts to call from a provincial town to the capital city's congested local network may, in turn, congest equipment and call traffic in the provincial town's network.

11. In this book, the World Bank refers to the International Bank for Reconstruction and Development and its affiliate, the International Development Association. Operations by the two other affiliates, the International Finance Corporation and the Multilateral Investment Guarantee Association, are not included.

12. For example, about 50 percent of the profits from telecommunications operations in India between 1979 and 1983 were passed on to the government, and 30 percent of telecommunications profits, or about $150 million annually, were used to meet losses incurred by the postal service. In 1977, about 20 percent of the postal service's net capital and operating expenses were financed by telecommunications surpluses. More recently, starting in 1989, the government of Uganda began to collect 60 percent of the overall income of the Uganda Posts and Telecommunications Organization. During the five-year period 1989–94, the government will receive about $10 million in direct tax revenues, along with payments from the Posts and Telecommunications Organization of 8 percent interest on loans totaling $4 million. The net flow of funds from telecommunications to the treasury (taxes, duties, and dividends minus government contributions to investment) was about $600 million in Brazil in 1984 and about $350 million in Mexico in 1987. Both amounts were equivalent to between 30 and 40 percent of operating revenues.

13. Wellenius (1978).

14. See, for example, Webber (1980).

15. The French government passed legislation that enabled special financing for telecommunications. The installation of new telephone lines reached 2 million a year in 1979, up from 350,000 lines in 1968. *Telecom France* (1981), pp. 29–32; and Thery (1977).

16. Wellenius and others (1989).

17. American Telephone and Telegraph Corporation (1990).

18. On occasion, variations of this view have been attributed to multilateral development banks and some sources of bilateral aid. In fact, between 1949 and 1974, lending for telecommunications by the Inter-American Development Bank amounted to $87.2 million, or only 1 percent of total lending. The amount of funds allocated for telecommunications lending by the World Bank averaged about 2 percent of total lending between 1960 and 1990. For the period 1983–89, telecommunications represented only 1 percent of total lending. Also, for a considerable period during the mid- and late 1970s, several major bilateral donors virtually ceased to support projects for telecommunications infrastructure.

19. Clippinger (1977).

20. Melody (1987).

21. The ITU is an intergovernmental agency representing virtually all countries in the world. It is responsible for setting international technical standards, coordinating the use of the radio spectrum, and handling development assistance. The CCITT (from the French acronym for International Consultative Committee on Telephone and Telegraph), one of the ITU's permanent bodies, became involved in the early 1960s in examining issues of telecommunications and economic development. The other permanent bodies of the ITU are the CCIR (International Consultative Committee for Radio Communications), the IFRB (International Frequency Registration Board), and the Secretariat. Following decisions of the 1989 plenipotentiary conference, a fourth permanent body, the BDT (Bureau for Telecommunications Development) was established, taking over and expanding functions that were earlier vested in the ITU's Technical Cooperation Department, as was a high-level committee to make recommendations for a broad reorganization of the ITU.

22. CCITT (1976), chap. 1, sec. 2.1; and ITU (1989) chap. 5.

23. Okundi and Evans (1975) and Okundi (1975). The idea of using satellites for regional communications in Africa revived in the late 1980s, and by 1990, the ITU had completed a feasibility study for RASCOM, a regional African satellite communication project.

24. Pool (1976).

25. Kochen (1982) pp. 230–58 and Center for Science and Technology for Development (1986).

26. See Block and others (1984); Hudson (1990); and Tietjen (1987). Despite a few outspoken advocates, the question of whether such satellite applications are now the least-cost way to provide such rural services in most developing countries has not been sorted out. Many contend that in most instances they are not. For discussion of more general approaches to the role of telecommunications in rural development, see Hudson (1984) and Parker and others (1989).

27. Webber (1980).

28. Key related decisions were (a) to open to private competitive supply all information services except the electronic directory; (b) to develop a national packet-switched data network for cost-effective connection of users and service providers; and (c) to adopt a simple tariff structure and billing procedure. A lively debate has been raging for years on the extent to which cross-subsidy is involved in getting Minitel off the ground, on whether France's Telecom is now recovering all costs, and on whether there have been benefits to the economy over and above what users pay for the new services. Whatever the answers, it is clear that the French videotex system is by far the largest and fastest growing in the world, with hundreds of profitable independent service providers. The French system stands in stark contrast to that of other industrial countries, such as the United States, where most videotex initiatives by individual businesses have failed to develop a sufficiently large market base.

29. This would be similar to one aspect of the development of railways, which in nineteenth-century Europe was a favorite sector of infrastructure for both public and private investment. One cause of the generally poor economic performance of several European countries in the nineteenth and early twentieth centuries is thought to have been the massive diversion of investment into the railway system, in imitation of countries such as England, before the rest of the economy was ready to make full productive use of it.

30. Except where explicitly noted, these examples are drawn from unpublished reports or memoranda on specific projects in telecommunications and other sectors, undertaken mostly in the decade from 1980 to 1990.

31. Pye and Lauder (1987).

32. Malgavkar and Chebbi (1988).

33. Castilla and others (1989).

34. Nettleton (1990).

35. Kojina, Hoken, and Saito (1984).

36. Malgavkar and Chebbi (1988).

37. Mayo and others (1987).

38. Examples are cited in Goldschmidt, Hudson, and Lynn (1980).

39. Goldschmidt (1978).

40. The example is cited in Hudson and others (1979).

41. Organization for African Unity (1980), pp. 10–11.

42. This view is not unique. See Moss (1981), which summed up by stating that, "overall, the papers in this volume suggest that we know relatively little about the influence of telecommunications on productivity.... We are still a long way from discerning the important links and their magnitude." Also see Baer (1981).

43. That this continues to be the case some twenty-five years after the first published

studies on telecommunications and development was noted by Jan Bjerningen of the Swedish International Development Authority during the consultative meeting with development financing agencies convened by the ITU in Geneva in January 1990.

44. Prest and Turvey (1965) p. 685.

45. Prest and Turvey (1965) pp. 683–90 and Weisbrod (1968) pp. 257–62. Benefits are also classified as tangible and intangible.

46. Wellenius and others (1989) and Wellenius and Stern (1991).

47. Although the sharp distinction between broadcasting and two-way communication media is rapidly breaking down with the advent of new types of telecommunications services, at present this is important mainly in limited segments of the more advanced economies of industrial countries.

48. All this is not to argue that innovative services may not be important in some developing countries. The very early stage of expansion reached by the telecommunications network in many developing countries and the correspondingly low degree of commitment to established sector structures present an opportunity for innovation. See Hobday (1986) and Mody (1989). To seize this opportunity successfully will require creative but realistic application of criteria for the suitability of technology. Of course, many of the telecommunications innovations currently appearing in the industrial countries are unlikely to deserve priority in many developing countries for several years. Video telephone and cable television might be put in that category, for example. Other innovative services and technologies, however, deserve more immediate consideration. These include modern electronic text communication services, which are superior to telex in most aspects and which may cost no more to implement in a new system; the use of data communications for such purposes as remote access to technical information; and the use of simple audio and graphic teleconferencing facilities as a partial substitute for and complement to travel. Data communications and specifically computer-based conferencing systems in developing areas are discussed in Balson, Drysdale, and Stanley (1982). See also National Research Council (1990); Oeffinger (1987); and Paine (1986).

Chapter 2

A Perspective on Technology and Costs

COSTS IN THE TELECOMMUNICATIONS INDUSTRY are declining throughout the world, and this global trend is projected to continue in the foreseeable future. The real costs of providing and maintaining telecommunications facilities and services in both industrial and developing countries have fallen in recent years. This is particularly important to developing countries, which, with their small base of existing facilities and potential for high rates of growth, could achieve rapidly declining unit costs. Two related factors account for this trend: technological change and economies of scale.[1]

Technological Change

Until the 1960s, the analog technologies used in telecommunications were evolving and improving at a fairly steady rate.[2] During the 1970s and 1980s, however, with the development of digital electronics, the increasing application of computer technologies to telecommunications, and the development of very wide band systems, the pace of change accelerated sharply. This lowered equipment costs, significantly increased the maximum capacity of individual equipment units, improved reliability, lowered the amount of power used by most types of equipment, and significantly reduced space requirements. New concepts in network management and structure allowed greater flexibility and more efficient use of assets. And finally, a wide range of new facilities and services was developed. Rapid changes are expected to continue, and further developments, especially new radio technologies, are likely to affect the cost and performance of systems.

Three interdependent series of events contributed to these changes. First, innovations altered basic electronic technology. Some of these originated within the telecommunications industry (for example, the transistor in 1950), while others were exogenous to it (for example, microprocessors for computers in the 1970s). Second, new industrial processes suited to large-scale mass production were introduced in the manufacturing of telecommunications equipment, increasing reliability and lowering costs. In particular, the advent of transistors followed by printed circuit boards and large-scale integrated circuits shifted the emphasis in production from mechanics to electronics and the emphasis in basic production elements from simple components to complex subsystems. The development and use of large-scale integrated circuits lowered power requirements and led to further miniaturization of discrete elements accompanied by sophisticated, automated printed circuit boards and system assembly. Third, a large spurt in demand for telecommunications services made innovations economically feasible in both basic telecommunications technology and equipment manufacturing processes. Hence, to a large extent, the effects of new technology and of scale are inseparable.

Equipment Costs

Innovations, especially in microwave, satellite, and fiber-optic systems, have dramatically reduced the unit costs of medium- and long-distance transmission. These cost reductions resulted from both lower equipment costs and greatly increased bandwidth (which determines the number of simultaneous voice and other signals that the equipment can carry). The cost of switching also decreased, but not as rapidly as that of transmission. Considerable improvements and some cost reductions have been achieved in junction and local cable networks.

Transmission equipment. During the 1970s, the cost per transmitted voice channel for analog microwave radio terminals (which accounts for two-thirds of a microwave route's costs) was cut in half in real terms.[3] This was mainly the result of increases in equipment capacity, which grew from 1,200 voice circuits per radio channel in the 1960s to 2,700 voice circuits in the 1970s and to 6,500 voice circuits in the 1980s. When microwave systems and other portions of the networks were digitalized in the 1980s, terminal costs per channel were further reduced by more than half again. Although these cost reductions were largely the result of increased bandwidth on major routes, important breakthroughs were also achieved in some of the low-capacity transmission systems used to extend service to small communities with a small volume of traffic. Compact and robust integrated analog or digital multiaccess, ultra high-frequency or very high-frequency radio systems now provide high-quality rural communications at a cost well below that of both conventional terrestrial transmission systems and satellites.[4] In the late 1980s, for example, digital multiaccess, ultra high-frequency systems cost about $3,000 per channel compared with $10,000 per channel for conventional analog ultra and very high-frequency radio systems.[5] Such systems are also increasingly cost-effective on routes with no access to electricity. In Australia, for example, the cost of solar power for a repeater station dropped from $A30 to $A2 over the past decade.[6] The merging of multiaccess radio with cellular technology, fixed use of equipment designed for mobile systems, and new generations of radio technology were developed for use in urban areas of industrial countries and promise to provide cost-effective solutions in rural areas of developing countries as well.

Improvements in coaxial cable equipment increased the capacity per pair of coaxial tubes from between 960 and 1,200 channels in the 1960s to 10,000 channels in the 1970s and thus reduced the unit costs of transmission at a pace comparable to that of microwaves. The introduction of optical fiber cable systems in the early 1980s further reduced costs. By the early 1990s, optical fibers that transmit from 560 to 2,000 megabits per second (Mb/s) were physically smaller and cost less per pair-kilometer than coaxial tubes with only a fraction of the capacity.[7] Furthermore, although the cost of coaxial cables is largely determined by the price of copper in commodity markets, innovations both in the technology for manufacturing optical fiber and in transmission equipment have steadily reduced the cost of optical fiber systems at about 70 percent annually.[8] Other features result in further savings. Unlike copper cables, optical fiber cables are impervious to electrical disturbances and can therefore be installed at low cost on existing power transmission lines and along electrified rail tracks. Optical fiber repeaters can be spaced at 50 kilometers or more compared with 10 kilometers for coaxial cables. Driven by all these cost advantages, optical fiber technology is gradually displacing copper wires and cables throughout telecommunications networks.[9] The first commercial optical fiber cable systems were installed in industrial countries in the late 1970s to serve routes with heavy long-distance traffic. Since then, rapidly decreasing costs in both industrial and developing countries have made these systems the standard choice for new long-distance cable transmission routes and for interexchange junction routes in multiexchange urban networks. In industrial countries, optical fiber cables are also being used increasingly in local loops for customers requiring broadband capability (such as high-speed data or video); by the end of the 1990s, they are expected to cost close to what copper cables do and thus to be widely used in local networks.

Satellite systems, in addition to the international telecommunications for which they were initially designed, now provide a variety of cost-effective solutions to domestic long-distance communications in a growing number of developing countries.[10] The unit cost of transmission by satellite has decreased markedly, largely because the capacities of satellites have increased. For example, the typical price for leasing international circuits from INTELSAT dropped from $32,000 per half-circuit a year in 1965 to $8,500 in 1975, $4,700 in 1981, and $4,400 in 1987.[11] In real terms, the price dropped three-fourths every ten years. The price of full-size earth stations with 30-meter antennas has remained roughly unchanged in current dollars, which reflects a further reduction in real price of about one-third every ten years. The increasing power of modern satellites has also allowed low-cost, small earth stations to be used for low-traffic routes. Improved encoding for voice channels (which reduced the required bandwidth per channel) and improved earth station amplifiers have helped reduce the size of earth stations. VSAT (very small aperture terminal) antennas that are 0.6 meter in diameter are now being used either for television and other signal reception only or for transmission and reception of data and voice by small users (the application of satellites for delivering social services is discussed in appendix A). As of the early 1990s, VSAT terminals are available for about $3,000 for receive-only and for between $6,000 and $10,000 for receive-and-transmit capabilities.[12]

Switching equipment. The costs of switching equipment declined from an average of $300–$400 per connected line in the early 1970s to $200–$300 per line in the late 1980s.[13] Thus the real cost per line was cut roughly in half each decade. The cost per line of electronic analog exchanges, initially higher than that of electromechanical exchanges, tended to fall, reflecting general trends in the electronics industry. Prices for older equipment remained roughly constant in real terms, being tied mainly to the price of metals and labor and to the limited scope for further gains in manufacturing productivity. The introduction of digital technology accelerated the decline in unit prices. By the early 1980s, digital long-distance telephone exchanges were available at about half (or less) the price of analog units with similar capacity, and prices continued to fall. Then digital local telephone exchanges became available at prices comparable to those of electromechanical and electronic analog equipment, while offering other significant cost and operational advantages. By the late 1980s, the global production of analog switching equipment had been largely phased out.[14]

Nonetheless, whereas the price of most advanced electronic equipment, especially computers, fell precipitously, that of digital telecommunications switches declined rather slowly. One important factor keeping costs up has been the high cost of developing specialized software. First introduced in the early 1970s with stored program control analog exchanges, software today accounts for a large and still growing proportion of the total cost of digital switching systems. For example, 45 percent of the initial cost of AT&T's ESS 1 exchanges was in software.[15] Another factor is that although switching systems are increasingly modular in structure, the modules of different systems are not interchangeable. Thus different manufacturers compete with one another only when selling complete systems. Having chosen a system, users can only obtain spares and expansions from the initial supplier. Also, since the modules are not standardized among systems and use proprietary technologies, the benefits of competitive global sourcing of subsystems (which plays such a major role in innovation and cost reduction in the field of computers) have not materialized for telecommunications switching.

Junction and local cables. During the 1970s and early 1980s, the introduction of pulse code modulation systems expanded the capacity of existing interexchange junction cables in large multiexchange local networks by a factor of six.[16] By the early 1990s, optical fiber systems had become the standard solution for expanding junction circuits in urban networks. These systems vastly increased the capacity of interexchange circuits in existing duct space, eliminating the need for expanding the cable duct networks and saving the attendant costs and disruption of the urban environment. When combined with digital switching equipment, which enables more efficient use of junction circuits, the use of optical fiber cables for junction circuits can also reduce the overall costs of local networks.

Technological improvements in telephone sets and other terminal equipment on the customer's premises, and the new switching equipment's capacity to work over higher resistance subscriber loops, have also reduced the cost of materials used in the local cable networks. For example, in the 1940s, standard telephone sets required the use of 6.5 or 10.0 pounds of copper conductors per mile between telephone exchanges and the customer's premises. By the 1970s, the norm was reduced to 2.5 to 4.0 pounds of copper conductors per mile. Since then, advances in telephone set technology and sophisticated loop extenders have enabled even longer subscriber loops. Tone dialing has extended signaling limits beyond those possible with dial pulse systems, increased the reliability of telephone sets, and enabled more facilities to be provided within sets.

Energy and Space Requirements and Network Performance

The technological changes that helped decrease the real costs of telecommunications equipment also had a significant impact on the costs of the wider system. The most important gains have been reduced consumption of energy and building space requirements, improved network reliability, and gains in efficiency that are the result of better management and positive changes in the structure of the network.

Energy and space requirements. Semiconductor technology and large-scale integration have reduced the amount of energy consumed by many types of equipment. In the 1970s, for example, a typical 960-channel microwave repeater required a power source of about 1,000 watts. By the 1980s, this had been reduced to about 100 watts. Currently, 300-channel repeaters are available that require less than 10 watts, which can be supplied by solar cells in sites without electrical power or access to fuel for diesel generators. In contrast, the advent of electronic switching has not improved the amount of energy required by telephone exchanges per line. Unlike electromechanical telephone exchanges, whose energy consumption was roughly proportional to the volume of traffic actually flowing through the exchange at a particular time—very high during the busy daytime hours, low during the night—energy used by digital exchanges is mostly defined by total exchange capacity rather than by actual traffic flow. As a result, power consumption is constantly high. Electronic components also have limited ranges of acceptable operating temperature, so that air conditioning is required, thus raising energy consumption.

Progressive miniaturization and integration of electronic components have produced large reductions in the space required by all tele-communications equipment: switching, transmission, and customer terminal equipment. For example, in the early 1950s, a standard multiplex transmission equipment bay held 24 channels; the bay capacity gradually increased to 240 channels in the 1960s, 600 channels in the 1970s, and 2,000 channels by the late 1980s. The amount of space required for digital switching equipment, excluding complementary items (such as the main distribution frame or a power plant), has also declined. A digital exchange of four to five times the line capacity of an old electromechanical exchange can be installed today in the same amount of space. Significant reductions are also being achieved by eliminating the amount of space occupied by maintenance personnel in each exchange. This follows from the enhanced reliability of equipment and the introduction of centralized maintenance and operational capabilities, including facilities for monitoring exchange performance and for extending failure alarms to distant, centralized maintenance locations.

Network performance. Technological changes have also improved reliability. The higher reliability and greater stability of solid-state components, which gradually replaced vacuum tubes in the late 1960s, virtually ended the need for preventive maintenance of most types of transmission equipment and also greatly reduced the incidence of repairs. For example, for a typical microwave repeater, the mean time between failures increased from about 15,000 hours in the 1950s to 400,000 hours or more in the 1980s. Use of ducts for major cables, and use of modern plastic-sheathed, jelly-filled cables that effectively prevent the ingress of water without costly gas pressurization, has reduced both the damage that can occur during transport and construction and the incidence of faults. However, equipment using advanced electronic technologies has also made local repair of some parts more difficult and more costly because specialized repair and test facilities are needed. Consequently, although advanced technology requires fewer maintenance staff, it also makes developing countries more dependent on suppliers. This increases turnaround time for some repairs and means that local stocks of costly spare subsystems must be maintained to overcome failures that could shut down the system altogether. Also, whereas the incidence of faults has been reduced, complex software and centralized operations make both individual pieces of equipment and total networks more vulnerable to extensive breakdowns of the system through just a single fault.[17]

The introduction of stored program control analog techniques in switching equipment, followed by digitalization, has made equipment more flexible and improved operational facilities so that the overall system is more efficient. Call metering facilities, which required additional equipment in electromechanical exchanges, is now included in digital exchanges at no extra cost, enabling the adoption of tariff systems that more accurately reflect the time and duration of calls (pricing is discussed in chapters 13 through 15). Likewise, the automatic generation of traffic data expedites and facilitates computerized customer billing. Centralized supervision and network management facilities may also make network management more feasible for developing countries. The management of modern digital networks has become so complex that many functions surpass human ability and have had to be assigned to computer robots. Although this equipment is very expensive, it enables fewer skilled managers to run a given system than would have been possible with analog technology.

Technological innovations and associated changes in relative costs are also altering the structure of networks. Digitalization has virtually eliminated the traditional boundary between switching and transmission, reduced the interface costs, and enabled more efficient and flexible use of equipment. In the days of all-analog equipment, transmission and switching equipment required hybrid connections. As digital transmission technology was introduced, analog-to-digital conversion required interface relay sets between switches and transmission equipment. Now that digital switching is becoming the norm, the system is becoming much more transparent. For example, in an analog system, the transmission system would premultiplex outgoing voice circuits in a digital format to at least the DS 1 level (that is, 24 voice channels at a total bit rate of 1.5 megabits per second); in a totally digital system, this function would occur within the switch itself. The very low cost of bandwidth now makes it possible to locate more of the switching function closer to the users, which would result in substantial savings in local loops that traditionally account for about 40 percent of the total cost of telecommunications investment in developing countries. The use of optical fiber in local cable networks, which are being installed initially for business users requiring large bandwidths, is likely to extend gradually to individual users. It will allow the introduction of integrated services digital networks (ISDN) providing end-to-end digital connectivity and universal network interfaces for a wide range of voice, video, and data communications.[18] New generations of cellular and other radio technologies (such as personal communication networks) are expected to offer increasingly cost-effective technical choices to substitute for, or at least complement, the expansion of wired subscriber loops.

Services Development

Technological changes have made it possible to extend basic telephone service in developing countries to small populations and remote areas, as well as to introduce new services increasingly required by the modern sectors of the economy. Satellite communication can reach otherwise inaccessible areas where use of cable or line-of-sight radio communications would otherwise be prohibitively expensive. Current applications include connecting widely separate points in large or archipelagic countries, distributing bulk data, and networking televisions.[19]

The introduction of stored program control analog techniques in switching equipment followed by digitalization made possible the provision of new facilities and advanced services that would have been too difficult or too costly using electromechanical switching equipment. Enhanced facilities and sophisticated new services designed to meet the requirements of business customers in industrial countries are now more or less built into the designs of new digital switching systems. They are thus inherently available in developing countries even though such a high level of sophistication may be required only in parts of the network. For example, although ISDN is only in the planning stage in most developing countries, privately leased circuits and data terminals connected to dedicated public data networks are already fairly widespread, and international packet-switched links are becoming available in major cities. Facsimile transmission has spread rapidly throughout the developing world. It is particularly relevant in countries that use a script other than the Roman, where use of telegraph and telex facilities becomes problematical unless costly special machines are designed and produced to suit the local script. Mobile communications are also growing. Many developing countries have licensed private companies to install and operate cellular telephone systems, which provide telephones primarily to business subscribers willing to pay the high costs of cellular telephones in order to overcome shortages of wired telephone service.

More broadly, technology has been the primary driving force behind the wave of change in the market structure, policy, and regulation of telecommunications that has swept the world since the mid-1980s. This is discussed briefly in chapter 16 and is covered in more depth in other publications.[20]

Economies of Scale

It is generally accepted that economies of scale apply to the costs of telecommunications systems. This phenomenon is particularly important to developing countries, which start with a small base of existing facilities and have large unmet demand, both of which offer an opportunity to reduce costs by accelerating growth of the telecommunications system.

Little explicit information is available, however, to help estimate the extent to which accelerated expansion can reduce unit costs in developing countries. Studies of economies of scale have been undertaken based on econometric analysis and on engineering cost data, but only in industrial countries with large, well-established networks, where the supply of telecommunications has kept pace with demand.[21]

In econometric analyses carried out mainly in Canada and the United States, aggregate production functions have been estimated using historical operating costs and the book value of existing systems taken as a whole. According to these studies, economies of scale have been present to the extent that a 1 percent increase in input to a large, highly industrial telecommunications system has been associated with an increase in output of an average 1.05–1.15 percent. This is, however, a very rough estimate, since results of econometric studies vary significantly (some are much higher, and a few are lower) and, in some instances, different conclusions are reached from the same data.[22] Furthermore, the effects of historical changes in scale and technology cannot be readily separated in these analyses.[23]

The engineering cost approach to quantifying economies of scale uses data on costs relating to current best-practice technology.[24] These studies have largely focused on the costs of investments in terrestrial long-distance transmission, where there are large indivisibilities of plant. Estimates show that a 1 percent increase in scale leads to a 0.6 percent decrease in the average unit cost (a 1 percent increase in input being associated with a 2.5 percent increase in output). However, for large multiplex systems and to some extent for switching systems, the potential economies of scale are much lower. When multiplexing is included in transmission systems, the effect of a 1 percent increase in scale is closer to a 0.3 percent decrease in unit costs (a 1 percent increase in input being associated with a 1.4 percent increase in output); when both multiplexing and switching are included, the estimates approach the upper end of the range of those obtained in the econometric studies.[25]

Overall, for countries with large, well-developed telecommunications systems, econometric and engineering studies both suggest that some economies of scale do exist in the provision of telecommunications services but that these are not uniform throughout the system. However, the potential for economies of scale is substantial in most developing countries, which have smaller networks, are usually less encumbered with outdated technology, and have significant scope for rapid expansion. These cost savings relate to increasing the efficiency of equipment and networks, procurement, and organization and management of the enterprise.

Equipment and Network Efficiency

Large economies of scale can be obtained in transmission systems. In long-distance radio transmission networks, the costs per channel and kilometer of investing in 1,200-channel microwave radio equipment (excluding multiplex) are about one-seventh the costs of investing in 120-channel equipment over an equivalent distance. For the multiplex equipment, in one African country, the average cost per channel was $960 for a 60-channel set, $460 for a 240-channel set, and $390 for a 600-channel set.[26] Similar economies of scale occur in optical fiber cables. An Australian study estimated that the cost per channel for a 30-kilometer optical fiber link was more than $600 for 34 Mb/s capacity, $200 for 140 Mb/s, and $100 for 565 Mb/s. Moreover, as the capacity increases, the cost increases less with distance. For example, the cost per circuit of a 565 Mb/s system covering 60 kilometers was only 10 percent more than the cost of the same system covering 10 kilometers.

Another Canadian study compared unit investment costs for three intercity transmission technologies: analog radio, digital radio, and fiber optics (see figure 2-1). In all three cases, high start-up costs made the cost per circuit and kilometer very high for systems with less than 5,000 voice channels. Costs continued to fall until each system's maximum capacity was reached, that is, the exhaust point; after this point, further expansion required an additional system. The researchers drew two conclusions from this comparison; first, for a transmission system of more than 10,000 voice channels, a fiber-optic system costs less today than any of the earlier transmission systems did in their day, and second, the exhaust point for fiber systems is considered to be virtually unlimited in today's context. Fixed costs, such as the costs of cable, cable installation, and the right of way, so domi-

Figure 2-1. *Comparative Unit Investment Costs for Three Intercity Transmission Technologies in Canada, 1988*

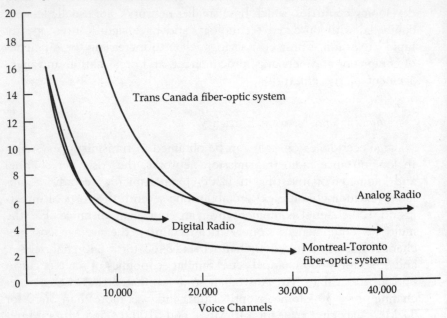

Dollars per circuit-kilometer

Trans Canada fiber-optic system

Analog Radio

Digital Radio

Montreal-Toronto fiber-optic system

Voice Channels

Note: All dollars are Canadian dollars. The analog radio curve is represented in dollars of the early 1960s and assumes that additional systems do not share an existing right-of-way. The digital radio curve is based on dollar values of the mid-1980s and ends at its exhaust point. Installing larger capacities would produce a curve similar to the analog radio curve. The Montreal-Toronto fiber-optics curve is in 1987 dollars. The Trans-Canada fiber-optic curve is still theoretical and is shown only for purposes of comparison.

Source: Federal-Provincial-Territorial Task Force on Telecommunications (1988).

nate the overall cost of a fiber-optic link such as the Montreal-Toronto link that the total cost of installing 12,000 voice channels was found to be less than 4 percent higher than the cost of installing the same system for 2,000 channels.[27]

The costs of transmission can be reduced further when the average use of circuits increases with route capacity. Consider, for example, subscriber-dialed, long-distance telephone calls made from city A to city B. Calls from A are generated at random. To ensure that all calls can be connected as they are generated, the number of circuits from

A to B should equal the number of subscribers in A. However, since these subscribers call city B only occasionally and for a limited time, the average use of the circuits is very low, resulting in high cost per call. The average circuit occupation can be increased by sharing a smaller number of circuits among all subscribers, who are automatically connected to any available circuit as they make a call. This, however, results in a certain proportion p of unsuccessful call attempts since a free circuit is not always available. Values of p between 0.001 and 0.005 (that is, between 1 and 5 calls out of every 1,000 fail at the first attempt for lack of circuits) are performance targets commonly used in telephone traffic engineering. For a given level of p, the average occupation of circuits increases with the number of circuits in the route. Figure 2-2 shows, for example, circuit occupancy that is 40 percent on a 10-circuit route but 81 percent on a 100-circuit route, both

Figure 2-2. *Average Occupation of Circuits as a Function of the Number of Circuits for Two Levels of Probability (p) of Lost Calls*

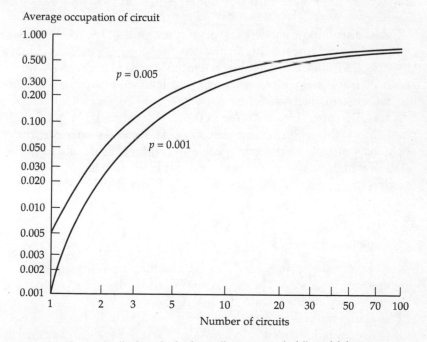

Note: Based on Erlang's formula for lost-call systems with full availability.

designed so that 0.5 percent ($p = 0.005$) of calls are lost. Figure 2-2 also shows a tradeoff between quality (the probability of losing calls) and cost: the lower the proportion of calls lost, the lower the circuit occupancy and, hence, the higher the investment required to cater to the same volume of traffic.[28]

Switching equipment also exhibits economies of scale, although they are less pronounced than those achieved in transmission. For example, in an African country in 1990, the prices paid per line for exchanges with a capacity of 5,000, 23,000, and 40,000 lines (including power plant, test equipment, and installation) were $420, $340, and $240, respectively. Because of their fundamental design and architecture, however, digital switching systems produce fewer economies of scale than electromechanical exchanges, and the prices paid per line for digital exchanges may not vary much with total capacity. For example, in a recent tender in a Latin American country, the bid prices offered for individual digital remote line units (small exchanges parented on main digital exchanges) with capacities between 500 and 2,000 lines ranged between $150 and $250 per line, while the bid prices for main local exchanges with larger capacities, between 3,000 and 10,000 lines, were within a similar range of $150 and $280 per line.[29]

Cable distribution networks, typically accounting for about 40 percent of the total investments in the telephone system in a developing country, also offer considerable potential economies of scale. In an Asian country, average bid price per pair-kilometer of cables, including the cost of ducts and installation, was $176 for 50-pair cables, $35 for 300-pair cables, $21 for 600-pair cables, and $11 for 2,400-pair cables. Although the unit cost of cable decreases with size, these economies of scale in the urban cable network result mainly from the relatively high costs of ducting and cable installation, which are mostly independent of the size of cables being installed.

Procurement Efficiency

Important cost savings can also be obtained by using appropriate procurement practices. These may overshadow the savings obtained from the economies of scale discussed above.

Packaging equipment into large orders of relatively few units normally reduces bid prices. For example, in a Latin American country in 1990, the price quoted for 22,000 lines for three telephone exchanges averaged $198 per line, while on a separate order of only 13,000 lines for nine exchanges of the same type, the price quoted was $349 per line.[30] At the same time, a total of 68,000 lines of equipment, similar to that of the larger Latin American order, were quoted in an African country at a much higher average of $298 per line, which is largely the result of the African purchase being fragmented into three, separately bid packages.

The price is also strongly affected by the modality of procurement and financing. Hence, a vendor may simultaneously quote very different prices to different customers. Procuring equipment through international competitive bidding or shopping extensively among suppliers can significantly reduce the costs of procurement compared with procuring equipment from a single source. A rule of thumb sometimes quoted is that prices under international competitive bidding are about 30 percent lower than those negotiated with a single supplier. Export development credits and most bilateral aid are tied to procurement from suppliers in a single country, which often means a single supplier for each type of equipment; this leads to negotiated prices that are often well in excess of those at which the same equipment is sold in competitive markets.[31]

There is also a tradeoff between cost and uniformity. In general, having a large variety of types of equipment from different manufacturers is not desirable, since this increases the cost of spares, training, and maintenance. However, locking onto a single type or source tends to escalate costs over time, despite declining costs worldwide. This occurs because a common marketing strategy is to offer a potential customer a very low price on the first order in anticipation of developing a captive customer who will later be forced to buy additional equipment, at much higher prices, from the same supplier. This is particularly true with switching equipment, which was discussed earlier in terms of technology. For example, in one developing country in the mid-1980s, unit prices for add-ons and new equipment were between 50 and 60 percent higher than prices for the initial order. Placing price adjustment clauses in the original contract sometimes helps control prices in the future. This approach is often used for cables; manufacturing costs are largely determined by prices in world commodity markets, so a price formula is relatively simple to establish and apply. Ultimately, however, a balance needs to be struck between maintaining a continuous supply from established sources and opening the market to competition, in order to lower prices as well as facilitate technological innovation.

Organizational Efficiency

Size also affects the efficiency of the operating enterprise. Every telecommunications entity, whether it serves 1,000 or 1 million telephone lines, has to carry out a similar number of operation and management functions.[32] It is not uncommon for large entities in industrial countries to have five or fewer employees per 1,000 telephone lines whereas those in developing countries with well-managed but much smaller networks have staff levels several times higher.[33] Because of the frequency and size of individual purchases, large entities also tend to develop greater strength as buyers in the international market and to improve their ability to mobilize financing; on both accounts, they can obtain lower costs than those incurred by smaller entities.

Organizational economies, as well as equipment and network economies, may also result from the integration of fragmented services within a given service area and, to some extent, among areas. Providing telex, data, or packet-switching services adds relatively minor additional administrative costs for a telephone company that already manages 80–90 percent of the total telecommunications investment in an area. Integration can reduce interconnection costs; equipment can be readily standardized, which reduces both the inventory of spare parts and the investment in training staff to maintain and operate it. Furthermore, facilities can be shared among services, thus reducing duplication of plant.[34] Facilities sharing is especially important, because the same digital electronic technologies are being used to provide an increasing variety of telecommunications services.[35]

Diseconomies of Scale

Networks can, however, become so large that they encounter diseconomies of scale.[36] Several factors account for this. First, the traffic generated per line may increase with the size of the network because more people can be called.[37] The higher the traffic per line, the greater is the quantity of switching equipment needed to process the traffic. Second, as the demand is met in the inner business districts of the main cities, further expansion takes place as new subscribers are added in the less dense outer areas, requiring longer and costlier local loops.[38] Third, growth in the number of subscribers increases the quantity of long-distance plant per subscriber.[39] This is because the ever-widening community of interests, both business and social, fostered by the expanded reach of communication and transport systems, results in significant increases in demand for long-distance service. This, combined with the fact that long-distance telephone traffic often grows faster than local traffic, results in increases in both circuit-kilometers and the quantity of long-distance switching equipment needed to process the increased flow of traffic. Finally, in most developing countries, a high proportion of subscribers is concentrated in large cities that require multiexchange networks. The requirement of interexchange circuits increases more than proportionally to the number of exchanges, since all exchanges must be interconnected with each other either directly or through intermediate (tandem) exchanges.

Operating entities can also become too large. For example, as networks expand, the existing organizational structures may become increasingly ineffective and unresponsive to the needs and circumstances of the regions or provinces served. Overgrown headquarters become unduly costly to house and service in the expensive urban centers where they are usually located, while regional staff have inadequate authority and incentives to deal with local matters and limited prospects for developing their careers. Limited authority to retire or reassign personnel made redundant by technological changes may slow the gains in labor productivity and aggravate problems of space, support facilities, management, and welfare. Managers preoccupied with meeting large outstanding demands for basic telephone service and extending the network throughout the country have little interest, energy, and resources to deal with emerging needs for more advanced services.

Given these potential diseconomies of scale, average cost per line may or may not decrease as the system grows very large.[40] The most likely situation is, however, that even if reductions in unit costs were to taper off along major routes and costs eventually began to increase in certain large centers, costs would continue to decrease in most parts of a developing country's telecommunications system (moreover, as networks become totally digital, the possibility that the costs per line will increase in the future becomes even less likely). The inherent potential for reducing the unit costs of individual pieces of equipment and network components through economies of scale can continue to be realized if this potential is kept in view while equipment and networks are being technically designed and engineered and while the procurement procedure is being determined for specific items. And, even if at some point the average cost per line begins to increase, the average cost per unit of traffic will almost certainly continue to decline.

Constraints on Achieving Economies of Technology and Scale

A number of factors constrain the pace of technological innovation and system growth and hence limit the extent to which related cost savings can be achieved. Lack of timely attention to developing human resources can more than offset the advantages of technological innovation. New technology is generally more complex than the technology in existing equipment. For example, digital telephone exchanges incorporate advanced electronic hardware and specialized software that require sophisticated skills to install and maintain and, consequently, need higher-caliber staff than did electromechanical equipment. Management of software and changes in network design have become more difficult and require higher levels of skill and training. While telecommunications enterprises in developing countries are often overstaffed with poorly trained, low-level personnel, these countries have a universal shortage of highly qualified technicians and engineers capable of specifying, installing, and servicing modern equipment. This has been a major constraint on system performance.

The adoption of modern technology is also often constrained by a shortage of capital. Insufficient funds for investment make it difficult for enterprises in developing countries to expand their facilities to meet large unmet demands and replace old equipment at the same time. Under pressure to extend services, managers often defer replacement of obsolete equipment, despite deteriorating performance and the increasing scarcity and high cost of spare parts. The accelerating pace of technological change in recent years, and the consequent reduction of the economic (as distinct from the physical) life of most new types of equipment, has exacerbated this dilemma.

A related issue is whether developing countries can realize substantial savings by purchasing the used equipment that is in good condition and available from industrial countries. To take advantage of rapid technological change, operating companies in industrial countries are replacing equipment well before the end of its physical service life. Some then attempt to sell the recovered equipment, particularly analog switches overtaken by digitalization, to developing countries, often at prices well below those of new equipment. Because such equipment appears to be less expensive than new modern equipment based on initial capital cost, countries short of capital funds may find procurement of such old-technology equipment attractive at first sight. However, the recovered equipment often needs to be reengineered, at considerable cost, to meet new, very different traffic patterns. It also requires much more building space than new equipment. Most important, old equipment imposes heavy penalties on future development of the network. The developing country will very quickly face exactly the same problems that caused the industrial country to replace the old technology: high costs of operation and maintenance, difficulties integrating into an expanding and technologically advancing network, inability to provide the modern services and facilities required by business and government customers, and difficulty and high cost of obtaining adequate spares for maintenance and components for expansion.

The timing of technological innovation can involve considerable uncertainties and risks, especially in the context of major shifts. Although the direction of technological change is foreseen years in advance, the pace at which such changes will materialize is harder to predict. The cost of telecommunications equipment embodying new technology is uncertain. This is partly because the cost of equipment and its components is highly sensitive to scale and to the rate at which new-technology equipment is produced, which depends, in turn, on demand. The price of initial supplies of equipment incorporating new technology is high. In the early stages of marketing such equipment, competition among suppliers is limited; also, to speed their recovery of the high costs of developing new technology and equipment, manufacturers attempt to load such costs onto the price of initial supplies of the equipment. The latest technology is often the best choice for developing countries choosing among the technology commercially available in world markets. However, there is a tradeoff between adopting the latest products quickly to avoid further investments in obsolescent equipment and waiting for these products to be proven in extensive field applications to avoid costly failures caused by equipment that has not been tested sufficiently.[41]

The factors that constrain technological innovation can also prevent developing countries from reaping the potential benefits of economies of scale. The extent to which economies of scale reduce the unit cost of providing telecommunications services depends on the speed with which the network expands. This is limited, in turn, by the availability of investment capital for large-scale development, the responsiveness of managers to customer demands, and the cost-effectiveness of procurement practices. In the least-developed countries, the ultimate constraint on the pace of expansion often is the operating entity's technical and managerial ability to implement large development programs and thereafter effectively maintain and operate the assets created.

Combined Effects of Technology and Scale on Cost and Price of Service

Technological change and economies of scale have produced sustained reductions in investment costs in developing countries. The average investment cost per line added under telecommunications programs partly financed by the World Bank declined, in current terms, from about $2,000 in the 1970s to between $1,400 and $1,700 in the 1980s. Thus the real costs were cut roughly in half every decade.

Cost reductions can be dramatic if the networks are properly designed and engineered and the procurement of required equipment is properly organized. Taiwan is an example of what an efficient telephone administration can achieve through rapid expansion, purchase of equipment at competitive prices in the world market, and adoption of correct least-cost technical and organizational solutions. In the twenty-one years between 1956 and 1977, the number of telephones increased from 37,000 to 1.2 million lines; by 1989, this figure had risen to 7.5 million. Between 1956 and 1977, the average annual growth rate was 19 percent; between 1967 and 1977, it was an astounding 25 percent a year. Strong growth has continued, registering over 10 percent a year for the period from 1985 to 1990. At the same time, labor productivity, measured as the number of employees per 1,000 lines, improved from 114 in 1956 to less than 8 in 1981 and then to less than 5 in 1989. Rapidly decreasing cost per line kept tariff increases well below the general level of inflation. The 1956 tariffs, which were already low by world standards, dropped by half in real terms by 1977, while the quality of service greatly improved and the number of subscribers who could be reached by telephone in the country increased by a factor of thirty.[42] Tariffs are still being reduced; in 1989, local service rates were effectively cut by about $12.5 million when the areas falling into the local basic rate were enlarged, local installation fees were reduced from $623 to $467, and prices charged for other services, such as international telex, fell also. Notwithstanding these low and falling tariffs, the rapidly growing telephone system generated internally more cash than the operating entity needed to invest, and the telecommunications sector transferred increasing proportions of revenue to the government in the form of taxes. In 1989, out of total net revenues of $862.7 million, about 63 percent, or $531.4 million, was returned to the national treasury.

Because of economies of scale and the adoption of new technology, there can be little doubt that an operating entity with a large and rapidly growing system should be able, all other things being equal, to offer its subscribers more service for a given amount of money than an entity with a smaller system and a network that is growing more slowly. Table 2-1 shows that in the 1980s the annual telephone bill of a hypothetical business subscriber (making 5,000 local calls and 200 long-distance calls a year) was roughly the same in Togo, which had only about 10,000 subscribers and where the longest call did not exceed 500 kilometers, as the bill of a similar subscriber in India who

Table 2-1. *Business Telephone Expenditures under Selected Model Assumptions in Selected Developing Countries*
(U.S. dollars unless otherwise indicated)

Country	Annualized connect and deposit charge[a]	Annual line rental charge	5,000 local calls[b]	200 trunk calls[c]	Total expenditure under the model	Thousands of main lines in service
Bangladesh	45.76	21.89	258.40	231.65	557.69	130[d]
China	84.82	38.88	41.04[e]	97.20	261.94	5,550
Colombia	67.06	18.00	70.00	42.00	197.06	2,070
Côte d'Ivoire	26.97	60.32	754.00	301.60	1,142.89	38
Ecuador	26.23	14.40	25.00	36.00	101.63	450
Guatemala	67.96	144.00	74.67[e]	90.00	376.62	138
Hungary	363.86	63.94	295.26	142.04	865.10	858
India	87.87	132.66	316.58[e]	753.75	1,290.85	3,488
Indonesia	40.91	23.06	205.88	370.58	640.42	829
Jordan	89.42	108.00	270.00[e]	180.00	647.42	203
Laos	242.92	1,200.00	5,000.00	2,400.00	8,842.92	7
Mexico	77.20	59.05	570.00	205.79	912.04	3,774
Morocco	6.29	29.97	277.50	133.20	446.96	286
Nepal	39.33	4.52	184.73	180.96	409.55	30
Pakistan	17.88	21.60	180.00	259.20	478.68	637
Papua New Guinea	8.57	74.75	805.00	386.40	1,274.72	31
Senegal	35.41	62.37	742.50	178.20	1,018.48	29
Togo	139.04	108.00	750.00	180.00	1,177.04	10
Uganda	3.78	10.80	350.00	504.00	868.58	28

Note: Tariff schedules date from 1982 to 1990.

a. Sum of connection charges and deposit, if any, amortized over a ten-year period at 8 percent interest.

b. Three minutes if timed.

c. Priced as three-minute calls of distances between about 200 and 300 kilometers.

d. Estimated.

e. Free calls: In Guatemala, rental includes 400 local call impulses a month; in India, 1,650 free calls a year; in Jordan, 2,000 free six-minute calls a year; in China, 60 free calls a month for residences and 100 for businesses.

Source: World Bank data and ITU *Yearbook of Public Telecommunication Statistics* (1990).

could reach 3.5 million subscribers and call places several thousand
kilometers away. Similarly, table 2-2 shows that a hypothetical residential subscriber (making 1,500 local and 10 long-distance calls each
year) would have to pay annually $870 in Uganda, with 28,000 lines,
less than $200 in various countries with several times as many lines,
and only about $100 in the People's Republic of China, with over 5
million lines.

Differences among countries in costs and tariffs generally cannot,
however, be attributed only to differences in technology and scale.

Table 2-2. *Residential Telephone Expenditures under Selected Model
Assumptions in Selected Developing Countries*
(U.S. dollars unless otherwise indicated)

Country	Annualized connect and deposit charge[a]	Annual line rental charge	1,500 local calls[b]	10 trunk calls[c]	Total expenditures under the model	Thousands of main lines in service
Bangladesh	45.76	21.89	77.52	11.58	156.75	130[d]
China	84.64	24.62	8.42[e]	4.86	122.55	5,550
Colombia	37.26	6.00	21.00	2.10	66.36	2,070
Côte d'Ivoire	26.97	60.32	226.20	15.08	328.57	38
Ecuador	13.11	3.60	4.50	1.08	22.29	450
Guatemala	67.96	48.00	0.00[e]	4.50	120.46	138
Hungary	36.33	21.31	66.60	5.33	129.57	858
India	87.87	132.66	65.66[e]	24.12	310.31	3,488
Indonesia	40.91	23.06	61.76	18.53	144.26	829
Jordan	44.71	72.00	0.00[e]	9.00	125.71	203
Laos	242.92	1,200.00	1,500.00	120.00	3,062.92	7
Mexico	77.20	55.04	166.50	10.29	309.03	3,774
Morocco	6.29	29.97	83.25	6.66	126.17	286
Nepal	39.33	4.52	52.78	9.05	105.69	30
Pakistan	17.88	21.60	54.00	12.96	106.44	637
Papua New Guinea	8.57	74.75	241.50	19.32	344.14	31
Senegal	24.34	62.37	222.75	8.91	318.37	29
Togo	60.80	108.00	225.00	9.00	402.80	10
Uganda	3.78	10.80	350.00	504.00	868.58	28

Note: Tariff schedules date from 1982 to 1990.

a. Sum of connection charges and deposit, if any, amortized over a ten-year period at 8
percent interest.

b. Three minutes if timed.

c. Priced as three-minute calls of distances between about 200 and 300 kilometers.

d. Estimated.

e. Free calls: In Guatemala, rental includes 400 local call impulses a month; in India,
1,650 free calls a year; in Jordan, 2,000 free six-minute calls a year; in China, 60 free calls a
month for residences and 100 for businesses.

Source: World Bank data and ITU, *Yearbook of Public Telecommunication Statistics* (1990).

Many other factors complicate cross-country comparisons, including organizational structure and management efficiency, configuration and stage of network development, effectiveness of operation and maintenance, economy and efficiency in procurement, cost and productivity of human resources, and government policies on tariffs, taxation, financing, and foreign exchange. In particular, no consistent relation exists between the number of lines in a given system and the price charged for service. For example, for eighteen countries analyzed in 1982, the correlation coefficients between the number of lines in service and the annual bill of hypothetical business and residential subscribers were not significantly different from zero.[43]

The following chapter discusses the major reasons for this surprising result: sector management, organization and financing, and government policy toward the sector.

Notes

1. Strictly speaking, economies of scale are said to exist when proportional increases in all inputs yield an increase in output that is larger than proportional.

2. It is assumed that the reader has at least an elementary knowledge of the technical characteristics of telecommunications systems and the main types of general solutions and equipment used. For an introduction to these topics, see Green (1986); Chapuis and Joel (1990); and Martin (1991). For an excellent history of the early evolution of telephone switching systems, see Chapuis (1982).

3. "Real terms" refers to prices expressed in a currency of constant value, for example, after discounting for world inflation. The real cost reductions cited in this chapter assume an average dollar rate of inflation of about 5 percent a year. For example, over a ten-year period, a price that remains unchanged in current dollars decreases in real terms to about 60 percent of the original price.

4. One advantage is that these systems permit customers to be served farther from exchanges, which can, in turn, be located in more populous areas with better access to power and skilled staff. Basic systems provide capacity for 100 or more subscribers, who share about twenty-eight speech channels. Larger systems, with up to sixty channels, allow even greater service coverage at lower costs per subscriber.

5. Implementation and Management Group Party, Ltd. (1980).

6. Mack and Lee (1989). Besides solar cells, low-cost power can be generated for rural telephony using alcohol-burning generators that convert energy and compact wind turbines.

7. At 2,000 Mb/s, a pair of fibers (one that transmits and one that receives) can accommodate about 30,000 voice frequency channels. The actual bandwidth of a particular cable installation depends on its length, since capacity is reduced as the cable is lengthened. See Green (1986).

8. Flamm, (1989) p. 20.

9. Fiber is more expensive than copper in, among others, the area of terminal equipment and installation. For a long, high-capacity trunk system, this disadvantage is offset by some cost advantages; for a small system with many users, such as a wired office building, using fiber still costs much more than using copper. See Simpson (1990).

10. For a comparative cost analysis of satellite, optic, and microwave transmission media, see Polishuk, Guenther, and Lawlor (1987).

11. A COMSAT study on the future of satellite communications predicts that the net annual costs of circuits for satellite use will continue to decline, falling to somewhere around $2,000 by the year 2000. See Crockett (1989).

12. A VSAT satellite distribution system also requires the installation of a hub; either electronic equipment costing approximately $250,000 must be added to an existing master satellite antenna or a new antenna costing at least $1 million must be installed.

13. These are rough ranges of prices paid for switches purchased through international competitive bidding. Prices vary considerably depending on the method of procurement and financing, the size of the exchanges (discussed later under economies of scale), and other factors.

14. Despite being obsolete, some electromechanical equipment was still being manufactured in the early 1990s in a few developing countries, including India and the People's Republic of China.

15. For a more detailed discussion of the effects of computerization of telecommunications switching on cost trends, see Flamm, (1989).

16. A digital technique invented in the 1940s, pulse code modulation only became commercially viable with the advent of integrated circuits.

17. This happens as well in industrial countries with sophisticated installations and highly skilled staff. In June of 1991, for example, a rash of large-scale breakdowns occurred in various locations throughout the United States; in one instance, 5 million business and residential telephone subscribers in a congested exchange in the Washington, D.C., metropolitan area were left for several hours without service due to a software malfunction (caused by a few lines of code buried within millions of lines of programming). Damaged fiber-optic cables have also shut down service recently, affecting increasingly large numbers of users in highly congested areas. For example, in January 1991, a fiber-optic cable was accidentally cut by a maintenance worker, and this cut all long-distance service provided by a major carrier between New York City and the rest of the country. This event nearly shut down all three New York City airports along with many other critical operations for an entire day.

18. ISDN is the result of an international agreement among cooperating countries, sponsored by CCITT in Geneva in 1984. It sets electronic parameters and standards and provides a guide for subsequent investment in existing national networks so that within the next ten to twenty years, national networks could become ISDN networks. The goal is to allow connected users to transmit and receive the full variety of communications services (ranging from voice telephone to cable television to data communications) over what used to be just a telephone network. The cost of converting to ISDN is enormous, and universal implementation for all customers in the public network is by no means certain or even feasible. Many feel that it would only benefit large users and affluent individuals, at the expense of those unable to pay its high costs. See Green (1986) chap. 25 and Krechmer (1989).

19. Despite the expectations of advocates, satellite technology has not become a cost-effective, general alternative to conventional wire, cable, and radio technologies for providing basic domestic services. Several studies have compared the costs of satellite systems in rural areas with those of other technologies, such as radio and cellular, that provide mobile service. In Alaska, Hills (1988) found that when compared with existing options of single sideband and microwave/very high-frequency radio systems, a proposed mobile satellite service was generally more costly, but offered functional advantages. Another study (Khadem, 1987) looked at the cost of cellular and mobile satellite systems for applications in developing countries and found that the cost per subscriber of a hypothetical cellular system would be about $2,500 (not counting administration and maintenance), while that of an equivalent satellite system would be $5,400. The main difference lies in the range of coverage, which was quite limited for the cellular system. Investment sufficient to give the cellular sys-

tem as much range as the satellite system would make the costs of the two systems comparable.

20. Wellenius and others (1989) and Wellenius and Stern (1991).

21. More detailed summaries of several relevant econometric studies can be found in Meyer and others (1980), pp. 125–34 and app. D and Littlechild (1979) pp. 49–52. The engineering studies are described in Meyer and others pp. 199–213 and Littlechild pp. 53–73.

22. For example, Meyer and others (1980) report that the Federal Communications Commission in the United States used a study prepared by a firm called T&E and concluded that the case for economies of scale had not been conclusively proven. Then, AT&T employed the Stanford Research Institute to review the T&E study and produce its own conclusions, which were that the economies of scale were probably in a range of between 1.1 and 1.25. See Meyer and others (1980) p. 131.

23. One of the problems encountered in econometric studies of scale economies is the difficulty of netting out the effects of technological change. Meyer notes that "studies where comparisons are possible suggest that about one-third of the economies of scale which would be estimated if the impacts of technological change were not taken into account could in fact be due to technological change." See Meyer and others (1980) p. 126. Another study on productivity gains at Bell Canada for the period 1958–80 found annual increases of about 3.5 percent overall. In this case, one-quarter of the gain was judged to be the result of technological change, and three-quarters were the result of economies of scale arising from growing demand. See Kiss (1983).

24. The numerical estimates presented are taken partly from Meyer and others (1980) and Littlechild (1979).

25. The allocation of costs to local and to long-distance facilities is an important factor in determining economies of scale where possibilities exist for bypass. The high costs of local exchange facilities are common to the supply of local and long-distance calls and value added services. Assigning these facilities entirely to local telephone services can raise the price so high that access is placed beyond the reach of many users. Conversely, assigning them entirely to long-distance and value added services can stimulate large users to set up private networks to bypass high tariffs. Bypass threatens the economies of common supply in a single system. See ITU (1989) p. 20.

26. Even larger economies of scale apply when the costs of initial civil works (such as roads and sites) and power supply are included.

27. Federal-Provincial-Territorial Task Force on Telecommunications, Canada (1988).

28. Figure 2-2 is based on Erlang's formula for lost-call systems with full availability.

29. Also, whereas the unit cost per termination of a purely long-distance electromechanical analog trunk exchange could be as much as three times higher than the cost per termination of a subscriber line in a local exchange, digital exchanges essentially do not differentiate among types of circuits.

30. This price difference is too large to be attributed wholly to the smaller average size of the exchanges in the second lot.

31. The cost of such financing can, however, be considerably lower than that of multilateral and commercial sources, so the net effect on total cost has to be examined case by case. One way to deal with this uncertainty is to invite bids for price (of equipment) and terms (of financing) and to award the contract to the bid that reflects the lowest present value of total cost including the cost of financing.

32. Many instances demonstrate the organizational economies that can result from increased size of the system. An East Asian telecommunications entity with about 300,000 main telephone lines has set up a fundamental planning unit, whose functions revolve around long-range planning, technology monitoring, land acquisition, and ascertaining that

telecommunications investment programs and pricing policies complement national development goals and policies in other sectors. The unit is staffed by about ten senior professionals (five engineers, two financial analysts, and three economists) and support personnel. This same unit will, however, be able to handle the entity's planning needs when the system increases to more than 1 million main lines.

33. Staff ratios as high as 100 and more employees per 1,000 telephone lines are found in some developing countries, including some with very large systems. Whereas this extremely low labor productivity relates to antiquated technology (for example, the number of operators and clerical staff is excessive because switching and billing are not automated), it also reflects employment obligations, constraints on relocating or terminating redundant personnel, lack of performance incentives, and other factors.

34. For example, part of the telephone cable network could connect telex and data subscribers, and long-distance circuits could carry telegraph, telex, data, radio, and television program signals.

35. The integration of services and facilities within and among areas involves economies of scope in addition to economies of scale. Although taken for granted in most cases, integration within areas is by no means universal. An extreme example existed in the late 1970s in an East Asian country and has subsequently been only partly remedied by policy reform. At that time, two of the country's more than sixty telephone companies overlapped in six exchange areas in the capital city, each providing completely separate exchanges and distribution networks, with only limited facilities to communicate among them; the same two entities had duplicate long-distance networks throughout part of the country, with no interconnections, and two other companies also operated long-distance service, partly duplicating these two networks.

36. Some evidence of diseconomies of scale has been observed in a comparison of the average total costs per telephone line in sixteen urban districts in India. Costs tended on average to decrease up to about 20,000 lines and then to level off up to about 40,000 or 45,000 lines. Of the four telephone districts with more than 45,000 lines, three had higher costs than those with between 20,000 and 45,000 lines, while the largest, Bombay, showed little difference. See Kaul (1980).

37. This does not apply, of course, to situations in which expanded investment in an underprovisioned and highly congested system actually reduces the amount of traffic per line.

38. In recent years this problem has been partly offset through the use of digital remote stand-alone line units, which can be located in the center of subscriber clusters located far from the main telephone exchanges.

39. It has also been argued that, at least in North American systems, full economics of scale on long-distance transmission routes are attained at a level of output that is small compared with the market. Waverman (1975) chap. 7.

40. Cross-country comparisons of the costs per line of telecommunications equipment yield meaningless results. Dividing estimated total gross investment in telecommunications plant by the total number of telephones for several countries yields values ranging from nearly $100 to well over $2,000, with a random spread in between that is unrelated to the size of the system. What these numbers reveal, in part, is nothing more than differences in accounting practices among countries, particularly with regard to depreciation and asset revaluation. For the raw data, see Fargo (1982). Of course, management, planning efficiency, and procurement practices also affect costs per line. For example, in one large developing country in Africa, the cost of equipment per installed line runs into many thousands of dollars.

41. The latest proven technology is, however, designed and manufactured mainly for the markets of industrial countries. Equipment is rarely aimed specifically at the service and traffic requirements, climate, power supply, human resource skills, and other features of de-

veloping countries. Thus developing countries may have to pay for unnecessarily sophisticated equipment and for facilities that will not be used extensively for quite some time. Furthermore, this equipment may require a supporting physical and human infrastructure that is not readily available. The high and still growing cost of research and development, however, and the relative size of markets make it unlikely that this situation will change for major equipment in the near future. Niches may be identified, and manufacturers in newly industrializing countries such as Brazil, India, and Korea are, however, beginning to turn out products especially addressed to the conditions of developing countries. See Ivanek, Nulty, and Holcer (1991) and Mody (1989).

42. For example, the monthly rental of a residential telephone in a large exchange area including a number of local calls, increased between 1956 and 1977 from NT$48 to NT$100 ($2.50). With a correction for inflation (the consumer price index rose more than four times above the 1956 level), the 1977 residential rental was actually half the 1956 rental. Unit call charges changed at roughly the same pace, and in 1977, a local call cost only $0.018, one of the world's lowest call charges.

43. The correlation coefficient measures the strength of association between changes of two (or more) variables. A correlation of 0.0 means that the variables move independently of one another, whereas a correlation coefficient of plus (or minus) 1.0 means that both variables move exactly together (or exactly in opposition). The level of significance measures the probability that a correlation coefficient as high as or higher than the coefficient measured could be obtained if the variables moved purely at random. Correlation analysis is an important tool when it is not self-evident whether the changes in variables are related or not.

Chapter 3

Sector Organization, Management, and Financing

IT IS DIFFICULT TO DISCUSS the organization and management of national or international telecommunications activities without reference to the organization or functioning of other communication activities and the rules and regulations under which they operate. Telephone, telegraph, telex, data and facsimile transmission, postal messages, postal packages, money orders, messenger and other services, and physical travel serve to varying degrees as both complements to and substitutes for each other. Numerous factors affect the combinations and types of communication networks that emerge both within and between countries, such as the extent to which these activities are regulated by governments; are undertaken primarily by governments or by private enterprise; are limited by shortages of foreign exchange; are supported by other local infrastructure (roads and electric power); are supported by national training or education programs, which result in literate or skilled local labor; and are encouraged or held back by direct or indirect government policies, duties, subsidies, competition, and so forth. Nevertheless, given the scope of this book, this chapter focuses primarily on only a portion of the highly diverse communication sector—the organization, management, and financing of telecommunications services.

Telecommunications Organization in Developing Countries

In most developing countries, the telecommunications sector comprises (a) one, two, or occasionally more operating entities, which provide monopoly public and leased telecommunications services; (b) various networks that meet the needs of large government and other public entities (such as the armed forces, police, railways, and public utilities); (c) private networks that have been set up to meet specialized user needs or to supply additional service because the primary telecommunications operating company cannot provide adequate coverage or quality of service; and (d) one or more bodies that perform technical and economic regulatory functions.[1]

Although a trend toward liberalization and privatization can be observed globally, most public telecommunications services in developing countries are still provided by entities that are partly or wholly owned by the state. In a number of instances, postal and telecommunications services are grouped under the same organizational structure. These government-owned telecommunications operating entities are organized in various ways; some form part of a conventional government department (as in Algeria, Cameroon, and India), some are semi-independent, state-owned enterprises (as in Ecuador, Indonesia, Jordan, Malaysia, Thailand, Togo, and Uganda), and some are state-owned corporations organized under company law (as was the case in Argentina, Chile, and Mexico before privatization occurred between 1987 and 1990).

Under these arrangements, three important constraints limit the development of telecommunications: the operating entities' lack of autonomy from the government, their inadequate internal organization and management, and a shortage of funds for investment.[2]

The Autonomy Issue

Telecommunications are a rapidly evolving sector, in which significant technological change is taking place, costs are falling, and the variety of options available for new and innovative services is increasing. Three additional factors are particularly relevant for developing countries: (a) the excess demand for even basic telephone service is large, (b) government policymakers are increasingly aware that the economic development of provincial and rural areas requires at least a minimum level of public access to telecommunications facilities, and (c) advanced data, facsimile, electronic mail, and information systems hold the potential for facilitating large productivity increases in the modern sectors of the economy. This situation means that the management of telecommunications entities must be professionally disciplined as well as flexible, responsive, and alert to evolving needs and opportunities. Hence, a certain degree of financial and management autonomy is essential for public telecommunications entities to perform well.

Independence from Government

Organizing such entities as corporations with their own board of directors and separating telecommunications from the postal service may appear to be a straightforward way to achieve significant autonomy. Experience shows, however, that adequate autonomy can sometimes also be attained even when telecommunications entities are closely tied to government, such as being part of a government department, provided appropriate organizational and financial measures are implemented.[3]

Such measures must free the telecommunications entity from day-to-day government interference and should provide for continuity of management despite political changes in the government. To operate efficiently, the entity should not have to secure government approval for normal technical, procurement, and expenditure decisions; top corporate management, following national development policy and regulatory guidelines, should be able to make these decisions. Specifically, this situation could be brought about by (a) streamlining the relations with the few government agencies that should be legitimately interested in influencing the sector's long-run plans; (b) expediting procedures for approving investment programs, which should have a multiyear horizon and be only generally specified to allow management flexibility in the details of implementation; (c) allowing freedom to operate under general guidelines, which specify tariff levels that allow full recovery of costs, generate substantial funds for new investment, and promote efficient allocation of resources; (d) simplifying the procedures for periodically revising tariffs where general price inflation requires frequent adjustments; (e) giving the operating entities authority to collect bills from all users and to disconnect nonpaying subscribers (this should explicitly include all government subscribers); (f) requiring that accounts be maintained on a commercial basis; (g) ensuring that internally generated funds are available for investment under approved medium-term development programs, with the operating companies either retaining these funds or easily recovering them when required;[4] and (h) allowing operating entities to set salaries, wages, and other benefits so that they can be competitive employers and attract and retain qualified staff at all levels.

The government should expect management to pursue the objectives and follow the policies laid down by the government through the autonomous operating entity's board of directors and to manage the operating company in a professional and cost-conscious way. Government should also expect the entity to pay normal taxes and duties, market rates of interest on all new debt (including government debt), and dividends on government equity at a rate that would attract equity capital to private enterprises with similar risk profiles.

Independence from the Postal Service

In many countries, the provision of telecommunications services is organizationally linked with that of postal services, because of the early history of the sector. In the middle and late 1800s, when telegraph and telephone services were first being developed, many European countries treated them as an extension of postal service communication. Several important colonial powers then passed on the combined Posts and Telecommunications (P&T) Department or Office des Posts et Telecommunications (OPT) type of organization to their colonial administrations.

In a more modern context, however, several arguments can be made for separating the management of postal and telecommunications services: (a) although advanced technology and management systems play a big role in the provision of postal services, in general, postal operations are highly labor intensive, whereas telecommunications operations are capital intensive with rapidly decreasing labor inputs; (b) the provision of telecommunications services requires a much higher proportion of skilled labor and professionals than does the provision of postal services—lumping them together in a developing country often contributes to telecommunications salaries that are too low to retain qualified staff or to personnel regulations that are overly rigid; and (c) the contrasts between the technologically dynamic and rapidly evolving telecommunications sector and the more mature labor- and physical-transport-dominated postal sector are so great that the same organizational structure and management style cannot plan for and implement both services efficiently. Given such considerations, a growing number of countries have partitioned, or are beginning to consider partitioning, their postal and telecommunications organization and are establishing each service as an independent entity. The breaking up of the British Post Office in 1981 into an independent British Telecom and the Post Office is a classical example.

Of course, in a developing country there may still be some advantages to an organization that combines posts and telecommunications, particularly in more isolated areas and small towns in which the local post office building may house not only postal mail services but also savings bank facilities and a telephone public call office (PCO). However, the P&T does not have to be combined organizationally for a combined facility to function effectively. Several countries provide PCO services in rural areas without undue problems by having the separate telecommunications entity pay the post office or small local shop or cafe a commission for housing and operating a PCO. It could be argued that in the future, when sophisticated electronic text systems are more widely available in developing countries, postal and telecommunications services should at least partly be administered by the same organization, since they will converge technically. Although this suggestion has some validity, it does not seem desirable to tolerate the major operational inefficiencies apparent in many developing countries in anticipation of an uncertain development in the future. Also, when electronic text is introduced outside large urban areas, at least initially, the electronic transmission function could be handled by the telecommunications entity and labor-intensive distribution handled by the postal service. It is even possible that some competition between the two entities providing text services could stimulate more efficient management and service responsiveness (for a more detailed discussion of electronic mail and the interaction of postal and telecommunications services, see appendix B). All in all, at least in developing countries, the case can be made that separating posts from telecommunications could, organizationally as well as financially, streamline the management structure, and hence increase the long-term efficiency, of both the postal and telecommunications sectors.[5] If using telecommunications revenue to subsidize postal services was thought to be desirable, this could still be done; it would just have to be done on a more explicit basis.

Internal Organization and Management of Operating Entities

A second major problem for the telecommunications sector in developing countries is inadequate organization, management, and staffing of the operating entities. The following weaknesses are often observed. The entity's organizational structure is frequently inadequate for the size of the telecommunications development effort required, or for the nature of the business (as noted above this is sometimes the result of combining the functions and control of postal and telecommunications and the inherent organizational and operational problems associated with such an arrangement). Job descriptions and requirements, service standards and staffing norms, separation of responsibilities, and lines of delegation of authority are not well defined. Administrative and financial controls are loose or nonexistent. Management is timid and lacks objectives, goals, and accountability. Financial management has little influence on planning and day-to-day decisions. Sufficiently comprehensive management information systems are not in place. No one unit is responsible for long-range planning and economic analysis. Commercial forms of accounting often are not used or do not produce timely signals for decisionmaking and for assessing performance. Billing and collection of receivables are slow and not well monitored, which places an unnecessary financial burden on the organization. Little thought is given to maintaining strong, central engineering planning or to coordinating and supervising projects. Finally, decentralization of administrative work and technical, operational, and billing and collection responsibilities is unduly delayed as the size and complexity of the organization increases.

Staffing Problems

The key to building a capable operating entity is adequate staffing. Many telecommunications organizations in developing countries have yet to develop and maintain an effective and expanding system for training all levels of engineering, financial, and administrative personnel. Effective training programs increase productivity and in the long term consolidate a stable and qualified cadre of senior staff who are supported by middle-level personnel from which future executives can be promoted.

Good telecommunications training centers exist in several developing countries, and many of them were set up with technical assistance from the ITU and financed partly by the United Nations Development Programme. Typically, technical personnel up to the level of technician are trained in national or regional centers and on the job. Junior professional engineers are also educated in these centers or in local university programs. Education abroad (other than in regional centers) is typically reserved for personnel at selected professional levels or with narrow technical specialties.[6] Training problems generally include inadequate coordination between training centers and the operating companies and lack of attention given to training in the important areas of administration, finance, and accounting.

An especially difficult problem is faced by some of the smaller, poorer developing countries, which depend on foreign personnel and have virtually no sources for recruiting new staff with the level of general education needed to undergo training. Nevertheless, it is crucial to transfer all posts to nationals as soon as sufficiently competent replacements are available. Dependence on foreign staff has on occasion led to unnecessary overstaffing at the most senior levels and to acute personnel crises in several countries, sometimes resulting in the virtual collapse of telecommunications operations. A comprehensive strategy for recruitment, training, further education, and career development of staff is needed in such countries, often with a ten- to fifteen-year time horizon. With such a plan the rapidly rising demand for middle-level personnel could soon be met with adequate local staff, who in turn gradually take over all posts held by temporary foreign experts.[7]

A Management Checklist

The following is a checklist of selected items to which management and policymakers might refer in identifying problem areas and evaluating steps that might be taken to improve the internal management of monopoly operating entities.

Organization. Does the entity have adequate autonomy? Is it separated from the postal service? Does it operate on a commercial basis? Does it have appropriate medium- and long-term objectives or guidelines relating to the services offered, access to services, quality of services, finances, tariff policy, and so forth?

Management and control. Does the operating entity have an adequate management information system that facilitates effective and responsible control of current operations, stores, maintenance problems, financial affairs, and long-term planning and programming? Are adequate statistics collected on traffic, faults, outages, equipment performance, management efficiency, and so forth? Does management have difficulty controlling or coordinating among departments, divisions, or units? Are subscribers connected at a reasonably high rate when exchanges are ready to be cut over? Do adequate procedures exist for maintaining, testing, and checking the quality of work performed? Are there enough qualified professionals, managers, technicians, and accounting and finance officers? Are job descriptions adequate at all levels? Can the existing staff cope with the additional responsibilities of implementing an expansion program and managing the expanded entity? Are salaries adequate to retain qualified staff? Are policies relating to staff advancement and promotion appropriate, are the accompanying rules and regulations sufficiently flexible, and are incentives for achievement built into the system? Is the level of general staff excessive? What use is made of expatriates, of consultants? Are present arrangements suitable for an expanded work load? Is technical assistance needed in the short run? Does management make an adequate effort to keep the public informed of major developments and to promote public support for the sector?

Planning. Are economic analyses of expressed and hidden demand and of the telecommunications needs of both urban and rural areas undertaken, and are the results of such analyses adequately conveyed to officials at the finance and planning ministries and to the public? Are the telecommunications development plans closely coordinated with government plans and programs in other sectors as well as with private plans? Do updated national numbering, charging, switching, signaling, and transmission plans exist? Are they judged to be sound in light of the country's requirements and its technological and financial resources? Do regional development plans exist? Do physical development plans exist for the capital city and major metropolitan areas? Are the service targets reasonable given current service levels and national goals for access and quality of service? How are the current (one to four years) programs prepared? Are the development of services and service priorities balanced properly? Is exchange equipment coordinated adequately with external plant works? What is the basis of costing in the works programs, and are there accurate and updated records of unit costs? Are the costs of various works reasonable? Is preparing civil works a bottleneck? Have the major works been subject to economic comparisons of alternatives (for example, coaxial cable versus microwave and optical fiber for a major trunk bearer)? Is the discount rate appropriate? Does it (or should it) reflect capital scarcity? Do plant practices specify the forward provisioning periods for the various classes of plant, for example, main, secondary, and tertiary distribution cable, junction cable, exchange equipment, multiplex equipment? Are the provisioning periods reasonable with regard to growth rates, availability of capital, and prevailing costs of equipment and labor?

Accounting system. Is the accounting system on a commercial basis, and, if relevant, are telecommunications accounts kept separate from postal accounts? Is the commercial accounting system adequate to keep management informed and able to foresee trends and possible problems? Are there adequate internal financial controls (control of cash receipts and payments, inventories, plant retirements, sale of property, related materials, and so forth)? Is the costing or cost control system adequate? What accounting practices are being followed, and are there any problems relating to depreciation, the transfer of plant investment from "under construction" to "in service," and the handling of foreign exchange losses, bonuses, and so forth? Are pension funding arrangements adequate? Are there sufficient procedures for writing off bad debts?

Billing and collection. Are billing arrangements satisfactory? What is the billing cycle, and is the accounts receivable position satisfactory for the length of the billing cycle? Could billing be expedited through increased computerization? How much would this cost, and what would it save in time and cash on hand? How are collections made? Are there disputed account problems, and how can these be resolved? Are there satisfactory procedures for disconnecting nonpaying subscribers, including all government and other public sector subscribers? For revenues not collected directly by the entity, how are charges set, and how do they reach the operating entity?

Inventories and stores. Is the inventory or stores control system satisfactory? Are inventory levels satisfactory? Is there an adequate system for replenishing inventories quickly, for checking whether turnover is reasonable in relation to the scale of operations (and of construction), and for identifying and disposing of obsolete material?

Training. Have personnel development policies and practices been recently evaluated in the light of current and forecast requirements of the operating entity and of the sector? How do training programs and facilities within the entity and within the sector compare to the needs? Has this topic been adequately examined by national or international agencies or experts (ITU or consulting firms)? What action might be initiated to deal with any personnel shortages likely to affect current or future development within the sector?

Audit. Are present arrangements for external and internal audits satisfactory? Are the nature and extent of the audits appropriate? Is there any reason to question the independence of the external auditors, the adequacy of the qualifications and experience of the personnel conducting the audit, or the adequacy of the procedures followed? Do the audit reports contain all the financial statements and supplementary information needed for analysis and comparisons? Do external auditors systematically advise management of weaknesses in the accounting system that should be improved? When used, is government audit adequate, or should it be supplemented by other audits? What is a reasonable period for submitting audited financial statements? Are internal audits and inventories done systematically? Are internal audits adequate? To whom does the internal audit section report?

Tariff policy and financial plans. Does the telecommunications operating entity generate sufficient resources from users of services to cover operation and maintenance costs as well as debt and interest payments, to generate a reasonable return on assets and pay dividends on equity, and to cover a reasonable proportion of the costs of expanding the system in the future? What proportion of cash is expected to be generated internally for the next three to five years? Is it satisfactory? Is the amount of net internal cash generation influenced significantly by the capital structure and terms of debt financing (unusually low or high debt-equity ratios, easy or harsh debt repayment terms, and use of supplier credits)? If the percentage of self-financing for capital requirements is below 35 or 40 percent, is this level acceptable in view of the size of the work program, recent tariff action, expected tariff revisions, or expected level of cash generated internally after the full revenues from the completed work program are realized? Do special factors, such as taxes, bonuses, dividends, or other payments to government or exceptional customs duties, depress the percentage of cash generated internally? If excess demand exists, is some form of price rationing used to encourage business and government priority access or use? Are peak-period pricing (local and long-distance calls) and toll ticketing practiced? Do long-distance charges increase too rapidly with distance? Do tariffs for specific types of service reflect the cost of those services? Are any deviations from pricing according to the cost of service justified on grounds of promoting a more efficient or equitable use of, or access to, service? Has the incremental cost of expanding or adding specific services been studied? Are telecommunications tariffs used partly to supplement general government taxes, to subsidize postal services, and so forth? What are the present implications of this, and what are the longer-term goals?

The Financial Constraint

Another major factor that directly constrains the organization and management of the telecommunications sector in many developing countries is the scarcity of funds for capital investment. This constraint is especially significant because the capital requirements of the sector are relatively high; the ratio of capital to output (measured by revenue) in telecommunications has been estimated to be about 3:1.[8] Likewise, the incremental capital-output ratio, which describes the relation between new investment (incremental requirements for additional capital) and expanded output, is large compared with the value of the same ratio for most other industries. Of course, such ratios give only the most general indications of capital requirements; capital requirements per unit of system expansion vary widely, depending on program composition, geographic characteristics, cost of local inputs, conditions of procurement, and a host of other factors.[9]

In the 1970s, developing countries invested about $3 billion annually. In the 1980s, this figure more than doubled to reach about $7 billion; by the late 1980s, it had reached some $12 billion (all figures are in 1988 U.S. dollars). As a tentative estimate, developing countries are likely to require about $25 billion a year in the 1990s if they are to catch up with the unmet demand for basic telephone service by the year 2000 (a rather modest target). In real terms, this is more than three times the investment level achieved in the 1980s and eight times that of the 1970s. Additional funds will be needed to meet the still relatively small but very fast-growing demand for the advanced services increasingly required by modern economic sectors.[10]

Figure 3-1 presents the approximate breakdown of telecommunications financing for developing countries during the 1980s by the main source of funds.[11] About 60 percent of investments were financed by the operating entities themselves through internally generated cash (mainly retained earnings and provisions for depreciation). It is clear, therefore, that despite the large capital investments required for telecommunications operations and expansion, a lack of local currency should not be a constraint on this process for developing countries. As outlined in chapter 1, monopoly telecommunications entities can easily generate large financial surpluses in the local currency. With correct pricing policies, the full costs of providing telecommunications services, including the cost of capital, can be recovered from tariffs, and a large proportion of the funds required for subsequent improvement and expansion can be generated internally (the pricing

Figure 3-1. *Sources of Funds for Telecommunications Investment in the Developing World, 1980s*

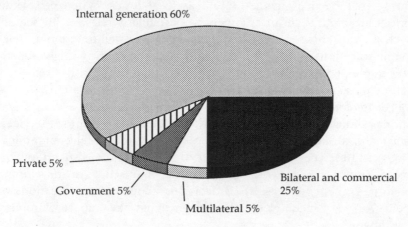

Internal generation 60%

Private 5%

Government 5%

Multilateral 5%

Bilateral and commercial 25%

Source: Wellenius (1990b).

of telecommunications services is discussed in chapters 13, 14, and 15).

In the absence of sufficient local capacity for manufacturing telecommunications equipment, however, an investment program will have large foreign exchange requirements. The foreign exchange components under fifteen recent telecommunications expansion programs partially financed by the World Bank, for example, ranged widely between 12 and 90 percent of the total investment requirements, with the norm being between 70 and 85 percent. Hence, although the telecommunications entity can generate large financial surpluses in local currency, these can rarely be reinvested within the sector as fast as they could be absorbed, because of shortages of foreign exchange.[12] This is the case even though the sector's foreign exchange requirements are usually a small proportion, in the range of 0.5 to 3.0 percent, of the total annual imports of a developing country.

Although telecommunications entities in developing countries usually do not directly generate enough foreign exchange to support rapid expansion programs, considerable foreign exchange earnings and savings are generated in other sectors as a result of improved or expanded telecommunications investment. For example, the foreign exchange requirements of a country may be reduced because access to and quality of telecommunications services are improved, which increases administrative and management efficiency and operational productivity in other sectors and, in turn, reduces investment requirements in foreign exchange as well as local funds. When petroleum prices are high, such an argument is particularly relevant for sectors such as transport. As discussed in chapter 7, a well-functioning telecommunications system in a typical developing country might substitute some petroleum-consuming transport and begin to use transportation more efficiently.

The problem, of course, is that the foreign exchange earnings and savings brought about by investments in telecommunications services, but realized outside the sector, do not accrue to the telecommunications entities. They are also difficult to identify and measure. Government finance and economic planning ministries in developing countries, therefore, base their decisions about sector allocations or policy on very incomplete information about the impact of foreign exchange.[13]

Aside from internal generation, other options include supplier credits, foreign commercial bank loans, and bilateral aid. These account for about 25 percent of total investment and often constitute the main source of finance for telecommunications imports in many developing countries. The drawbacks that sometimes limit the extent to which these options are used include the normally high cost of supplier credits and commercial bank loans, the relatively short repayment periods, and the relatively high price of equipment procured through supplier credits and other forms of tied or restricted procurement. Also, less well-off developing nations may lack the financial strength to be eligible for these sources of funds. Multilateral development banks are another source of foreign exchange for telecommunications development.

Multilateral development banks, such as the World Bank and the Inter-American, Asian, and African development banks, accounted for about 5 percent of telecommunications funds invested in developing countries in the 1980s. From the early 1960s to 1990, the World Bank lent about $4 billion to support over 100 projects, costing almost $20 billion, in some forty countries. This makes it the largest multilateral source of financing for telecommunications. However, since the Bank generally lends for telecommunications only when lending from other sources is not available at reasonable terms, telecommunications amount to only 2 percent of all Bank operations; this proportion varies considerably from year to year, depending on country priorities and the availability of other sources of funds (see table 3-1). For example, in the fiscal year ending June 30, 1987, the

Bank extended new telecommunications loans and credits totaling more than $682 million; the following year, the figure dropped to $36 million, and in 1990, it reached more than $616 million.

World Bank telecommunications loans and credits traditionally finance a three- to five-year slice of the country's total public telecommunications investment program. These operations support rehabilitation, expansion, and modernization of local, long-distance, and international facilities in urban and rural areas. They also finance technical assistance needed to improve the operating entities' organization, management, and human resources. Increasingly, however, World Bank telecommunications lending focuses on selected aspects of particular complexity or urgency, such as the rehabilitation of rundown plant, the development of cable networks, or the strengthening of financial management or project planning and implementation. Further details showing the amount of loans and credit granted by the World Bank to specific countries for telecommunications projects between 1962 and 1989 are given in appendix E.

In addition to loans and credits to the telecommunications sector, the World Bank supports telecommunications development through a variety of lending operations not included in the above figures. In fiscal 1988–89, about 20 percent of Bank-financed projects in other sectors (mainly railways, power, agriculture, rural development, and earthquake and other reconstruction projects) had telecommunications components. These components were estimated to cost about $161.6 million, of which about $64.8 million was financed by the Bank. A major area of growth since the mid-1980s has been that of policy adjustment. Telecommunications enterprises have been included in Bank-financed public sector management projects, especially in Sub-Saharan African countries. Loans of up to $400 million, to help offset the costs of wide-ranging public sector reforms, were made in the early 1990s to several Latin American countries subject to progress in implementing the governments' plans to overhaul telecommunications policy, restructure and privatize telecommunications operating entities, and develop competition and public regulation (these aspects are discussed briefly here and in chapter 16).[14]

A brief comment is in order on the two remaining sources of financing shown in figure 3-1. Government loans and equity contributions, although apparently substantial in many developing countries, are often more than offset by the transfer of operating surpluses from the operating entities to the national treasuries or to meet postal and other deficits. Overall, it is estimated that net government contributions do not exceed more than 5 percent of total funding requirements. Until about 1990, private investment was limited to the very

Table 3-1. Trends in Telecommunications Lending by the World Bank, Fiscal Years 1986-90

Sector	1986 Millions of dollars	1986 As a percent of total lending	1987 Millions of dollars	1987 As a percent of total lending	1988 Millions of dollars	1988 As a percent of total lending	1989 Millions of dollars	1989 As a percent of total lending	1990 Millions of dollars	1990 As a percent of total lending
Telecommunications	50.4	0.3	682.3	3.9	36.0	0.2	161.0	0.8	616.7	3.0
Agriculture and rural development	4,777.4	29.3	2,930.3	16.6	4,493.9	23.4	3,490.0	16.3	3,656.1	17.7
Development finance companies	1,449.2	8.9	2,297.9	13.0	1,712.5	8.9	2,366.7	11.1	1,271.7	6.1
Education	829.2	5.1	439.8	2.5	864.0	4.5	890.7	4.2	1,486.6	7.2
Energy	3,018.0	18.5	3,704.3	21.0	2,395.0	12.4	3,863.6	18.1	3,304.3	15.9
Industry	821.1	5.0	418.4	2.4	2,224.6	11.6	1,982.5	9.3	795.6	3.8
Nonproject	1,321.0	8.1	2,437.1	13.8	1,687.0	8.8	3,418.5	16.0	3,044.0	14.7
Population, health, and nutrition	419.5	2.6	54.1	0.3	304.9	1.6	623.0	2.9	933.4	4.5
Public sector management[a]	—	—	—	—	—	—	—	—	525.6	2.5
Small-scale enterprises	274.5	1.7	421.5	2.4	513.0	2.7	585.0	2.7	207.5	1.0
Technical assistance	137.9	0.8	103.9	0.6	95.7	0.5	175.3	0.8	141.0	0.7
Transportation	1,498.2	9.2	1,745.9	9.9	2,642.5	13.7	1,830.8	8.6	2,785.3	13.5
Urban development	1,117.5	6.8	1,469.1	8.3	1,716.3	8.9	1,188.5	5.6	1,002.1	4.8
Water supply and sewerage	604.8	3.7	969.4	5.5	535.3	2.8	791.2	3.7	931.8	4.5
Total	16,318.7	100.0	17,674.0	100.0	19,220.7	100.0	21,366.8	100.0	20,701.7	100.0

— Not available.

a. First introduced in 1990.

Source: World Bank Annual Report (various years).

few countries that allowed the private sector to provide telecommunications services in any way. Private investment in telecommunications was largely ruled out in the 1950s and 1960s in many countries, as the emerging Asian and African nations became independent and many Latin American countries nationalized the foreign-owned telecommunications companies. Only a few countries (notably the Philippines) retained service in private hands. Some private investment has occurred in a small number of state companies that were partly owned by subscribers (mainly through mandatory purchase of shares, which occurred in Brazil), were publicly traded in domestic stock markets (which occurred in Chile), or, even more rarely, were publicly traded in foreign markets (American Depository Receipts of shares of Teléfonos de México were traded over the counter in the United States in the 1980s; until 1990, 51 percent of Teléfonos de México had been owned by the state). Altogether, private investment (voluntary and mandatory) probably did not account for more than 5 percent of total telecommunications investment in the developing world during the 1980s.[15]

The following chapters examine some of the facts and contentions surrounding two other constraints perceived to limit telecommunications investment in developing countries: the benefits of telecommunications investment are not enumerated and quantified as fully as are those of other sectors, and telecommunications investments, although financially profitable, are perceived to benefit directly only a relatively narrow—and privileged—portion of the population.

Notes

1. Some countries also have a significant domestic industry that manufactures telecommunications equipment and cables. These activities, although closely linked to the provision of telecommunications services through the procurement policies of the operating entities, must be treated as a subsector under the rubric of manufacturing industry, not telecommunications.

2. The following three sections draw on Saunders (1982).

3. In some countries, telecommunications groups within government departments have, in practice, more management autonomy than so-called government corporations in other countries, which on paper appear to be more independent.

4. From a longer-term development point of view, one problem is that government can easily reduce planned investment in the sector when budget deficits occur or unexpected national fiscal crises develop. With large monolithic projects such as dams, airports, or power stations, either the facilities are built or they are not; they are never left half done since that would serve no useful function. A telecommunications investment program can, however, be cut piece by piece—fewer subscribers are connected, fewer towns are served, and less long-distance capacity is offered in the short run—without totally eliminating the

effectiveness of much of the investment already completed. Hence, the sector is more vulnerable than many to last-minute cuts in its investment program brought about by exogenous national fiscal problems.

5. Related to this, a study of forty-three countries showed that government organizations with joint responsibility for both posts and telecommunications tend to be the least responsive in setting tariffs that reflect costs and market demand. Government entities responsible for only telecommunications tend to be somewhat more responsive, whereas privately owned entities tend to be the most responsive. See Littlechild (1980).

6. The ITU recommends a pyramidal approach to technical training, in which more employees are given preliminary training at national centers, fewer well-prepared trainees are sent to subregional centers, and the best of these trainees are sent for specialized training to higher-level regional centers in industrial countries. See ITU (1980b).

7. Ethiopia and Burkina Faso are good examples of countries with very modest educational resources that were able to avert staffing problems for several years by timely awareness and action. They also illustrate the long lead times required. For example, in the 1960s, the telecommunications sector in Burkina Faso (then called Upper Volta) depended entirely on foreign staff to fill senior posts. By 1976, the sector was staffed exclusively by nationals. This successful transformation mainly resulted from a decision in 1960 to engage promising school students in long-term commitments by offering them extensive education and training in foreign universities and telecommunications administrations (in Europe and, later, in African regional centers). By 1976, although the OPT in Burkina Faso still had to overcome limitations in its telecommunications, engineering, and management capabilities, as well as some political interference, the operating entity had become a viable organization staffed entirely by nationals, with some of its key positions held by individuals who had reached high standards of education and competence.

8. Chapuis (1975); Huntly (1967). Given technological advances and declining costs, the ratio in 1990 was probably less than it was in 1967; in the specific case of AT&T (now long-distance only) in the United States, statistics of the Federal Communications Commission show that the ratio of total assets to total operating revenues in 1988 was only 0.64:1.00 but that for local exchanges the ratio was 2.14:1.00.

9. For example, in recent telecommunications programs partly financed by the World Bank, the investment cost per telephone line to be added ranged between about $1,200 (for a program mainly confined to a large metropolis and its environs) and about $4,000 (for a country whose national infrastructure had to be built from a very incomplete base).

10. These figures are based on actual and projected growth of telephone lines in countries for which statistics were available in 1990, assuming an average investment of $2,000 per additional line connected. Although the numbers are subject to considerable error, they are probably accurate enough for the limited purposes of this discussion. These are updated calculations of the figures given in World Bank (1989a).

11. Figure 3-1 is based on estimates of total telecommunications investment in developing countries, on the financing plans of telecommunications programs supported by the World Bank in the 1980s, corrected to reflect that many developing countries do not borrow from the World Bank, and on rough figures for telecommunications lending by other multilateral agencies. Again, although these figures are subject to considerable error, they are probably accurate enough for purposes of this discussion. See Wellenius (1990b).

12. Limited foreign exchange particularly affects investment in rural areas, where overall revenues are likely to be much lower than in urban areas. See Goldschmidt (1984).

13. The limited information available is confined to literature surveys of transport and energy substitution, such as the survey presented in chapter 7; to isolated cost-benefit exercises, such as several of those outlined in chapters 8 and 9, which show savings in the costs of transport and gasoline in specific cases; and to a few individual cost-benefit exercises,

which outline specific potential for direct savings in foreign exchange (see example 26 in chapter 9).

14. Wellenius (1990b).

15. The situation changed quite rapidly in the late 1980s and early 1990s, as sweeping telecommunications reforms in a growing number of developing countries, especially in Latin America (for example, Chile in 1988, Mexico and Argentina in 1990), attracted foreign private operators and investors to undertake large expansion and modernization programs. The limits to overcoming telecommunications constraints under the traditional scheme of public sector monopoly, and the trends toward privatization and competition, are outlined in chapter 16 and treated at length in other publications. See Wellenius and others (1989) and Bruce, Cunard, and Director (1988).

Part II
Macroeconomic Analysis of Benefits

Chapter 4

Aggregate Correlation Analysis

THE MOST WIDELY CITED EVIDENCE on the benefits of telecommunications investment compares measures of availability and use with various measures of aggregate national economic activity, such as gross domestic product (GDP). This chapter and the next review such macroeconomic evidence and conclude that, although the evidence could be improved by more rigorous econometric analysis and increased disaggregation, it provides only general and descriptive insights into the question of the benefits and development priorities of telecommunications. Nevertheless, this aggregate evidence helps set the scene for the more detailed sector analysis and project studies discussed in subsequent chapters.

Telecommunications infrastructure may be viewed as an input to a productive process, a "factor of production" like petroleum or electricity.[1] Consequently, most economic empirical work assesses the effect of telecommunications at the macroeconomic or country level in one of two ways. The first is statistical correlation or regression analysis, which usually specifies a macroeconomic country-level model (often with only one equation embodying supposedly causal relations) and then estimates the parameters of that model from data on the provision or use of telecommunications and from one or more indicators of the level of national economic activity. This approach includes both cross-sectional studies, in which the variables are compared for different countries and regions at a single time, and time-series studies, in which the values of variables are traced over time for a single country or region. Several such studies are reviewed in this chapter.

The second general method, which can be referred to as structural economic analysis, focuses on the structure of the economy as re-

vealed by the levels of activity in different sectors (agriculture, manufacturing, services, and so forth). This approach, which is reviewed in
the following chapter, relies primarily on the classic tool of input-
output analysis, which generally describes an economy in terms of
more or less stable coefficients, relating the outputs of particular sectors to their requirements for inputs. One such input is the use of
telecommunications services.

Correlation and Regression Studies

In 1963, Jipp brought to public attention the strong correlation between telephone density (the number of telephones per 100 persons)
and what he called the "wealth of nations."[2] Since then, single-
equation representations of this relation have been formulated many
times, using different groups of countries and different periods as well
as GDP per capita or related indicators as proxy measures for the
wealth of a country. Representative examples are examined below, including references to the benchmark work carried out under the aus-
pices of the International Telecommunication Union's CCITT.[3]

In retrospect, some of the early cross-country comparisons done by
the CCITT may seem naive. It is important to remember, however, that
when its work began in 1964, few relevant economic studies existed
on which to draw, and almost no qualified economists or econometricians were working in the field of telecommunications. The somewhat
simplistic first efforts did throw some light on the association between
telecommunications and aggregate economic activity and gradually
made planners aware of the complexity of the problems at hand.

Example 1. CCITT's Cross-Sectional Analysis

In the mid-1960s the CCITT used cross-sectional data from thirty industrial and developing countries to examine the correlation between the density of telephone lines
(d) and GDP per capita (g).

Figure 4-1 shows the data and model fitted for 1965. A scatter diagram using loga-
rithmic scales for both variables showed that most of the data clustered along a line,
which can be roughly represented by the equation

$$d = ag^b$$

or equivalently

$$\log d = \log a + b \log g$$

where a is the intercept, and b is the slope.

Figure 4-1. *Density of Telephone Lines as a Function of* GDP *per Capita for a Cross-Section of Countries, 1965*

Telephone lines per 100 persons (*d*)

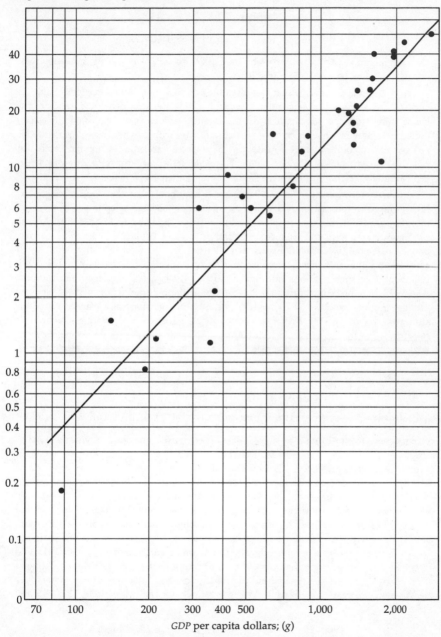

GDP per capita dollars; (*g*)

Source: CCITT, GAS-5 *Handbook: Economic Studies at the National Level in the Field of Tele-communications* (1968). Reproduced with permission.

The parameters a and b were estimated separately for 1955, 1960, and 1965 by ordinary least-squares regression of log d and log g:

$$1955 \qquad \log d = -3.0932 + 1.444 \log g$$

$$1960 \qquad \log d = -3.1171 + 1.432 \log g$$

$$1965 \qquad \log d = -3.1329 + 1.405 \log g$$

High correlation coefficients were obtained (0.91 to 0.92), indicating a strong cross-country relation between the variables.

The slope b is positive, which indicates that density increases as GDP per capita increases. The slope is also larger than 1.0, which indicates that density, d, tends to increase faster than GDP per capita, g. To the extent that a cross-section of countries can indicate the way in which telephone density will vary with GDP per capita over time for any particular country, then density can be expected to grow at approximately 1.4 times the growth rate of GDP per capita. For example, if GDP per capita grows at 5 percent annually, telephone density will, on average, increase 7 percent.[4]

The CCITT recommended in the 1968 GAS-5 handbook that the slope of the fitted lines be used to forecast both demand for and supply of main telephones, in terms of expected growth in the country's GDP and population, and also to forecast initial levels of demand and supply. This recommendation did not explicitly claim that the observed cross-country correlations implied a causal link between the provision of telephones and the growth of GDP, although it was suggested elsewhere in the 1968 handbook that such relations can indeed indicate the benefits of telecommunications. The 1972 handbook advocated such a method only in combination with other forecasting methods, and the 1976 version omitted all suggestions of a causal interpretation and the use of the estimated equation as evidence of consequent economic benefits.

Example 2. CCITT's Time-Series Analysis for Individual Countries

An alternative to examining a cross-section of many countries at a given time is to examine the density of telephone lines for a single country as its GDP per capita increases through time. Figure 4-2 shows an exercise of this type for Sweden, reported by the CCITT in its 1968 GAS-5 handbook. Unfortunately, reliable long-term data were not then, and are not now, available for most developing countries.

The Swedish data show two trends, which represent different rates of exponential growth of telephone density in relation to the corresponding growth of GDP per capita. The CCITT interpreted the first trend, starting in about 1900, as reflecting the period in which telephone service was introduced. The second trend, starting between 1915 and 1920, shows the more gradual process of connecting large proportions of the population to the system. The parameters of the exponential equations were estimated by least-squares regression of log d and log g separately for the two periods, 1900–15 and 1920–65:

$$1900\text{–}15 \qquad \log d = -10.4106 + 3.1935 \log g$$

$$1920\text{–}65 \qquad \log d = -4.6445 + 1.5476 \log g$$

Figure 4-2. *Density of Telephone Lines as a Function of* GDP *per Capita in Sweden, 1900–65*

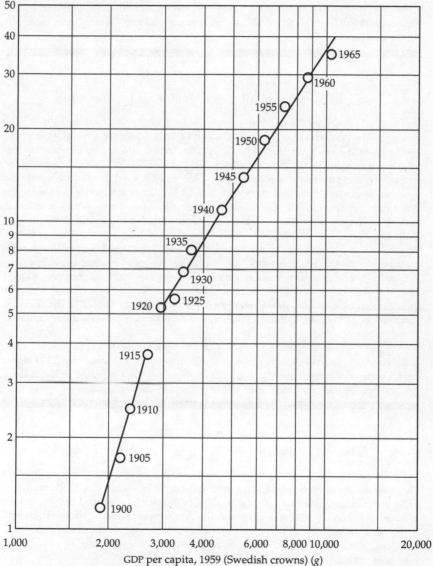

Telephone lines per 100 persons (*d*)

GDP per capita, 1959 (Swedish crowns) (*g*)

Source: CCITT, GAS-5 Handbook: Economic Studies at the National Level in the Field of Telecommunications (1968). Reproduced with permission.

High coefficients of correlation were obtained ($r = 0.99$), indicating a strong relation between the transformed variables.

The coefficient of log g for 1920–65 (1.55) is similar in magnitude to those for the cross-country models discussed in the previous example (1.41 to 1.44) and in this case implied for Sweden that telephone density increased more rapidly than GDP. Perhaps encouraged by this similarity, the CCITT also carried out correlations using mixed time-series and cross-sectional data (1954 to 1968) from both developing and industrial countries. This analysis yielded no findings significantly different from those observed previously.

Example 3. CCITT's Utilization Factor and Other Studies

The CCITT tried several other correlations involving GDP, which were thought to hold some promise, but which in fact proved to be of little additional value. One exercise used a so-called telecommunications "utilization factor," defined by the CCITT as the number of telephones (all stations, presumably) per $100,000 of GDP. Examining the utilization factor for fifty-five economies in 1961 and 1971 led to the conclusion that whereas, in general, industrial economies had high utilization factors (10 or more for the United States, Canada, and most of Western Europe), lower factors were found among developing economies with a strong industrial sector (5 to 8 for Brazil, the Republic of Korea, Mexico, Singapore, and Taiwan). The very lowest utilization factors (typically 1 to 3) tended to be associated with the poorest developing economies. As in other CCITT exercises, this simply suggests that economies become increasingly intensive in telecommunications services as they grow. The CCITT did, however, suggest that the observed disparities in utilization factors can be used as a general argument that developing countries should raise their priorities for investment in the telecommunications sector. This implies an assumption of causation between telecommunications investment and GDP.

The CCITT also attempted to find some simple trend in the proportion of GDP or of gross fixed capital formation that goes into telecommunications investment. No meaningful correlations were found, although developing countries appeared to invest lower proportions of both in telecommunications than do industrial nations.[5]

Example 4. Proxies for GDP and Telephone Density

Other exercises based on statistical correlations have been undertaken to improve the representation of economic and telecommunications activities. Gellerman and Ling, for example, observed a strong relation between the number of telephone subscribers in Panama's two main cities and the total consumption of electricity, as well as between the ratio of telephone to electricity subscribers and annual electricity consumption per subscriber.[6] Another study, based on a cross-section of twenty-nine countries, was also published in 1976 by Bebee and Gilling.[7]

Since there is evidence that the secondary and tertiary sectors of an economy are the most intensive users of telecommunications, Bebee and Gilling selected the GDP per capita produced by these two sectors as the relevant indicator of a country's level of economic development. They then defined a "development support index" in terms of measures of capital expansion and quality of manpower, both of which they noted

Figure 4-3. *Telephone Index as a Function of the Development Support Index*

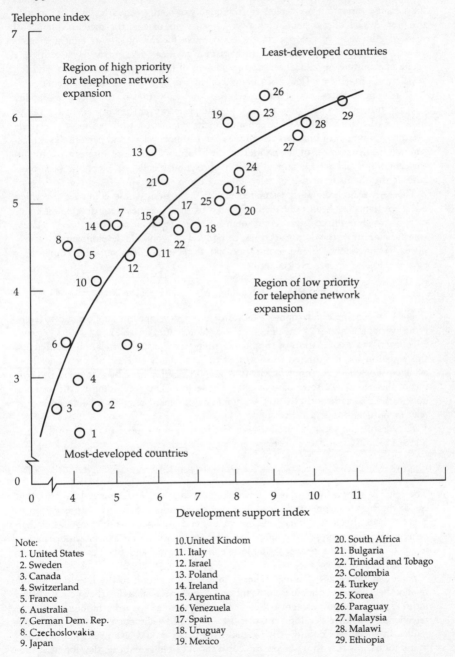

Note:
1. United States
2. Sweden
3. Canada
4. Switzerland
5. France
6. Australia
7. German Dem. Rep.
8. Czechoslovakia
9. Japan
10. United Kindom
11. Italy
12. Israel
13. Poland
14. Ireland
15. Argentina
16. Venezuela
17. Spain
18. Uruguay
19. Mexico
20. South Africa
21. Bulgaria
22. Trinidad and Tobago
23. Colombia
24. Turkey
25. Korea
26. Paraguay
27. Malaysia
28. Malawi
29. Ethiopia

Source: Bebee and Gilling (1976). Reproduced with permission.

were essential ingredients of development. Under the capital expansion category they included the proportion of GDP used in fixed capital formation and the value of gross fixed capital formation per capita. For the quality of manpower variables they included the literacy rate of the population over fifteen years of age, the median number of years of education of the population (as a proxy for the quality of educated manpower), per capita daily intake of protein and calories (as a proxy for health), per capita expenditure on education, the proportion of population that is urban, and the population growth rate (low rates being considered favorable to development).

A "telephone index" was constructed from three measures of telephone availability and use: number of telephones per 100 literate persons over fifteen years of age (as an indicator of the availability of telephones to the population who can most effectively use them), number of business telephones per 100 nonagricultural persons (as a measure of the penetration of telephones in the telecommunications-intensive sectors of the economy), and average annual number of telephone calls per telephone (as a measure of the intensity of use of telecommunications services).

Those three indexes were calculated for 1970 for twenty-nine countries for which adequate data had been published. Thirteen of the countries were developing countries. Through multiple linear regression, several analytical models were fitted to linear transformations of the data categories, using level of economic development (GDP produced by the secondary and tertiary sectors) as the dependent variable. The authors gave the following as a typical result:

$$Y = 5.928 - 9.078 \ (1 \ / \ X_1) - 7.093 \ (1 \ / \ X_2)$$

where Y = economic development indicator, X_1 = development support index (composite variable), and X_2 = telephone index (composite variable).

Following an examination of the related simple and partial correlation coefficients, this equation was interpreted to show that the availability and use of telephones have an important effect on other factors that support development by increasing the output of the economy's secondary and tertiary sectors. This, according to the authors, upheld their basic hypothesis that economic development requires a balanced mix of telecommunications and other support factors.

To examine the specific mix, least-squares regression was used to correlate the telephone index (X_2) and the development support index:

$$X_2 = 7.8 - 18.0 \ (1 \ / \ X_1).$$

The result is shown in figure 4-3. Since the curvature is greatest near the origin, that is, for the higher-income industrial nations, the authors concluded that as countries become more developed, telephone availability and use appear to increase faster than other development support factors. This is consistent with what was shown by the earlier CCITT correlations between simple telephone density and GDP per capita and in CCITT utilization factor exercises.

Although Bebee and Gilling make an interesting addition to the literature on the relation between economic development and telephone availability, they go a little too far in arguing that government policies should be directed toward attaining the mix of telephone and other development support described by the curve derived from their model. In their view, countries lying above the curve should give priority to telecommunications, whereas those below the curve should emphasize other development support factors. They also considered that the derived curve offered a method for ranking countries by priority for telecommunications expansion. It is not at all clear that such prescriptive conclusions can be legitimately drawn from this kind of approach.

Example 5. Telex Density

Telex density has also been compared with several economic variables.[8] A cross-section of eighty-seven industrial and developing countries was used to estimate parameters for a model to explain variations in telex density (number of telex lines per 1,000 inhabitants). Seven variables were initially specified to reflect various proxies that could be associated with the use of telex service in a country: GDP, value of exports, value of imports, bank deposits, telephone lines, telephones, and energy consumption, all on a per capita basis.

Since the independent variables showed significant correlation among themselves, stepwise regression was used to select a simple model with high explanatory power. The result was the following:

$$d = 8.9324 \cdot 10^5 g^{0.70560} i^{0.44062}$$

where d = telex lines per 1,000 persons, g = GDP per capita in 1974 dollars, and i = value of imports per capita, in 1974 dollars. The correlation coefficient was $r = 0.92$. Hence, as would be expected, telex density, like telephone density, was positively correlated with GDP per capita across countries. In addition, since telex is used extensively in international trade, it is no surprise that telex density was also correlated with imports.

Example 6. Telecommunications Traffic

Correlation analysis has also been used to explain variations in telecommunications traffic across countries and through time. For instance, in the 1968 and 1972 editions of the GAS-5 handbook, the CCITT reported on several traffic studies based on time-series data for individual countries, which essentially followed well-established techniques used to forecast traffic in industrial countries.[9] The exercises consisted of correlations between telephone calling rates and GDP, GDP per capita, value of imports, value of exports, and other measures of economic activity. Results were mixed. The CCITT concluded that, with the data available, local telephone call traffic could not be shown to be influenced by the normal course of economic development, whereas long-distance telephone traffic increased at roughly double the rate of increases in real GDP. Since wide variations occurred among countries and from one year to the next, however, this rate could not be used as a universal benchmark. Not surprisingly, international telephone traffic through time was most closely associated with changes in the volume of international trade.

With regard to cross-sectional exercises, Yatrakis published a study based on an examination of forty-six industrial and developing countries.[10] He analyzed the correlations between international telephone, telex, and telegram traffic and twenty economic indicators specified as dependent variables. The number of tourists per year, GDP per capita, and size of the country's population were found to be the main variables explaining international telephone traffic among countries; all three were positively associated with traffic. Telex traffic was the most closely correlated with tourism and trade. The CCITT, in reviewing this study in the 1976 GAS-5 handbook, concluded that it showed that good and reliable telecommunications are a stimulus to a viable tourist trade, which in turn has a positive effect on the host country's economy.

Limitations of Aggregate Analysis

As outlined above, since the early 1960s, statistical regression and correlation analysis at the aggregate level has frequently been used to show how telecommunications services are linked to economic development. The simplicity of the approach and its modest requirements for data made it an attractive tool in the initial stages of economic analysis of the effect of telecommunications in developing countries. However, such analysis is subject to important limitations. Among the most constraining of these are the inadequacy of using a single equation to represent highly complex relations, the heterogeneous nature of the data used, and the inability to attribute causation to any of the observed correlations.

With regard to the first point, the relations between telecommunications and economic activity are far too complex to be usefully represented by a single-equation model with only one or a few independent variables. The large aggregation of factors involved in the variables specified in these single-equation exercises by definition obscures most of the relations that could be meaningful to decision-makers who are concerned with investment needs in many sectors, and who probably intuitively understand at least some of the interdependencies and tradeoffs involved.

Second, the data used for both the cross-sectional and time-series correlation exercises are generally not comparable. The highly developed countries at the upper end of the telephone density range tend to have efficient, reliable, and modern telephone service, wide telephone coverage throughout the country, and only minimal (if any) waiting lists and waiting times. These countries are compared with those at the lowest end of the telephone density spectrum, which have very limited national coverage, a large unsatisfied demand, limited access to subscriber-dialed, long-distance calls,[11] and major call traffic congestion during business hours on both local and long-distance networks.[12]

Also, in the center of the telephone density spectrum there are certain anomalies.[13] The OPEC countries (Organization of Petroleum Exporting Countries) tended to have lower telephone densities than would have been expected on the basis of a cross-country correlation between income and telephone density. The reason, of course, is that a telephone infrastructure takes time to develop, and telephone services in the OPEC countries were generally unable to keep up with the rapid increases in GDP. Also, telephone densities for the former Eastern European countries tended to fall below the aggregate cross-country income-density regression lines. In these centrally planned countries, with tightly controlled markets, the planners had evidently concluded that some types of telephone communication (most prominently residential) were not necessary or desirable or, alternatively, that in many instances telephones were not necessarily the least expensive way to communicate, as they were in market economies with similar income levels.

The third and perhaps most significant problem with the GDP–telephone density correlations is that, even if the data were homogeneous and the aggregation problems less severe, the movement of two variables through time, or their association across a set of countries, does not imply that changes in one of the variables causes changes in the other; correlation is a test for joint variation, not causation. This was often forgotten in the applications reviewed.[14]

For instance, using GDP per capita to predict telephone density (as in example 1) assumes that a change in economic activity results in a change in telephones, which is simple one-way causation. One could also expect, however, that the economy's output to some extent reflects the fact that telephone service is available as a production input.[15] In example 6, the high correlation between telex traffic and tourism was interpreted by the authors as showing that telecommunications stimulate tourism and that this, in turn, has a positive effect on the economy. Although these statements are compatible with the statistical evidence presented (which is a necessary condition for them to be true), they do not necessarily follow from it. Example 4 illustrates the temptation to draw unwarranted policy implications. Not only do the authors suggest that their hypothesis is proven by the statistical regression analysis (which it cannot be), they also interpret what in fact is a poor fit of the model to the data as having a normative value: they contend that countries placed above the curve should give priority to telecommunications investment, whereas those below the curve should expand other development inputs.

Value of Aggregate Analysis

Given these problems, what, if any, is the value of using correlations at the aggregate level to indicate the effect of telecommunications in a developing economy? First, such exercises orient the thinking of analysts and planners by providing simple quantitative descriptions of existing patterns of telecommunications at various levels of economic development. The findings also hint that the combina-

tion of market prices, physical and managerial constraints on investment, and government preferences seems to generate a somewhat
consistent—although not necessarily optimal—pattern of evolution
across a wide variety of countries.[16] This has some illustrative value.
Finally, by throwing some light on the complexity of the phenomenon
being examined and by providing some limited empirical results with
which more elaborate analyses must be compatible, such studies
could direct efforts to yield more useful results.

There would seem to be some, albeit limited, scope for improving
the aggregate statistical modeling approach. Better techniques are
available.[17] Econometricians often face the task of disentangling several simultaneous causal relations of varying directions and between
different variables—the identification problem.[18] Also, a growing
body of research on the roles of communication in the family, in the
organization, and within regions is beginning to throw some light on
relations between telecommunications and society. This might provide
a starting point for an analysis of telecommunications in terms of specific models and strategies of development. One outcome of such an
analysis could be better criteria for selecting the variables to be included in the models and the shapes of the analytic functions to be
filled.[19]

In addition to the analytic limitations of aggregate correlation analysis, other shortcomings must be addressed: the quality of the data
needs to be improved, and, equally important, the data need to be analyzed on a more disaggregated basis. Nevertheless, the approaches
outlined above at the national level are in many instances not suitable
for analysis of smaller territorial units, since even the simple data
specified become increasingly unreliable—and often are not available—for provinces, regions, and individual cities, towns, and villages.
A more manageable way to approach disaggregation might be to examine the association between telecommunications infrastructure and
economic activity in various sectors. This is the subject of the following chapter.

Notes

1. This is particularly true in most developing countries in which commercial and industrial uses of telecommunications by far outweigh residential uses.
2. Jipp (1963) pp. 199–201.
3. The International Consultative Committee on Telephone and Telegraph (CCITT, from
the French name) is one of the ITU's permanent divisions. From 1964, the CCITT included a

work group on economic studies of telecommunications at the national level, designated GAS-5. See CCITT (1976, 1984a, and 1984b).

4. Let g increase by Δg, and d by Δd. Then,

$$d + \Delta d = a(g + \Delta g)^b$$

and

$$d = ag^b$$

Dividing the first line by the second gives

$$1 + (\Delta d / d) = [1 + (\Delta g / g)]^b$$

For small values of $\Delta g / g$ (say, less than 0.1 or 10 percent), this is approximately equal to

$$1 + (\Delta d / d) \approx 1 + b(\Delta g / g)$$

For example, if the relative increase of gross domestic product (GDP), $\Delta g / g$, equals 0.05 (or 5 percent), it can be expected to be accompanied by a relative increase in density $\Delta d / d$ equal to about 1.4 times 0.05, or 0.07 (7 percent). The coefficient b remained fairly constant (at about 1.4) during the period examined by the CCITT.

5. That developing countries have tended to invest lower proportions of their GNP in telecommunications than industrial countries was also discussed in chapter 1.

6. Gellerman and Ling (1976). The authors also noted that the demand for both telephone and electricity connections was fairly well met during the period of analysis (1963–74), and thus connections could be considered generally to reflect demand.

7. Bebee and Gilling (1976).

8. Wellenius, Budinich, and Moral (1979).

9. See, for example, Taylor (1980) chaps. 3 and 4; Drewer (1973); and Fox (1973).

10. Yatrakis (1972).

11. In several telecommunications projects that the World Bank helped to fund in the 1960s and 1970s, the introduction of subscriber trunk dialing service almost immediately increased long-distance call traffic between two and six times.

12. Whereas in industrial countries the number of telephone lines is a fairly good measure of demand, in developing countries it represents only the level of supply—which often falls far short of demand—and therefore is not a good indication of telephone requirements consistent with existing economic activity.

13. The consistencies in deviations of OPEC and Eastern European countries were observed by the authors and found to be statistically significant at the 0.05 level in several correlation exercises using data from the mid-1970s for a cross-section of eighty countries.

14. In the United States, there has been a very strong positive correlation between teachers' salaries and total national alcohol consumption. No one has yet, however, argued that a way to reduce national alcohol consumption would be to reduce teachers' salaries.

15. For one of the very few applications reviewed, which mentions that this should give rise to two regression lines, not one, see Shapiro (1976).

16. It is likely that over several years, as an economy develops and becomes more complex, and as time becomes more important in production and distribution processes, some form of telecommunications gradually becomes the least expensive way to communicate. The correlation exercises examined above produce results that are consistent with this hypothesis but, of course, do not "prove" it.

17. One intercountry study ran regressions that pooled cross-sectional data into a time series, lagged the regressors by one year, and examined the results in terms of standardized regression coefficients. Unfortunately, the results of the numerous regressions seemed on occasion inconsistent and provided little insight into causality. The overall conclusions of the study—that "the first role of the telephone is as a channel of information" and "the

second role ... is in allowing a number of alternative communication patterns to occur in business and social activity"—did not necessarily follow from the numerous regressions. See Hardy (1980), pp. 1–146. Subsequently, another study drew on Hardy's correlations and actually claimed (incorrectly) to be able to estimate the effect on national GDP of telephone installations in regions of low telephone density. See Hudson, Hardy, and Parker (1982).

18. An introduction to the problem and approaches to its solution can be found in Walters (1968) and in the multitude of more recent econometric texts.

19. An interesting study initiated by O'Brien uses principal component analysis. See O'Brien and others (1977). See also Thomas (1984). The latter examined the temporal priority of communications in relation to other factors commonly associated with development; these included a spectrum of social modernization indicators, such as literacy and school attendance, and were subject to a number of control variables, such as type of colonial heritage and political regime. Although this analysis produced some interesting findings, it did not convincingly resolve the issue of causality. More recently, Heymann (1987) used correlations between national products and selected development indicators (established using regression calculations) to determine quantitative relationships between telephone density and other aspects of the national economy but explicitly avoided the entire issue of causality.

Chapter 5

Structural
Economic Analysis

AS INPUTS TO PRODUCTION PROCESSES, telecommunications services are not equally important to all productive sectors of an economy. Primary sectors make relatively little use of telecommunications, tertiary sectors are the most intensive users, and secondary sectors lie somewhere in between.[1] Several studies have illustrated this.

The CCITT reported that in 1950 in the former Federal Republic of Germany, where 25 percent of the economically active people were employed in agriculture, the agricultural sector had only 7 percent of the country's telephone lines and accounted for a bare 4 percent of telephone revenues. In contrast, commerce and transport, which employed 16 percent of the work force, accounted for 39 percent of telephone lines and 41 percent of revenues. Data from sixty-nine German cities with more than 100,000 inhabitants in 1962 showed that cities predominately engaged in mining or heavy industrial activities had far lower telephone densities than cities of similar population that were mainly active in other sectors of the economy.[2] Lönnström, Marklund, and Moo found that Sweden's telephone density has varied closely through time with the proportion of the work force in industry, commerce, transport, and communications.[3] Finally, a cross-sectional correlation analysis of twenty-two states in Brazil showed telephone density to be much more closely associated with the gross product of the service sector than with either industrial or agricultural output.[4]

The case of France referred to in chapter 1 also illustrates the relevance of questions about the sector's structure. Until the early 1970s, telecommunications development in France lagged significantly behind that of most other advanced industrial countries, including those with a similar level of per capita income. This had little obvious adverse effect on the aggregate rate of economic growth, which was more than 6 percent a year in real terms for most of the period from 1955 to 1970. It has been suggested, however, that France was able to enjoy rapid economic growth in the 1960s despite a highly deficient telecommunications infrastructure partly because of certain structural features. During that period particularly strong growth occurred in the primary and secondary sectors, with emphasis on agriculture, construction, petrochemicals, and steel. At the same time, growth in the tertiary sector clustered in only a few dominant centers, led by Paris. The massive effort first launched in 1974 to overcome the relative backwardness of France's telecommunications system reflected judgments made by the government that a structural shift in the growth process would occur in the late 1970s and in the 1980s, with emphasis on communications-intensive sectors and with increased regional dispersion of economic activity.[5]

If the use or benefit potential of telecommunications services is likely to be sector specific in developing countries, a further step in examining the relation between telecommunications and economic development would be to disaggregate an economy's structure, to the extent available data permit, and to examine the use of telecommunications services sector by sector. A useful starting point for such an exercise would be to consider how much of the output of the telecommunications service sector is sold to final consumers, whether households or public agencies, that are not themselves production units, and how much is sold as an intermediate good that contributes to the production of other goods and services in the primary, secondary, and tertiary sectors.

Very general information relating to this question can be obtained by examining telephone revenues by class of subscriber. For example, in the 1976 version of its GAS-5 handbook, the CCITT noted that nearly 90 percent of the expenditure on telephone services in developing countries was accounted for by subscribers in industry, banking, transport, and government. The proportion was lower among the more developed of these countries, which had a higher proportion of residential subscribers.[6] Similar but more detailed evidence on sector-specific usage is available from a study that examined the 1968–69 input-output transaction table for the Indian economy.[7] In that exercise, the consumption of postal and telecommunications services was partitioned roughly into three groups: 43 percent of the consumption was by households and nonprofit institutions, 42 percent by intermediate business consumers, and 15 percent by government departments and agencies.[8] Of the intermediate business consumption of postal and telecommunications services, 95 percent fell in the tertiary sector; of that amount, trade accounted for 68 percent; banking and insurance, 14 percent; education and research, 9 percent; and transport, 8 percent.

A more recent study on Korea in 1980 found that 6 percent of communications output went to exports, 24 percent went to private consumption, and 71 percent went to intermediate consumption. Of the output for intermediate consumption, only 1 percent went to agriculture, forestry, and fisheries (15 percent of GDP), 23 percent went to mining, power, and industry (33 percent of GDP), 4 percent went to building and public works (8 percent of GDP), and 72 percent went to commerce and services (44 percent of GDP).[9] Although this example closely conforms to traditional expectations about how telecommunications input and output are distributed among sectors, recent evidence from Italy and Greece does not follow the expected pattern. In Greece, for example, a large survey undertaken by the Greek telecommunications authority, OTE, found that 80 percent of telephone lines were connected to households, which would indicate a mature system with high residential penetration, but that only 7 percent of communications output was consumed by intermediate demand, which would indicate a poorly developed system in the areas of business and administration.[10]

A Cross-Country Comparison of Input-Output Coefficients

Suitable data for more elaborate analysis for both developing and industrial countries are limited, primarily because, as in the Indian example, readily available national input-output data rarely separate telecommunications from other communications services (mainly the postal services).[11] Nevertheless, input-output tables still provide useful insights. For example, in one comparative analysis of input-output tables for several developing and industrial countries, sets of data were collected and grouped so that three general types of indicators could be calculated.[12]

a. Communications input coefficients, that is, the amount of communications services purchased by each sector per unit of output (sales) of that sector

    b. Communications inputs to each sector as a proportion of total purchases by that sector

    c. Communications output distribution coefficients, that is, the proportion of total output from the communications sector purchased by each other sector.

Communications input coefficients and output distribution coefficients were derived from a United Nations input-output data base organized around twenty standardized sector definitions for 1960 or the nearest year available (data set A), and tables of interindustry flows were obtained primarily from World Bank files for seven countries varying from 1961 to 1970 (data set B).[13] As noted above, data for postal services and telecommunications are rarely separated; hence the data sets examined contain only a single communications category. Indeed, even the United Nations interindustry tables did not distinguish between transport and communication; in a few instances, however, a separation was approximated by going back to the original raw data.

Despite reservations about the suitability and accuracy of the data available for interindustry input-output comparisons, some significant patterns were discerned as outlined in the following three examples.

Example 7. Communications Output Distribution Coefficients

The communications industry is one of the few whose product serves as an input to nearly every other industry in the economy. That is, in most input-output tables, few cells in the communications row are empty. Table 5-1 shows communications output distribution coefficients for the countries in data set B.

It is apparent from the table that most intermediate communications output is consumed by service industries, whereas manufacturing and mining generally follow in second place. Classification of the major communications-intensive industries according to the nature of their inputs and the destination of their outputs highlights the tendency of these industries to have relatively high value added and to produce goods primarily for final demand markets.[14] In all countries examined, the agricultural sector uses relatively little communications; the highest agricultural coefficient was 2.7 percent of total communications output in the Philippines for 1961.[15]

Example 8. Communications Input Coefficients

Many differences in the consumption of communications among countries are partly determined by the different proportions of output that each sector contributes to the total output of the economy. For example, the high share of communications

Table 5-1. *Communications Output Distribution Coefficients, by Sector, for Data Set B*

(percentage of total sales of telecommunications and postal services purchased by each sector)

Output sector	Colombia (1970)	Japan (1965)	Korea (1966)	Philippines (1961)	Taiwan (1964)	Turkey (1963)	United States (1967)
Agriculture	1.84	0.26	0.33	2.71	0.67	0.22	0.55
Mining and manufacturing	17.78	34.82	18.23	48.47	15.88	11.05	15.33
Services	51.59	47.43	49.88	16.60	71.54	43.92	47.30
Other	—	5.09	1.49	—	2.39	—	3.97
Household consumption	28.79	12.40	30.07	32.22	9.52	44.81	32.83

— Not available.
Source: World Bank data.

output sold to the agricultural sector in Colombia (1.8 percent), relative to that in the United States (0.6 percent) and Japan (0.3 percent), partly reflects agriculture's contribution of more than 22 percent of gross output of the economy in Colombia compared with 3.9 percent and 6.6 percent, respectively, in the other two countries.

Such distortions can be reduced by examining the consumption of communications services per unit of output in other sectors. Table 5-2, for example, shows the average expenditure on communications services by each sector per $1,000 of sales for countries in data set A. A sharp contrast is found in the intensity of the use of communications inputs among primary, secondary, and tertiary sectors. Likewise, large differences are observed in the countries' level of communications services and broad economic development.

For a more detailed analysis, table 5-3 shows the expenditure on communications services incurred by aggregate sectors per $1,000 of output for the more disaggregated data available for the countries in data set B. These are the coefficients of *direct* input

Table 5-2. *Communications Input Coefficients, by Sector, for Data Set A*

(dollars of expenditure on communications services per $1,000 of sector output)

Economy	Agriculture	Mining and manufacturing	Services
Belgium, Finland, France, Fed. Rep. of Germany, Netherlands, United States	1.0	4.4	11.2
Greece, Italy, Japan, Yugoslavia	0.4	3.7	9.5
Korea, Philippines, Turkey	0.3	1.7	5.8

Source: World Bank data.

Table 5-3. *Direct Input Coefficients for Each Aggregate Sector's Use of Communications Services, for Data Set B*
(dollars of expenditure on communications services per $1,000 of sector output)

Input sector	Colombia (1970)	Japan (1965)	Korea (1966)	Philippines (1961)	Taiwan (1964)	Turkey (1963)	United States (1967)
Agriculture	0.64	0.34	0.08	0.28	0.16	0.03	2.55
Mining	2.37	8.67	2.79	5.33	1.36	1.89	1.85
Food	2.28	1.26	2.27	2.08	1.72	1.26	4.00
Textiles	4.50	4.06	2.82	0.39	0.95	1.44	2.99
Clothes	a	4.95	2.20	2.20	3.65	1.35	6.13
Wood, paper, printing	6.99	7.12	5.84	2.07	1.19	4.15	11.18
Rubber	a	5.11	1.89	2.70	2.18	1.44	5.46
Chemicals	3.14	4.12	4.03	1.37	2.10	1.41	4.63
Petroleum and products	0.67	3.22	1.44	0.30	1.28	0.83	1.58
Cement and minerals	4.28	4.56	3.25	0.75	1.95	—	5.87
Metals	5.20	3.81	1.99	0.43	1.11	2.13	3.43
Transport equipment	1.75	4.08	2.22	0.74	3.00	1.95	5.29
Machinery	7.05	6.75	3.36	0.89	2.82	2.08	7.43
Utilities	3.96	4.52	1.59	0.60	1.25	—	5.82
Construction	1.75	6.69	1.17	0.56	2.26	—	3.59
Trade	7.48	15.07	15.84	—	—	5.32	15.31
Transport	8.35	9.11	3.60	0.05	4.89	2.13	11.26
Communication[b]	99.13	3.69	0.27	—	0.67	37.53	16.24
Services	9.60	12.11	8.60	0.87	9.81	5.10	25.46
Other	—	29.27	4.80	c	0.28	—	—
Households	5.26	4.14	4.05	1.08	0.09	3.93	18.46

— Not available.

a. Included in textiles.

b. This row has little significance, since the entries vary according to the P&T administrations' practices in accounting for their own use of communications services. These tend to be arbitrary or nonexistent.

c. This result is highly anomalous and is not thought to be reliable.

Source: World Bank data.

for each sector's use of communications services. The patterns that can be observed in table 5-3 roughly correspond with those in table 5-2.

In addition to the direct purchases of communications services, each sector of an economy uses other inputs, whose production also requires the use of communications. Considering all relations of this kind produces coefficients of *direct and indirect* input for each aggregate sector's use of communications services. Such coefficients are shown in table 5-4.

As can be seen, among developing countries agriculture directly uses relatively few communications inputs. Even though the agricultural sector may contribute from 20 to 30 percent of the gross output in many developing economies, the very small size of the communications input coefficients for this sector reflects a low total expenditure on communications by the agricultural sector.[16] This was apparent from table 5-1. This is in stark contrast to the highly productive agricultural sector in the United States, where the direct use of communications is generally from ten to twenty times as high as it is in all but one of the developing countries considered (the exception being Colombia, which in 1970 used about 25 percent as much communications per unit of agricultural output as the United States).

A comparison between the direct inputs on the one hand and the direct and indirect inputs on the other shows a greater overall dependence of agriculture on the communications infrastructure than is apparent from the direct coefficient alone. This effect, however, is less apparent in the least developed of the countries considered, presumably because they use relatively few inputs, such as fertilizers, machinery, and technical assistance, and have relatively undeveloped markets and transport and telecommunications infrastructure.[17] In connection with nonagricultural activities, the use of communications inputs seems to follow a pattern related to overall development of the economy.[18]

For manufacturing sectors such as chemicals and machinery, the values for communications input coefficients are fairly widely dispersed among countries. Coefficients are lower for the developing countries, but there appears to be some lower limit for each sector, so that extremely low values (less than $1 per $1,000 of sector output) are rare. On average, the coefficients for the developing countries are about one-third to one-half the size of the coefficients for industrial countries. Also, the indirect communications requirement is important throughout. In Japan, for example (1965 data), the metals industry directly consumed ¥ 3.8 of communications services per ¥ 1,000 of output. It also consumed ¥ 26.6 of utilities per ¥ 1,000 of output. The utilities sector, in turn, consumed ¥ 0.12 of communications to produce ¥ 26.6 of output. Summing all such indirect demands for communications by the industries supplying inputs to metals gives a total indirect communications requirement of ¥ 14.6 per ¥ 1,000 of metals output. This is, of course, far larger than the direct requirement alone.

For nonagricultural, nonmanufacturing sectors such as services or trade, the differences between coefficients for industrial and developing countries are somewhat less pronounced; few direct and indirect coefficients are below $5 per $1,000 of output. For trade, few coefficients for developing countries are less than half the size for industrial countries.

Inspection of table 5-4 suggests that the structure of the coefficients—that is, their relative size for the various sectors—is similar in all countries examined. This, in turn, may imply that differences among countries result from factors of more general relevance (management and administrative practices, substitution of low-cost travel time for telecommunications, scarcity or low quality of telecommunications services), rather than from factors specific to individual sectors.

Table 5-4. *Direct and Indirect Input Coefficients for Each Aggregate Sector's Use of Communications Services, for Data Set B*

(dollars of expenditure on communications services per $1,000 of sector output)

Input sector	Colombia (1970)	Japan (1965)	Korea (1966)	Philippines (1961)	Taiwan (1964)	Turkey (1963)	United States (1967)
Agriculture	1.9	4.3	1.1	0.7	1.7	0.4	14.6
Mining	5.4	14.6	4.8	5.6	3.9	3.1	11.2
Food	5.2	7.7	4.9	2.4	4.7	2.3	16.5
Textiles	10.9	13.6	6.7	0.8	5.1	2.9	14.4
Clothes	a	16.2	7.7	3.0	8.6	3.1	17.3
Wood, paper, printing	14.9	16.5	10.6	2.7	5.4	6.7	24.0
Rubber	a	14.7	6.2	3.1	6.9	3.2	15.8
Chemicals	10.5	14.1	2.0	2.2	6.4	3.4	17.3
Petroleum and products	4.2	11.6	3.9	0.8	4.1	1.4	12.4
Cement and minerals	9.1	13.3	7.1	1.5	5.7	1.5	15.5
Metals	13.8	18.4	8.2	0.8	6.6	4.2	13.7
Transport equipment	8.5	14.8	8.0	1.1	7.3	4.4	16.4
Machinery	15.8	17.2	8.4	1.3	8.4	3.9	17.7
Utilities	7.7	12.0	3.5	0.9	3.4	1.6	14.8
Construction	7.5	17.2	6.0	1.0	6.7	1.8	15.0
Trade	12.5	18.2	16.8	—	—	6.2	22.9
Transport	12.7	14.3	6.2	0.2	8.1	3.6	17.1
Communication[b]	—	—	—	—	—	—	—
Services	12.8	15.7	10.6	0.9	11.7	5.7	34.9
Other	—	38.4	7.6	c	5.3	—	17.0

— Not available.

a. Included in textiles.

b. This row has little significance, since the entries vary according to the P&T administrations' practices in accounting for their own use of communications services. These tend to be arbitrary or nonexistent.

c. This result is highly anomalous and is not thought to be reliable.

Note: Indirect communications requirements were not calculated for the households sector.

Source: World Bank data.

Example 9. Backward Linkages

Direct and indirect input coefficients for the communications sector itself indicate the extent to which growth in that sector will stimulate growth in other sectors through the demand by communications enterprises for the outputs of other sectors. Table 5-5 shows production functions of the communications sector as they appear in the input-output tables for data set B. In terms of dollar input per dollar of communications output, the most important inputs to communications are generally households (labor), services, and transport. Machinery, trade, utilities, and construction are also important inputs in some countries. This concentration of inputs in the labor-intensive service sector and the high labor requirement (primarily for postal services) indicate that demand for an extra unit of communications output will feed back primarily into household income, with little direct effect on the major processing sectors of the economy.

Also of interest is the total increase in household income that results from a one-unit expansion of the demand for an industry's output. This includes direct income (additional labor used as input to produce the additional output), indirect income (income generated in the supplying industries), and induced income (income generated through direct and indirect income spent by households). The ratio between these estimates of total income generated and their direct income components gives industry income multipliers, which can be normalized by dividing each by the average income multiplier for each country.

Table 5-6 shows that the normalized income multipliers are generally lower for the communications sector than for all industries in the countries examined.[19] This means that if the growth of a sector is being constrained by demand, the expansion of final demand for the products of most other sectors gives rise to greater indirect income-generating effects than does the expansion of demand for communications. A general complication in interpreting such results in developing countries is that the output, not only of the communications sector but also of many others, is mainly constrained not by demand, but by insufficient supply.

Information Sector Analysis

Input-output tables, or some of the raw data that were originally collected to construct such tables and national income accounts, have also been subject to a different form of analysis and aggregation. Such work has addressed the concerns of some researchers that much existing economic literature does not adequately consider the information-related function of administration and that therefore much of the traditional economic theory of the firm is perhaps too far removed from reality.[20] The objective of this information sector analysis has been to identify the extent or the importance of directly productive activities that mainly involve handling information (as distinct from goods and materials) and processes of controlling, coordinating, monitoring, recording, or more generally organizing directly productive activities. These two groups of activities are referred to as the primary

Table 5-5. Purchase of Inputs in Dollars per $1,000 of Output from the Communications Sector, for Data Set B

Input sector	Colombia (1970)	Japan (1965)	Korea (1966)	Philippines (1961)	Taiwan (1964)	Turkey (1963)	United States (1967)
Agriculture	—	—	0.04	—	—	—	—
Mining	—	0.12	0.14	—	—	4.42	0.03
Food	—	—	—	—	—	—	0.08
Textiles	1.87	0.50	0.42	—	35.17	—	—
Clothes	—	2.55	0.50	—	—	—	0.75
Wood, paper, printing	3.74	12.33	30.13	—	—	8.83	3.37
Rubber	—	0.11	0.37	—	—	—	0.11
Chemicals	1.25	—	2.76	—	6.98	2.21	0.23
Petroleum	6.23	2.53	7.33	2.55	1.77	—	3.74
Cement and minerals	—	0.01	16.42	—	—	4.42	—
Metals	1.87	—	3.30	—	—	26.49	0.37
Transport equipment	19.33	2.51	1.64	18.75	3.63	4.42	5.52
Machinery	9.98	4.49	25.36	—	20.55	35.32	10.78
Utilities	2.49	7.99	5.83	13.60	4.19	4.42	7.90
Construction	—	5.99	3.80	13.10	8.20	—	21.30
Trade	—	2.13	18.62	—	—	11.04	8.10
Transport	31.17	29.16	18.97	—	38.74	24.28	29.40
Communications[a]	99.13	3.69	0.27	206.07	0.61	37.53	16.24
Services	53.62	12.14	9.90	—	28.06	15.45	106.81
Other	—	4.00	4.82	—	38.13	—	—
Households	521.20	475.65	348.28	537.55	642.34	737.31[b]	775.92[b]

— Not available.

a. The figures in this row represent communications input to the communications industry itself. They have little meaning since their size depends on the accounting practices used by the communications administrations of each country.

b. These household input figures include value added that is not household labor and so are overstated.

Source: World Bank data.

Table 5-6. *Normalized Income Multipliers, for Data Set B*

Input sector	Colombia (1970)	Japan (1965)	Korea (1966)	Philippines (1961)	Taiwan (1964)	Turkey (1963)	United States (1967)
Agriculture	0.62	0.81	0.78	0.66	0.47	0.81	1.13
Mining	0.66	0.51	0.60	0.70	0.48	0.68	0.72
Food	1.72	1.35	1.09	1.46	1.37	2.31	1.44
Textiles	1.05	1.09	1.11	1.14	1.53	1.17	1.38
Clothes	—	1.09	1.12	0.96	1.19	1.18	1.17
Wood, paper, printing	1.02	0.91	0.94	1.01	2.21	1.30	0.98
Rubber	—	0.89	0.99	1.09	1.14	0.98	0.93
Chemicals	1.28	1.31	1.01	1.31	1.07	1.16	1.19
Petroleum	1.77	3.02	2.39	2.97	0.80	0.70	1.62
Cement and minerals	0.82	0.78	1.05	0.98	0.85	0.82	0.88
Metals	1.07	2.18	1.65	1.06	2.46	1.02	1.16
Transport equipment	1.77	0.94	1.08	0.82	0.65	1.59	1.11
Machinery	1.14	0.81	0.98	0.94	1.08	1.00	0.95
Utilities	0.60	0.66	0.80	0.71	0.63	0.89	0.84
Construction	0.68	0.82	0.84	0.79	0.71	0.89	0.98
Trade	1.05	0.50	0.65	0.57	—	0.57	0.60
Transport	0.66	0.50	0.69	0.66	0.57	0.78	0.69
Communications	0.54	0.39	0.58	0.59	0.39	0.55	0.55
Services	0.55	0.44	0.65	0.59	0.40	0.55	0.69
Other	—	—	—	—	—	—	—
Average income multipliers	3.94	4.48	2.45	2.17	16.60	20.61	74.19

— Not available.
Source: World Bank data.

and secondary information sectors, respectively. Thus defined, the information sector includes many productive activities that are traditionally counted under the tertiary economic sectors (for example, services and government) as well as organizational activities that are lumped with the organization's main outputs in the primary (agriculture, mining) and secondary (manufacturing, construction, transportation) sectors.

Empirical work relating to this objective has grown out of a literature, starting in the late 1950s, on the role of information at both the market level and within individual economic units. The former is generally known as "economics of information,"[21] whereas the latter falls within the fields of "decision theory" and "modern organization theory."[22] A third body of related literature has also gained some prominence since the mid-1960s. It focuses on the broader role of all communications in national development and uses concepts from several disciplines, traditionally sociology, but increasingly also anthropology, psychology, political science, and geography. Until relatively recently such "communications" research focused almost entirely on mass media, although in recent years two-way communication has gained some attention.[23]

Modern organization theory primarily relates to the need to reconcile the somewhat diverse objectives of individuals and special interest groups within organizations. In so doing it emphasizes the processes by which expectations are formed and decisions are made under conditions of uncertainty. The economics of information, however, is concerned at a more aggregate level with types of information-related market failures and analyzes supply and demand decisions made under conditions of uncertainty. Both explicitly recognize that incomplete information and uncertainty are prime factors facing individual decisionmakers. Hence, the greater the extent to which communication can be used to reduce uncertainty, and thus increase the probability of a correct decision, the better the chance organizations, economic units, or individuals within those units will have of achieving their goals.[24]

The implications for the telecommunications needs of developing countries are not trivial. Communication emerges as a crucial factor in the performance of both individual economic units and of markets. Also, communication is necessary for any division of labor to occur, since the process of development implies specialization and greater interdependence. In a developing country, a subsistence farmer does not have to interact significantly with persons outside his own family. If, however, his output is to be raised above the subsistence level, communication must take place to facilitate the division of labor—the specialization by function—that increases productivity.[25] For a change to occur, communication, no matter how slight, is necessary for the initial coordination, reorganization, and perhaps technical innovation.

In a more urban setting it has been documented that some firms and individuals use more up-to-date techniques and are more ready to innovate than are others. The innovators have, among other things, knowledge about techniques, procedures, and opportunities that the others lack. Some literature in the communications research field examines the question of diffusion of innovations.[26] Early neoclassical economists pointed out the innovative role that entrepreneurs play in economic development, but they were never particularly interested in the process through which such innovative entrepreneurs arose. Communications literature, however, examines the place of communication in the process of technology transfers and investigates the characteristics of innovators and successful allocators, who in fact tend to be early adopters of new procedures and technologies.[27] The presence or absence of a communications infrastructure is an important factor in their findings.

In the broader context of a national economy—whether free market or centrally planned—information is required to coordinate economic activities. In a free market, producers and consumers attempt in their own self-interest to improve their decisions by reducing uncertainty, while in a centrally directed economy, where the division of labor between producers and decisionmakers is more marked, a rapid feedback of information can reduce the uncertainty associated with the myriad variables facing each group.[28] Following from this, several information theory analysts have argued that, from a communications perspective, an economic system consists primarily of two functions: first, the technical maximization of production (given resources and resource costs) and the technical minimization of the costs of distributing goods and services and, second, the exchange of information associated with organizing and coordinating production and distribution.[29] They note that the former is associated with production and distribution costs that depend on resource costs and technology, whereas the latter encompasses transactions costs that depend partly on the mode of resource allocation and on the communications infrastructure.

Empirical work based on a reaggregation of data previously collected for more traditional input-output or national income analysis has in fact shown that the organization and coordination of economic activity consume a substantial and increasing proportion of the resources of industrial countries. For example, for the United States, Porat estimated that about 46 percent of 1967 GNP was associated with the information-related processes of producing, processing, and distributing information goods and services. For 1970 he found that information-related activities accounted for more than 40 percent of the labor force and 53 percent of all labor income.[30] Subsequently, Jonscher attempted to estimate the proportion of U.S. resources devoted specifically to "organizing the economy—to directing, coordinating, monitoring, and recording economic activity."[31] Using a reaggregated version of Porat's original data, he found that roughly 43 percent of labor income in 1967 was associated with an organizing activity, with the remaining 57 percent being associated with doing the things that were being organized.

Similar empirical analyses of the information sector have been undertaken in other industrial countries. Halina, using Porat's classifications and definitions, estimated that 47 percent of Canada's GNP is associated with information sector activities.[32] Exercises undertaken in Australia, the United Kingdom, and the former Federal Republic of Germany indicated the same broad pattern for the importance of the information sector labor force; the percentage of population employed in the information sector was estimated as 28 percent in Australia (1971), 37 percent in the United Kingdom (1971), and 31 percent in Germany (1970).[33] For Japan, Uno has shown that "knowledge workers" encompassed 12 percent of the labor force in 1975 and have been increasing since 1960 at an annual rate of almost 5 percent, while other types of workers have been increasing at only 1 percent annually.[34]

It is therefore evident that the information sector or information-related portions of industrial economies are large and, at least until the early 1970s, were growing relatively rapidly. In recent years, growth in the information sector appears to have leveled off. In both the United States and Japan, the proportion of total GDP comprised by the information sector has remained at about 45 and 35 percent, respectively, since the early 1970s, although the proportion of "knowledge workers" in the total work force continued to grow into the 1970s.[35] Information activities in the primary and tertiary sectors have also continued to grow in industrial countries, as shown by data from a 1980 study on what was then the Federal Republic of Germany and from studies on rural regions of four European countries undertaken by the Special Telecommunications Action for Regional Development (STAR) program.[36]

During the 1980s, similar studies were undertaken in a growing number of developing and newly industrialized countries. Studies by Jussawalla and others in the Pacific Rim area have focused on backward and forward multipliers and linkages between the information sector (limited to the primary information sector because nontraded, internal information on organizational activities is difficult to include in the data) and all other sectors of a country's economy.[37]

The Pacific Rim is a natural focus of interest for this kind of research. It is an area where considerable investment in information-intensive industry, infrastructure, and services has occurred in developing and newly industrialized countries during the past ten to fifteen years and where the information sector is expected to provide the best opportunity for strengthening economic growth. In these studies, broad definitions of the information sector and subsectors were applied to each country's standard classification of the industry, and national interindustry accounts were then recast to highlight the information sector as separate from the traditional divisions of primary, secondary, and tertiary sectors. On the basis of the recast input-output tables, calculations were made to determine the information sector's value added, its total output, and the coefficients that measure its relation to other sectors.

In most of the countries studied, the primary information sector was found to have relatively high backward-linkage effects compared with other sectors in terms of derived demand for other goods and services produced in the economy and in terms of lowered input costs for other sectors of the economy.[38]

However, in the four largest countries studied—Indonesia, Malaysia, the Philippines, and Thailand—the information sector was found to be a smaller part of GDP than all other sectors, despite the recent, strong growth of their information infrastructure. For Singapore and Taiwan, where growth of the information sector is widely considered to have been spectacular, the information sector is still smaller than either the secondary or the tertiary sectors, although larger than the primary sector. Furthermore, in the case of Singapore, the information sector's share of GDP remained roughly constant at about 25 percent between 1973 and 1978, and labor productivity appeared to be no higher than average for the economy.

Overall, much of what is expected about the information sector is neither supported nor refuted by the data, and some key hypotheses still need to be tested. In particular, the presumed desirability of shifting the economic structure from agriculture and industry to information in the developing world should be examined in terms of broader macroeconomic considerations, and the assumption that the information sector poses exceptional opportunities for economic development must be substantiated.[39]

Where does telecommunications infrastructure fit into this picture? In a macroeconomic sense, the large unsatisfied demand for telecommunications services in developing countries and the high returns on new investment are evidence that the perceived communications needs of both producers and consumers are not being met. Not meeting this demand may worsen the unequal distribution of information between the parties and diminish the opportunity for transferring information.[40] The extent to which this retards increases in productivity by slowing functional specialization and the division of labor and the extent to which communication facilities are inadequate for the efficient organization and coordination of economic activity are not explicitly known.

In the context of increasing globalization and information intensity of economic activities, however, a plausible hypothesis is that although the gap between the telecommunications infrastructure of industrial and developing countries is narrowing (see chapter 1), a veritable information gap is building. It is still not clear how important this gap might be nor how policymakers could put in practice the broader concept of informatics.

Value of Input-Output Analysis

Input-output studies convey a general picture of the role of communications services in the overall structure of an economy. Of greater interest, however, is the contribution that such analysis can make to assessing or quantifying the benefits of telecommunications investment and to planning the scale and nature of services that should be provided.

Assessment of benefits depends on the functional relation between the observed communications inputs used by enterprises or organizations and their economic outputs, that is, on the nature of the production function. If the observed input coefficients varied little across countries, this could indicate a stable production function with fixed coefficients. In other words, one could conclude that managerial and technological considerations make it essential to use a fixed amount of telecommunications services and that if this amount is not used, the output of other sectors will drop. From the limited data available, however, this does not appear to be the case. Nevertheless, at least in the trade and services sectors, there is less difference among countries than in the other sectors examined, which suggests that a finer breakdown of sectors might identify some activities (banking, foreign trade, or tourism) where a fixed-coefficients model would, in fact, fit.

For sectors other than services and trade, the observed results can be interpreted in several ways. One possibility is that the variation in coefficients reflects some wasteful use of communications, particularly in higher-income countries. Another more likely possibility is that the quality of communications output varies greatly among countries; sometimes supply is less timely or the quality is relatively poor.[41] Finally, there may be systematic differences in sector organization, management, technology, work organization, benefits, and so forth between higher- and lower-income countries, including some substitution of other factors of production, such as increased use of labor or transport, for the intensive use of communications services.[42]

Whatever forces determine their size, input coefficients for communications might nevertheless indicate the minimum levels of demand from different portions of the economy that must be met if the existing allocation of resources is to remain unchanged. Consequently, input-output analysis might help to forecast demand, not so much in cases where the economic structure is changing slowly (where econometric methods might be more appropriate), but in cases where relatively large structural changes are expected. Examples that are not uncommon in developing countries would be the initial development of a large-scale tourism industry, a major expansion in mineral extraction, the creation of a new manufacturing sector, or the transformation of agriculture from subsistence to commercial forms.

At least one attempt has been made to estimate input-output demand for the United States. Using the 1963 U.S. input-output table and projections of future gross outputs by each sector, Bower estimated aggregate investment requirements for telecommunications.[43] The estimated gross outputs of each sector were multiplied by telecommunications input coefficients to arrive at the total industrial demand for telecommunications in a specified year. The final demand for telecommunications was estimated by assuming it to be, as a proportion of total demand, a function of the relative share of residential telephone connections in total connections. Once estimated, total demand was converted to a capital investment requirement by using an estimated capital-output ratio for the telecommunications sector.[44]

The results observed above in which input coefficients were shown to vary significantly across countries do, of course, imply that any such forecasting method would have to be based on local input-output data for each country. Such a method cannot be applied in the many cases where such data are not available nor in the many developing countries where levels of unmet telecommunications demand are high and where current investment in the sector is clearly suboptimal. Of course, it is not known if Bower's estimates are significantly more reliable than those obtained using other, much simpler estimating procedures.

Using input-output analysis to allocate telecommunications investment implies that the cost of not meeting the communications requirement indicated by the input coefficients is likely to be significant. In terms of economic theory, this assumes that the marginal revenue product of telecommunications services as an input to the productive process in each of the other sectors is significant and that telecommunications services are not readily substituted by other factors of production; the elasticity of substitution is low. Although these appear to be reasonable assumptions, their validity can be settled only at a more microeconomic level of analysis.

Problems with Input-Output Analysis

Other difficulties arise in using input-output techniques to analyze the relation between telecommunications and other economic activity. First, the usual industrial classification schemes used to compile input-output tables are inadequate for examining relations between telecommunications and other sectors. As most input-output tables are used primarily to study the flow of material goods, the highest level of detail is shown in the agricultural and manufacturing sectors. Postal services and telecommunications are rarely separated; this is a serious deficiency since their input requirements (especially capital and labor) are very different, as are the nature of the demands for their output.[45] Also, the service sector is often treated as a single group. Since this sector consumes 50 percent of all communications services in most countries, while manufacturing consumes about 20 percent and agriculture less than 1 percent, a more comprehensive analysis of communications use would require a more detailed classification within the service sector.

A second problem is a deficiency inherent in all cross-country input-output comparisons. Since prices across countries vary widely and in some instances bear little relation to costs, conventional input-output data expressed in terms of value of transactions may not accurately reflect the actual level of activity expressed, for example, in millions of telephone calls or tons of steel.

A third problem also inherent in all cross-country input-output comparisons is that the original raw data are subject to inaccuracies. Input-output tables are only as accurate as the data on which they are based, and this accuracy varies widely, particularly among developing countries. Analysis of a sector such as communications, which in most countries is relatively small, is especially liable to statistical errors.

A fourth problem is that data on the size of the information sector do not reveal much qualitative information about the sector; for instance, they do not address the extent to which a particular sector is information-intensive. The numbers can thus be interpreted many ways, leading to different conclusions and policy prescriptions. If an information sector were found to exhibit larger-than-average economies of scale, a high proportion of GDP in this sector could signal a small economy, rather than an advanced one, and the growth of services (which are partly counted in the information sector) could be either a positive sign of economic modernization or a negative sign of a bloated public sector, which is a major target of efforts to reform economic policy in many developing countries.[46]

A fifth problem, which in many ways is the most limiting, would impose a major roadblock on the usefulness of cross-country or through-time input-output analysis for telecommunications, even with accurate data, comparable price levels, and finely disaggregated activities. As has been noted frequently throughout this book, in developing countries the volume of purchases of telecommunications services is limited in most cases by inadequate supplies. An equilibrium between demand and supply cannot be assumed; developing countries typically experience acute and persistent shortages in the supply of telecommunications services, and the quality of service is often poor. Hence, existing expenditures that various sectors or economic activities make on telecommunications may not have much normative significance.

Notes

1. Examples of primary sectors are agriculture, fishing, forestry, and mining and quarrying. Secondary sectors include construction and manufacturing. Tertiary sectors include commerce (wholesale and retail), communication (posts, telecommunications, press, and

broadcasting), finances (banking, insurance, and real estate), public administration, transport (including storage), utilities (gas, electricity, water, and sewerage), and other services (education, health, personal, and professional).

2. CCITT (1965) pp. 32–33.

3. Lönnström, Marklund, and Moo (1975).

4. This study used TELEBRAS data for all of Brazil covering the years 1975 to 1980. More recently, this work was enlarged by a follow-up study, which found that the impact of telecommunications services on economic growth varies according to the development stage of the region involved. Telecommunications were compared with two other basic elements of infrastructure, energy and transportation, and with regional levels of revenue to determine the role that each played as economic growth occurred. The study found that the demand for telecommunications, as measured by density of installed telephone lines, was relatively strong in the early stages of development. As industrialization began, investment in energy was given higher priority than investment in telecommunications, but when industrialization reached a certain threshold, the increase in commercial activity stimulated a second surge in demand for telecommunications services. See Guiscard Ferraz (1987). A similar relation between growth in telecommunications infrastructure and economic growth was found in Korea, as measured by the ratio of elasticity of main lines to GDP over time. See Gille (1986).

5. Nora and Minc (1980).

6. To the extent that the data are based on the distinction that telecommunications companies make between business and residential lines, the results may be somewhat in error. Particularly in a developing country, the borderline between home and workplace is sometimes unclear, and in some instances a tariff provides an incentive to classify telephones as residential. See chapter 10 for further discussion of this point.

7. Kaul (1979).

8. The proportion of postal expenditures to total postal and telecommunications expenditures was 30 percent in 1976–77.

9. Gille (1986).

10. Pye and Lauder (1987).

11. Input-output is used here to denote national data detailing transactions among industrial sectors (inputs and outputs) and final consumption purchases made by households and government according to the linear input-output models developed by Leontief and others. For an introductory description of input-output analysis and some of its uses, see Miernyk (1965). For a specific application of input-output analysis to the information sector, see Karunaratne (1988).

12. World Bank staff member Colin Warren carried out much of the initial work on the input-output analysis discussed in the following pages. To build up a comparable data set for a larger number of countries, relatively old data had to be used. It would be possible to argue that dependence on communications services has increased since the time of the studies examined.

13. United Nations (1969).

14. Value added refers to the difference between the value of output of the industry and the cost of inputs not produced in the industry.

15. As noted in chapter 1, the small towns and rural areas of most developing countries have a large amount of unsatisfied demand for telephone access. Hence, observed usage by the agricultural sector could be considerably less than an efficiency-related optimum.

16. Of course, a small farmer making a call to the local market in a developing country is probably not counted as part of the agricultural sector. Also, the low coefficients reflect the lack of access to telephones in rural areas of developing countries. However, as noted in chapter 11, a survey in Kenya showed that although the agricultural sector may not purchase extensive quantities of telecommunications services, the variety of communications

contacts that it does maintain is extensive and covers a relatively large number of other productive sectors.

17. It is important to bear in mind the detail that may be lost in statistical aggregation. Whereas in developing countries telecommunications inputs are small for agriculture in general, they may be large for certain activities important for general economic strategy (such as plantation agriculture for supplying export markets or food to nearby cities).

18. Some of the data from the Philippines are thought to be unreliable.

19. Except for trade in the Philippines and trade and services in Turkey.

20. Chandler (1977), p. 490.

21. One of the first papers written was Stigler (1961). Among the more recent are Jonscher (1980; Jonscher (1984); and Lamberton (1990/91).

22. See, for example, Simon (1959); Cyert and March (1963); and March and Simon (1958).

23. One of the better-known pieces of this general communications literature is Schramm (1964). For other examples, see McAnany (1980) and Hudson (1984).

24. Arrow (1980) has argued that because information costs are generally independent of the scale upon which information is used, organizations that rely relatively more heavily on information can derive economic benefits disproportionately large in relation to their size.

25. For example, in East Africa, although the smallholder agricultural sector traditionally has not had a high potential use for telephones, the increasing production of cash crops is steadily changing this. Okundi, Ogwayo, and Kibombo (1977) p. 51.

26. Ashby and others (1980)

27. Contreras (1980), chap. 5.

28. Arrow has characterized market failure as "the particular case where transactions costs are so high that the existence of the market is no longer worthwhile." See Arrow (1970) p. 17.

29. Porat (1977) stated the same concept as functions involving the transformation of matter and energy from one form into another and the transformation of information from one pattern to another. Jonscher (1980) subsequently extended the analysis by explicitly specifying a model that includes both a production process and an information process.

30. Porat (1977).

31. Jonscher (1980) p. A.4.

32. Halina (1980).

33. Lamberton (1982), pp. 36–59. The U.K. and German data were cited from, respectively, Wall (1977) and Lange and Rempp (1977).

34. Uno defines knowledge workers as persons who produce and distribute knowledge. Empirically he counted them as natural science researchers, engineers, judicial workers, accountants, educators, physicians, dentists, pharmacists, writers, reporters, editors, fine artists, designers, photographers, government officials, managers, and clerical workers with college degrees. See Uno (1982) pp. 144–58. In the United States, Machlup (1962) first attempted to measure the share of the U.S. GNP connected with "knowledge" as opposed to other kinds of activities.

35. Wellenius (1988).

36. Deutches Institut für Wirtschaftforschung (1984) and Pye and Lauder (1987).

37. Jussawalla, Lamberton, and Karunaratne (1988).

38. Jussawalla (1986).

39. This section is based on material published in Wellenius (1988).

40. Spence (1975) has suggested that the organizational problem is to allocate resources under conditions of "imperfect and asymmetrically located information."

41. The wide discrepancies among countries in the quality and timeliness of telecommunications output were discussed briefly in chapter 4.

42. In the United States, for example, the input coefficient data show that the radio and television broadcasting industry spends $4 on telecommunications services for every $100 of gross output—a result of its heavy use of wideband transmission channels. This would be a poor basis for planning in a country where television broadcasting is not widespread, where physical distribution of audio and video tapes (with increased use of labor and vehicles) may be cheaper than real-time transmission, and where the corresponding delay may not be considered important.

43. Bower (1972).

44. Bower's suggested method goes a step farther than the CCITT 1976 GAS-5 handbook's section on input-output analysis, which notes that input-output tables might be used for telecommunications planning; Bower suggests estimating telecommunications investment requirements using projected outputs in all sectors of the economy.

45. Little is known about substitutability or complementarity of telephone and postal services. Although some rather descriptive information relating to the subject is presented in chapters 11 and 12 and appendix B, few attempts have been made to estimate the cross-elasticity parameters econometrically: one such attempt was a time-series study for the United Kingdom carried out by the Statistics and Business Research Department of the British Post Office. Despite the use of acceptable econometric techniques, it failed to produce statistically significant estimates of the coefficients.

46. Wellenius (1988).

Chapter 6

Location and Communication

THE PREVIOUS TWO CHAPTERS discussed statistical correlation and regression analyses of cross-sectional and time-series country data as well as cross-country comparisons of input-output sector indicators. Such exercises by definition examine telecommunications services and economic activity at a rather aggregate macroeconomic level. Several of the following chapters examine the other end of the spectrum by reviewing selected microeconomic project- and program-related attempts to identify and quantify the benefits of telecommunications as individuals and groups perceive them. This chapter and the next, however, examine some of the middle ground between the global analysis of previous chapters and the microeconomic exercises of later chapters. This is done primarily by, first, reviewing several issues pertaining to spatial aspects of telecommunications investment and, second, in a related sense, examining some of the existing evidence about general linkages between telecommunications investment and the transport and energy sectors.

The Spatial Framework

A voluminous literature addresses the question of why economic activity or specific forms of economic activity locate where they do.[1] The literature recognizes that an inevitable consequence of economic development is increasingly complex interdependency among individuals and organizations. In general, the various models formulated by

121

economists and regional scientists assume that businesses, workers, and consumers tend to locate in places that minimize the net costs of production and consumption. Various factors affect these costs, such as population skills and density, topography and natural resources, climate, availability of modern technology, existing infrastructures, pricing policies, government controls and regulations, and political boundaries. The availability of rapid and reliable communication is, of course, another important consideration in the decision governing location, that is, a reliable telecommunications system has significant space-bridging qualities.[2]

In a country or a region within a country, a basic spatial order is often clearly recognizable. Centers (cities, towns, or villages) can be ordered or ranked according to the groups of central functions found in them.[3] Such a ranking defines a hierarchy of central places, which is closely related to a hierarchy of population size, of economic functions, of economic units, of government administrative centers, and of ways of life.

Several analyses of the relation between central places and their hinterland have examined the number and distribution of existing telephones. Christaller, in pursuing the hypothesis that a hierarchy of places depends on the relative size and economic importance of central places, used the number of telephones as a proxy for a common measure of centrality in his examination of the size, number, and distribution of central places in southern Germany in 1933.[4] In 1982, Kilgour examined central places in Costa Rica and, using Christaller's methodology, concluded that the geographic placement of the telephone in national space elucidates a country's spatial structure and identifies efficiently organized areas as well as underorganized ones. She found that the spatial structure in Costa Rica conforms to the interlocking structure predicted by Christaller and argued that the telephone could be used in any country to indicate centrality and that this indicator would be comparable across countries.[5]

Other telephone-related analyses also support the contention that the flow of people, communication, energy, and goods that produces the interdependences among a region's centers to a large extent reflects a basic spatial order. The following two examples illustrate this in the context of telephone traffic.

Example 10. Spatial Order and Regional Telephone Traffic (Chile 1971)

The relation between long-distance telephone traffic and the structure of central places was partially examined by using 1969 data on twenty-two cities and towns in the Aconcagua Valley region in central Chile.[6] A high concentration of long-distance traffic was observed: the average small city or town exchanged 50 percent of all calls with just one other place, 71 percent with two places, and 81 percent with three places.[7] Traffic also tended to follow the lines of the population hierarchy: as many as 90 percent of the region's smaller centers exchanged at least 60 percent of their calls with larger places.[8]

Given this high concentration of long-distance traffic and its tendency to follow the population hierarchy, it was hypothesized that it should be possible to estimate independently the ordinal distribution of a large proportion of each place's total long-distance telephone traffic from a relative measurement of the interdependence among pairs of centers.[9] To this end cultural, economic, administrative, and locational interdependence indicators were combined to rank overall interdependence between a center and each other center. Postulating that traffic between one center and all others is distributed in the same order as overall interdependences, the ordinal distribution of 80 percent or more of each place's long-distance telephone calls was estimated accurately.[10]

Example 11. Hierarchical Level of Contact (Kenya 1980)

Another study, in two regions in western and south-central Kenya, also examined the level in the hierarchy of towns or villages among which contacts were taking place.[11] Telephone calls into and out of four general types of towns within the two regions were examined; the towns from largest to smallest were denoted as principal town, urban centers, rural centers, and market centers. In all cases more calls were received than were initiated, and, as would be expected, telephones were used most intensively in the principal town. Presumably this last result is partly because access to service was greater in the principal town, which therefore had more subscribers to call. Also, most calls were terminated within the region in which they originated, and the proportions were relatively higher for smaller places.

The most notable result is that the called parties tended to be at either the same hierarchical level or a level above the calling parties. This higher level was generally the principal town at the center of the region. Furthermore, contact among places at the same level decreased as the level of hierarchy decreased. This effect spilled over into telephone communication patterns with the rest of the country. Hence, principal towns and urban centers generally had a much wider horizon of contacts.

As the exercises for Chile and Kenya illustrate, a spatial analysis of the interactions between telecommunications and economic activities at the regional level can be used to describe and perhaps even predict the patterns of telecommunications traffic as a function of regional structure and change. Such studies also suggest that development tends to compel more specialization and less self-sufficiency.[12] In essence, a development process generates the need for greater connectivity: the more development, the more specialization, the more interdependence, the more need for connectivity.[13]

For national development policy, however, the above descriptive material is of limited value. It would be of greater interest to understand the extent to which telecommunications services can be used to influence locational decisions, rather than merely to reflect them. Three of the most common spatial problems experienced by developing countries are the following:

a. The heavy concentration of employment, especially in modern sectors, in one or a few urban centers, and the difficulty of inducing modern enterprises to locate in smaller centers or the hinterland
b. The rapid pace of migration from the countryside to major urban centers—one cause of the very rapid growth of urban populations—and the corresponding growth of the largely unplanned squatter settlements sometimes referred to by planners as the "transitional urban sector"
c. The apparent lack of major success so far in implementing, on a large scale, the multicentered urban structures that city planning studies frequently recommend for developing countries.

Work in locational analysis explains these patterns, in part, by examining communications linkages among organizations, and empirical research in the industrial countries suggests that the enhancement of telecommunications infrastructure can have a significant effect on the location of economic units.[14] In recent years, several studies have examined the role that telecommunications has played in the economic development of rural and less favored regions of Europe and North America. Interest in this area arose primarily in response to the structural changes that accompanied the dramatic increase in information-intensive activities taking place in the growing service sector during the postindustrial phase of economic development.

These changes present rural regions with new opportunities for growth and development but also threaten to increase economic marginalization if such opportunities are not aggressively pursued. In particular, the traditional objective of universal service may not be practicable in an environment of technical change and liberalized telecommunications policy, and structural changes may increase the gap between information-rich and information-poor regions.

In North America, investment in advanced telecommunications infrastructure in rural areas is seen as an important catalyst for local economic growth, and failure to invest is predicted to cut such regions off completely from the economic mainstream. Furthermore, investment in telecommunications alone is not sufficient; complementary investment in such areas as transport are needed if the potential gains from increased productivity are to be realized.[15]

For rural Europe, attention has focused on the supply of and demand for new telecommunications services in what the Commission of the European Council has termed less favored areas, paying special attention to the limited access afforded by private networks that are embedded within large organizations.[16] Expressed demand has generally been found to be low; one study on Greece found that of 425 provincial manufacturing firms surveyed, only 19 percent declared that "adequate infrastructure" affected their locational decisions, compared with over 30 percent that specified government incentives, low-cost land, and availability of labor.[17] Yet the potential benefits of increased investment are very high, particularly for the intermediate rural areas (as distinct from the extremely rural areas). One model that was developed to quantify the estimated aggregate gains in employment from investment in new information technologies in rural areas of Europe calculated that if all the rural and very rural areas of the European Community invested 1 percent of their GDP in new communications infrastructure, between 700,000 and 900,000 new jobs would be created, at a cost of about ECU11,600 per job.[18] In developing countries, however, little empirical work has yet demonstrated the extent to which suitably designed telecommunications policies and investment programs can help promote a specific pattern of settlement and estimate the resulting benefits.[19]

Regional Development and Balance

In developing countries, there is a widely held perception that the existing high degree of spatial concentration of modern sector activity is undesirable.[20] Following from this, government goals and policies are often aimed at dispersing development. This generally occurs at two levels: regional planning on a national scale, which seeks to achieve some form of balance between the levels of development in different regions or different cities, and, for the very large population centers, planning at the intraregional or metropolitan level, which typically calls for a pattern of several urban centers to relieve the heavy concentration in and around a congested central business district.

When looking at the problem of regional imbalance it is helpful to distinguish between structuralist and locationalist views. The former attributes slow growth to the unfavorable mix of economic activities in particular areas (a preponderance of slow-growth sectors or activities). In contrast, the locationalist view emphasizes spatial and infrastructural factors, such as the physical accessibility of particular locations, the absence of services such as water supply and power, and so on. It has been argued that for effective change, the structuralist point of view may be more relevant for more mature industrial countries, whereas the locationalist view may have greater relevance, at least in the near term, for many developing countries.

However, some arguments attempt to integrate elements of both the structuralist and locationalist approaches.[21] Nicol, for example, seeks to analyze the type and nature of effects that improvements in communications may have on economic performance and structure, and subsequently on the location of economic activities, in developing countries. This argument differs from traditional economic approaches, which do not treat information and communications as decision variables, in that it posits that access to information is a fundamental determinant of economic and spatial structures and that communications weakens the ties of proximity in the distribution and organization of economic activity in space.

Consequently, improved interregional communication should reduce information and transaction costs to businesses, thereby reducing the relative benefits of clustering in large urban areas. If firms have less need to cluster to capture agglomeration economies, they may relocate to smaller secondary centers where the costs of land and resources tend to be lower and where there is less congestion and fewer environmental constraints. It has also been suggested that improved telecommunications can ameliorate unequal income distribution among regions within a country by functioning as a partial substitute for physical proximity, if it can operate in tandem with an efficient physical transportation network.[22]

However, the evidence is ambiguous. The opposite point of view has also been argued in the case of rural areas of Europe, where the costs of transmitting information remain exceedingly high over distances, in spite of technological advances, and agglomeration economies have been strengthened by the introduction of new patterns of handling information.[23]

If it is accepted that improved telecommunications might facilitate achievement of locational objectives along with other incentives to locational change, relevant policies might take one of two forms: more generalized locational inducements, which have been familiar policy instruments in several industrial countries, or a more focused approach, which concentrates public investment and inducements to private investments at specific locations, sometimes designated as growth centers or growth poles.

A regional type of general inducement incentive was used in Sweden in the 1970s, where the relatively disadvantaged northern regions (Norrland) got priority in service improvements and public investment, and there was a large financial cross-subsidy from the central regions through telephone rates.[24] In 1973, the Swedish government reduced telephone charges for firms and individuals in Norrland by 20 to 50 percent. Calls made within the Norrland counties to the primary centers—usually the administrative headquarters—were often reduced 35 to 50 percent, and local calls made within the same community, 50 to 70 percent.[25] The resulting financial subsidies for telephone service in Norrland were relatively larger than the corresponding subsidies adopted for freight transport.

A telecommunications investment policy to consider explicitly the problems of disadvantaged regions and a shift in pricing policy to discriminate less against long-distance calls both seem on the surface to be relatively attractive ways to promote regional development. Also, the relatively high prices charged for medium-distance calls, which cross the boundaries of charge areas for local calls, might on occasion be mitigated by expanding local call areas to the maximum extent possible, while attempting to maintain a reasonably direct relation between price and costs.[26] The latter could encourage economic interaction and the development of local markets within the region, possibly even a more important goal than improving interregional linkages. In fact, it can be argued that effective intraregional policies should be implemented first, since without them, improved interregional linkages might only increase the dependence of the disadvantaged regions on, and their propensity to import from, more strongly developed centers.

Another approach is to stimulate economic development by more focused measures such as investing in communications or other facilities at designated growth poles.[27] For example, evidence exists that within a developing country or region, the larger-income growth pole villages tend to derive the most benefits from the introduction of a public call office (PCO) telephone. The following example illustrates this.

Example 12. Rural Village PCO Benefit Regression Analysis
(Costa Rica 1976)

For a cross-section of sixty-four villages in rural Costa Rica data were gathered on public telephone use and on selected economic and social characteristics for each village.[28] The data came from three sources: official government census data on housing, population, and agriculture; rural telephone traffic data collected by the national telecommunications entity; and survey data about individual telephone users compiled by telephone concessionaires.

Several models were specified and statistical regressions estimated in which telephone traffic or telephone use variables were used as the dependent variable. It was hypothesized that if differences in telephone use could be explained on the basis of the different economic and social characteristics of the villages, then it would be possible to predict which villages, then without service, would benefit most from gaining service. Those villages with a high potential to benefit would then be placed high on a priority list for new telephone investment.

The results of the regressions suggested that PCO benefits tended to be greatest in rural Costa Rican villages that possessed one or more of the following (sometimes collinear) characteristics:

a. Per capita village income was higher than the average of all villages
b. The village had a relatively large population
c. The village was located relatively far from the major economic, social, and government center of San José
d. The educational level of the population was above average
e. The population tended to be clustered more closely around the site at which the telephone was located.

It was also observed that the mix of calls made for agriculture and business purposes was clearly correlated with the corresponding economic base and demographic characteristics of the villages.

Migration between Rural and Urban Areas

Migration on a large scale from rural areas to large cities, sometimes through smaller regional centers, is generally acknowledged to be one of the most difficult problems facing developing countries—although some observers note that it also creates opportunities for economic and social development. This rapid flow of population often strains the social and economic overhead capital of the big urban areas, and it is widely assumed that if population flow were slowed, the big cities would be better able to absorb and generate employment for new immigrants and to cope with internal development programs. Investment in rural infrastructure, including roads, water supply, and telephones, has been suggested as one way to slow rural to urban migration because these improvements can reduce some of the basic disadvantages of living in a rural location.[29]

It is difficult to disentangle the negative and positive effects on migration of a PCO investment program in rural areas. Access to current agricultural prices, more efficient use of transport and movement of supplies, rapid flow of information in times of medical need or natural disaster, and so forth could improve conditions in rural areas of developing countries and may in fact create employment opportunities. People migrate for various reasons, however, and the relative short- or long-run lure of jobs, higher incomes, and educational and other opportunities in urban areas would not be greatly changed in the short run by providing a rural telephone connection. In fact, it might even be argued that a rural telephone program would encourage migration since the migrants could more easily keep in touch with their home village and family and would not have to endure long periods without access to local information.[30]

Metropolitan Development and Balance

Regional policy at the metropolitan level tends to mirror policies for planning location and land use at the national level. Metropolitan areas clearly have locales where the private benefits of a central location are sufficient to induce firms to pay large premiums, in spite of problems of congestion—problems that have led many planners to contend that the social costs of such spatial concentration outweigh the private benefits.[31]

Improved telecommunications may reduce such costs. For any given size, a city with a better system of two-way communication will enjoy a higher level of connectivity and is likely to achieve more agglomeration economies than one with a worse system. Also, with a fixed level of communication, larger cities will achieve higher agglomeration economies than smaller ones because they offer larger communities of interest.[32]

Nevertheless, in many developing countries, governments are planning more dispersed patterns of settlement within the urban region. These policymakers contend that the large and rapidly growing cities of poorer countries cannot afford to mirror the high-density spatial patterns of cities in richer countries. They argue that in developing countries jobs and residential locations should be more contiguous spatially, partly to reduce transport costs so that more family members in the lower- and middle-income groups can have access to employment opportunities.[33]

There are numerous reasons for the inability of developing countries to disperse economic activity significantly. One of the prominent reasons for this failure is the overdependence on relatively slow and costly face-to-face communication and messengers and the related absence of reliable and widespread telephone service. Webber argued that "a major contributor to transport congestion in such concentrated centers as Caracas is the thousands of messengers who race around the city carrying pieces of paper from one office to another ... Rather than sinking its investment capital into an underground rail system, Caracas might have realized a far greater return had it invested that same capital in the telephone and postal systems."[34]

Some evidence on the relation between telecommunications access and transport is discussed in the following chapter. With particular regard to the use of metropolitan space, the form of the metropolitan region is largely determined by the interaction of workplace location, residential location, and commuting. Well-focused telecommunications investment might modify the demand for travel, particularly regarding the journey to work, by altering two key locational processes: the choice of location by employers and the choice of residential location by households. The latter decision could, in turn, affect the location of workplaces by changing the supply of available labor.

Pool, in fact, noted that the telephone has had a profound effect on the spatial patterns of cities in the United States. He outlined the process by which, even as the centralization of business activities led to the growth of high-density centers in cities, the telephone encouraged the scattering of particular business districts. Once the telephone was available, businesses could move to cheaper quarters; a firm could move out from city centers or even up to the tenth or twentieth floor of new buildings. As early as 1902 a writer in *Telephony* calculated that if business continued to rely on messengers, the elevator wells of skyscrapers in big cities in the United States would have to be about double the size they were, which would make such buildings uneconomical.[35]

The general tendency throughout the world is still for business and industrial activity to be concentrated spatially, whereas residential areas are widespread. As cities grow in size, the journey to work becomes longer. This problem is particularly acute for developing countries. In a city of 1 million persons, the average journey to work is approximately 3 miles; in a city of 5 million, 7 miles. Some low-income groups in Asia and Africa reportedly walk two to three hours every day to reach their place of employment.[36]

Webber contended that for developing countries a feasible way to reverse such a trend is to undertake a large and focused investment in postal and telecommunications systems. He noted that "demand for connectivity generates city expansion; indeed it is the sole source of urban growth. If connections through geographic space were not costly, there would be no cities; and there would be no central concentrations within them either. People and organizations concentrate in cities only because the costs of connectivity are least where linked establishments are close to each other. That's also the reason they tend to concentrate in city centers, because each of them is striving for lower costs of being connected."[37] Following from this, he argued that greatly increased investment in modern telecommunications and postal infrastructure would "permit spatial relocation of many establishments for whom a central location is presently the only viable option. Phones, post, and the parallel data-transmission and video communications channels all are inherently random-access technologies. They do not rely on spatial concentration for their effectiveness. When they work (technically) well, they actually work best in spatially dispersed settings. These are systems that foster connectivity when partners to transactions are not spatially contiguous."[38]

There is a growing body of industrial-country literature on office location, such as Nilles's in-depth study of the possibilities for dispersing work from a major California insurance office.[39] Even for industrial countries with much of their infrastructure in place, improved telecommunications might promote locational changes that are thought to be economically, socially, or environmentally advantageous. Representative research on industrial countries was reviewed by Pye and Goddard, and a simplified framework for assessing the costs, benefits, and effects of such locational changes was presented by Harkness.[40] In general, however, these issues have not been sufficiently researched, and in a developing country where the greatest potential for innovative change may lie, little research has been undertaken.

Nevertheless, even if policies to increase the attractiveness of life in small villages and towns and to promote land use planning in metropolitan areas are strikingly successful in the foreseeable future, the large unplanned or transitional urban squatter settlements typical of most cities in developing countries will continue to grow.[41] Much of the population of these urban squatter settlements works in an informal economy—an economy that largely provides services and that has been characterized as having unregulated competitive markets, small-scale operations, ease of entry, family ownership, reliance on indigenous resources, labor-intensive and adaptive technology, and reliance on skills acquired outside the formal educational system.[42]

The mechanism to enhance the economic contribution and resulting welfare of urban squatter settlements cannot be totally divorced from communication. Production linkages between firms or group associations of different sizes were a striking feature of the economic structure in countries that became industrialized in the past century. Large enterprises subcontracted work heavily and efficiently to small ones. The same pattern was followed later by Japan, Korea, and Taiwan.[43] Such linkages might also be facilitated in developing countries by improving the access to public telecommunications services throughout the country.

Notes

1. For summaries of the substance of this literature, see Greenhut (1963 and 1970); Richardson (1969); Nourse (1968); Isard (1969); and Hagget (1965).

2. For a discussion of factors affecting the population density of cities in the context of the transport sector, see Linn (1983), pp. 90–105. The concept of the space-bridging qualities of telephones is examined in Kilgour (1982).

3. For instance, in the region of example 10, the main cities are seats of municipal authority and have a standard set of government services (for example, treasury, inland revenue office, and local court of justice), elementary and high schools and sometimes also technical or vocational schools and occasionally university campuses, one or more hospitals, police stations under the charge of senior officers, main lines of road and railway public interurban transport, branches of one or more national banks, and a fairly developed telephone exchange service. In less important places, education is limited to the elementary level, national health service is available only at clinics staffed by paramedics, police stations seldom include officers with rank above lieutenant, and banks are rare. Small towns often have only an elementary school, a small police station, a health station, a public long-distance telephone, and intercity transport only to the next larger town. About one-fourth of the region's settlements have none of these services.

4. Christaller (1966).

5. Kilgour (1982) p. 139.

6. Empirical evidence is presented in Klein (1971).

7. All places outside the region were counted as one.

8. This is consistent with findings in other developing countries. In Costa Rica in 1976, for example, an average of 60 percent of calls from surveyed rural PCOs were addressed to the capital city. Likewise, in El Salvador, in 1977, approximately 70 percent of telephone calls from rural places were directed to the capital city. More recently, in Peru during the two-year period of 1984–85, 70 percent of calls originating from five small rural communities in the province of San Martín went either to Peru's two major cities (Lima or Trujillo) or to the commercial center of the province, Tarapoto. See Mayo and others (1987).

9. For example, for each place the arc carrying the highest number of calls, the second highest, and so forth is identified until a large proportion of traffic is accounted for.

10. Klein also found that the proportions of a center's traffic that were carried over its highest traffic arc, second highest, and so forth were roughly constant among centers of like characteristics. Hence, he proposed that the proportion of each center's traffic that was carried over each of its major arcs could be predicted by mapping standardized distribution curves on the ranking of arcs obtained from measures of interdependence. With this, the problem of predicting long-distance telephone traffic in a region would be reduced to forecasting the total volume exchanged by each center. This might be useful, for example, for designing a regional network in a newly developing area or for predicting the changes in traffic patterns likely to follow changes in regional structure.

11. Cleevely and Walsham (1980).

12. The contention that development tends to compel more specialization is of course a form of Adam Smith's principle that division of labor is limited by the extent of the market.

13. For an elaboration of this point, see Webber (1980).

14. Thorngren (1970); Pye and Goddard (1977).

15. Parker and others (1989) and Bradshaw (1990).

16. See, for example, Gillespie and others (1989) and Mansell (1988).

17. Kottis (1986).

18. Hansen and others (1990) and Analysys, Ltd. (1989).

19. One limited but interesting study used a theory of hierarchies of places to examine the interactions between telecommunications and economic activities in Kenya's Rift valley and central provinces. See Cleevely (1979).

20. For example, costs of transport congestion in the form of high vehicle operating costs and wasted time, environmental damage, accidents, high energy consumption, and high costs for other infrastructure systems, notably water and sewerage. Research relating to the general contention that large urban areas are too large is, however, inconclusive. For a summary of the optimum city size argument, see Renaud (1981) and Saunders and Warford (1976), pp. 80–82.

21. Nicol (1983a and 1983b).

22. Preece (1987).

23. See, for example, Hall (1990).

24. Ministry of Labour and Housing and Ministry of Physical Planning of Local Government, Sweden, (1973).

25. By contrast, policies to provide and price telecommunications services in many developing countries probably increase the disadvantage to the less-developed regions. This is discussed in chapter 14.

26. This can have a major effect on traffic. For example, in the study of the Aconcagua Valley in Chile (example 10), extension of local area charges to some links normally priced as long-distance resulted in call traffic two to fourteen times as high as would be expected from the analysis of the region's economic and previous traffic patterns.

27. For a capsule discussion of some of the primary arguments associated with creating growth poles in rural areas of developing countries, see Saunders and Warford (1976) pp. 102–03.

28. This study is described in more detail in appendix C.

29. Other rural problems include lack of job opportunities; inferior social, economic, and health care amenities; and relatively poor-quality educational opportunities. For a summary of some of the literature on migration in developing countries, see Levy and Wadycki (1972) and Yap (1975).

30. Where migrants go is partly determined by distance and contacts, and it has been argued that improvements in communications have been more important in stimulating migration than has the reduction in the costs of physical movement. See Beier and others (1975).

31. One such premium is, of course, a high black market price for a telephone. It is not

uncommon in cities with long telephone waiting lists to find black market telephone prices in the central business district exceeding those in outlying areas by a factor of ten or more.

32. Nicol (1983a and 1983b).

33. Beier and others (1975) p. 63. This is in contrast to the case of several more industrial countries, which have achieved a more dispersed settlement within the urban region. Indeed, in some countries, notably the United States, a problem in several older industrial areas is how to stop businesses from leaving the city centers and to prevent those central areas from decaying.

34. Webber (1980) p. 9.

35. Pool (1979) p. 182.

36. Beier and others (1975) p. 62.

37. Webber (1980) p. 2.

38. Webber (1980) p. 9.

39. Nilles (1974) and Nilles and others (1976).

40. Goddard and Pye (1978) and Harkness. (1971).

41. In the late 1970s, an estimated 60 to 80 percent of the population in some Asian and Latin American cities (and perhaps 40 to 60 percent in African cities) lived in these settlements.

42. ILO (1972) and Beier and others (1975). Another study estimated that this informal economy employs between 50 and 70 percent of the labor force in Asian and Latin American cities, and between 30 and 50 percent in African cities. See Mazumdar (1975 and 1976) Of course by no means all of the urban poor are employed in informal sector activities. For a discussion of this point, see Linn (1983) pp. 37–42.

43. ILO (1972).

Chapter 7

Transport, Communication, and Energy Consumption

THE DISCUSSION IN THE PAST CHAPTER focused on some of the spatial issues of developing countries from a relatively wide, multisectoral point of view. In this chapter the general contention that the spatial characteristics of a developing economy are relevant to the relation between telecommunications and development is extended further. This chapter examines the one sector that, on a spatial basis, is particularly affected by distances between communicating parties, the time required for communication, and the alternative forms of communication available. This sector, which is both complementary to and competitive with telecommunications, is transport.

Clearly there are strong interactions between the telephone and alternative modes of transport, such as roads, railways, waterways, the postal system, telegraph, and telex. Both complementarity and substitution are present: a smoothly running transport system will necessitate fewer telephone calls to overcome disorganization in freight, business, and personal travel but will stimulate telecommunications demand by promoting trade and other interaction. An efficient postal system depends on efficient transport and may be a substitute for certain uses of telecommunications, especially in the case of overnight package delivery services. Efficient road and rail travel depends to a considerable extent on telecommunications, and good telecommunications facilities enable some trips to be avoided and others to be organized on short notice.

Two questions arise concerning the relation between the transport sector and telecommunications services:

135

a. How far can *complementarity* of telecommunications and transport be demonstrated? To what extent can gains in the efficiency or output of the transport sector resulting from an increased use of telecommunications be identified and measured?

b. Where telecommunications and travel are likely to be substitutes, can such *substitutability* be demonstrated? Can cases be identified where the use of telecommunications is the least-cost solution in meeting a specific communication need, and can the cost savings and economic rates of return on the investment associated with the use of this least-cost solution be estimated?

In addressing these two questions it is a simple step to move slightly beyond the association between telecommunications and transport to consider the important relation between telecommunications, transport, and energy consumption. There are essentially two reasons for doing this. The first is that the transport sector typically accounts for a significant proportion of the petroleum consumption in a country. In Canada, for example, which has the world's second highest level of energy use per capita, the transport sector accounts for 58 percent of petroleum used, and 60 percent of this is used to transport people.[1]

The second and more important reason is that because of changes in relative costs over the past decade, telecommunications services have become much less costly relative to transport. Transport costs rose significantly in the 1970s as a result of increases in the price of petroleum and the cost of petroleum-induced steel and other materials. Although these prices have more or less stabilized, they remain a significant component of transport costs for many countries where petroleum must be imported and paid for with hard currency. In contrast, telecommunications has been experiencing changes in technology (electronic switching, network digitalization, multiple-access subscriber radiotelephone, satellite communications, fiber optics, solar-powered repeaters, and so forth), which have brought about, and should continue to bring about, a decline in real terms in the cost of providing communications services. (Several such changes were outlined in chapter 2.) Hence, opportunities for realizing savings in the costs and use of transport through telecommunications continue to exist.

Compounding its potential to facilitate change in the general location of economic activity, which was discussed in the last chapter, telecommunications can help to conserve energy resources in at least two other ways: by reducing waste and improving efficiency in the use of fleets of road vehicles and by substituting telecommunications for certain types of travel. In addition to other ecological benefits, using petroleum more efficiently also creates less air pollution, which is a significant benefit in large urban areas in developing countries.

Research on the above topics relates primarily to industrial countries, although many of the studies have some relevance to developing countries.[2] The energy saving potential of both the telecommunications and transport systems in the more industrial countries is, however, reasonably well established, and substitution among telecommunications, transport, and energy, at least in the short run, is likely to be marginal. In developing countries, however, telecommunications networks are grossly inadequate in terms of unmet demand, low national penetration, and poor quality and reliability of service. Hence, improvements in telephone access, quality, and reliability could dramatically improve the efficient use of transport as well as energy.

Improved Use of Vehicles

In industrial countries it is generally accepted that good management practices together with the use of telecommunications can facilitate the efficient use of vehicle fleets. This is true for a wide range of applications, from taxi operations to delivery truck fleets or vehicles used to maintain equipment in the field. It is also apparent that benefits can be reaped at various levels of technology, including a one-vehicle truck business using a public telephone to locate a destination or secure a return load and a large business using a sophisticated system of radios to locate and identify vehicles automatically and to transmit posting instructions from a central control location. Some of the relevant technologies, all well known and tested in numerous field applications, are public telephones (PCOS and public coin telephones), radio paging, private mobile radio, open shared channel (citizen band) radio, switched radiotelephone service (with or without connection to the public switched telephone network), and data communications by radio (manual or automatic data acquisition, automatic identification of vehicles and location).[3]

Freight vehicles are costly capital assets, and replacement parts, driver time, fuel, and tires are important costs—for both the firm and the country, where imported petroleum and, sometimes, vehicles often contribute significantly to balance-of-payments deficits. Improved vehicle use entails full loading and minimizing idle time, misrouting, and return journeys made with an empty vehicle. Such improvements require organization, and efficient organization requires communication. The relevant questions, however, are on what scale and how significant are the benefits?

Evidence from Industrial Countries

Although fragmentary, the evidence suggests that even where telephone, telex, and two-way data and radio systems are used intensively to organize vehicle movements—as in the United States—further investments in complementary telecommunications systems remain attractive. One U.S. study suggested that using mobile radios to control the movement of vehicles for local pickup and delivery can increase output per vehicle by from 15 to 25 percent, so that four vehicles could do the work of five.[4] In Sweden in early 1983 more than 3,000 standard taxicab radios were replaced with computerized mobile radios linked to a central computer at headquarters. The central computer ran a constant check on the availability and location of every taxi, so the dispatcher could instantly locate the closest taxi to a fare and relay the address through the vehicle's printer. By cutting down the time that drivers spent roaming the streets and dispatchers spent locating them, it was claimed that Swedish taxi companies would increase their profits 30 percent by lowering petroleum consumption and using the vehicle fleet more efficiently.[5] During the 1974 energy crisis, the GTE Corporation allowed employees to take company vehicles home to be "home-dispatched" to their next task in the morning, saving 2.4 million gallons of fuel compared with 1973. In a 1975 study of the costs of operating freight and service vehicles in four U.S. cities, Lathey suggested that employing radio control could save 1,300 million gallons of fuel by 1980.[6]

Corresponding findings were reported in the United Kingdom and the former U.S.S.R.[7] The Soviet studies are of particular interest since they are fairly elaborate and come from an environment rather different from the industrial western market economies examined in most studies. One 1974 study analyzed the benefits of a project designed to promote the use of a telephone and mobile radio dispatch system to improve shared use of agricultural machinery in sixty farms in the Rostov area.[8] The study concluded that the system

a. Reduced idle times of machines by a factor of between two and three, improving the productivity of tractors and other agricultural machines between 20 and 25 percent

b. Reduced the time spent transporting machines between working locations from 30 to 40 percent
c. Reduced the time required for basic agricultural operations, such as harvesting, from 20 to 25 percent
d. Saved from 40 to 50 percent of the time spent by managers and specialists in machine operations.

By also using the dispatch system to monitor animal husbandry, feeding, and care, cattle productivity (output per ruble of expenditure on inputs of all kinds) increased 5 to 10 percent.

In a second study in the former U.S.S.R., a survey of managers in various industrial enterprises showed that increased use of long-distance telephone calls for managing vehicle fleets—made possible by recent improvements in service—had improved their use of vehicles. For example, improved long-distance telephone service reportedly increased the efficiency of the Sedin lathe-building works in Krasnodar 10 percent.[9]

In addition to using mobile radios and improving long-distance service in industrial countries, savings can also be achieved by using electronic devices that display up-to-date advice on traffic conditions, routes, and parking. Computer-based scheduling and data communications for coordinating and documenting freight are also available. Since much unnecessary vehicle mileage results from imperfect information about routing, traffic conditions, and parking, drivers could probably make trips between the same origins and destinations with fewer vehicle miles and a consequent savings of energy, time, and capital if they were better informed and advised.

Several studies in the United States sought to estimate the amount of wasted vehicle mileage that might be avoided by such systems. Schoppert and others found that 17 percent of the motorists at ramps of major highways were taking an indirect route.[10] Gordon and Wood estimated that the use of electronic route guidance and parking information technologies could reduce total vehicle mileage more than 3 percent.[11] System concepts for electronic route guidance were described by, among others, General Motors, the French PTT, and Stephens and others, for both the general user and for special purposes such as bus or truck fleets.[12] Such systems are obviously technically feasible, and if they included a parking advisory capability, a 5 percent reduction in vehicular energy consumption would not be unrealistic in urban areas of industrial countries.[13]

Evidence from Developing Countries

A few studies in developing countries showed that poor communication causes much unnecessary movement of empty vehicles. Kaul reported from a sample survey of fertilizer distribution in India that out of all vehicle trips made by farmers to fertilizer distribution points, the proportion of trips that failed (because the intermittent pattern of supply and demand had, unknown to the farmer, resulted in the required fertilizer being unavailable) was typically 10 to 25 percent, and for some locations it was more than 50 percent of all trips.[14] The trips typically covered more than 6 kilometers. Kaul's results are summarized in table 7-1.

It would seem that the impressive returns on investment in fleet-control telecommunications in industrial countries could at a minimum be paralleled by gains obtainable from the use of more efficient

Table 7-1. *Successful and Unsuccessful Trips by Farmers to Major Fertilizer Distribution Centers in India, 1975–76*

Region and state	Distance from farm fertilizer distribution location (kilometers)	Number of trips	Number of unsuccessful trips	Unsuccessful trips as a percent of total
North				
Haryana	11.4	2.3	0.5	22
Himachal	13.1	1.4	0.1	7
Punjab	14.7	3.1	1.7	55
East				
Assam	8.4	2.4	0.4	17
Bihar	9.9	2.1	0.6	29
Orissa	6.6	3.0	0.4	13
West Bengal	15.5	3.3	0.8	24
West and Central				
Gujarat	5.9	2.7	1.3	48
Maharashtra	8.7	1.8	0.5	28
Rajasthan	13.6	1.9	0.4	21
Uttar Pradesh	7.7	2.2	1.0	45
Madhya Pradesh	7.9	1.5	0.9	60
South				
Andhra	16.0	2.5	1.1	44
Karnataka	5.8	1.5	0.2	13
Kerala	2.7	3.4	0.3	9
Tamil Nadu	6.3	3.3	2.1	64

Source: Adapted from National Council of Applied Economic Research, India, "Fertilizer Survey, 1975–76." Summarized in Kaul (1978).

telephone service—including public telephones accessible by the roadside and mobile radio applications—in developing countries.[15] The data from India together with the limited data from the former U.S.S.R. suggest that a transition from the nonuse or very limited use of telecommunications facilities in developing countries to the use of basic facilities such as telephones in public call offices in both urban and rural areas could substantially reduce transport costs, even before considering other benefits that might accrue from selected applications of mobile radio facilities.

More recently, the ITU sponsored a detailed analysis of the transportation sector in the People's Democratic Republic of Yemen in order to examine the potential savings that could be obtained if telephones were installed along the country's major truck routes.[16] The study aimed to provide a quantitative basis for determining how much should be invested in telecommunications that support a transportation system and to estimate how much savings might accrue if emergency breakdown assistance and scheduling were improved. The study interviewed many transport officials throughout the country and transport workers on one route, the Aden-Mukalla-Seiyun road. The resulting data were used to calculate the value of losses incurred due to lack of communication (but only losses that could be reduced with better communication). These figures were then extrapolated to the country as a whole (see table 7-2).

Table 7-2. Annual Losses due to Lack of Adequate Communications Infrastructure for Transportation Routes in the People's Democratic Republic of Yemen, 1986–87
(U.S. dollars)

Type of loss	Aden-Mukalla-Seiyun road	People's Democratic Republic of Yemen[a]
Idle time due to transport breakdowns	883,503	11,666,094
Perished goods due to transport breakdowns	11,760	155,292
Unused return loads	17,096,606	—

a. The losses for the country were obtained by extrapolating data on losses for the route on the basis of information known about the volume transported: 363,136 tons (according to field survey figures) for the route compared with 4,861,000 tons for the country. An adequate coefficient for extrapolation could not be determined for unused return loads.
Source: ITU (1988).

The study then examined possible reductions in losses that could be obtained if a telephone system were installed at 20-kilometer intervals along the Aden-Mukalla-Seiyun road, where no telephones currently exist. The number of full return loads was not expected to increase more than 15 to 25 percent, since the problem of empty return loads was strongly linked with factors not related to communications. However, improved communications were expected to reduce the time needed by management to respond to reported breakdowns, which would reduce the amount of time wasted by loaded trucks while waiting for repairs and spare parts. It was estimated that providing quick access to telephones along the road would reduce the time needed for a truck driver to report a breakdown from as long as several days to just one hour. The total cost of implementing the system nationally was an estimated $600,000 a year, based on an estimated cost per terminal of $3,000, annualized with 5 percent depreciation and 6 percent servicing, maintenance, and attendance costs. The analytical model showed total potential transport losses avoided annually through the use of this system to be $687,480 (YD229,160) for the Aden-Mukalla-Seiyun road; when extrapolated for the entire country, the resulting losses avoided were $9,013,962 (YD3,004,654). This gives a benefit to cost ratio for the project of 15:1; other benefits would include a gain of $1,710,000 a year from the 15 to 25 percent increase in full loads for return journeys, a reduction in the size of the fleet and related costs and expenditures for the same volume of goods transported, better scheduling that could improve performance 10 to 15 percent, and avoidance of $156,000 in losses from spoiled perishables for the country as a whole.

Substitution of Telecommunications for Travel

Pool reported that within three years after the telephone was invented in 1876, the London Spectator newspaper predicted that the new device would replace personal meetings.[17] Predictions of this kind are still being made but are related, of course, to increasingly sophisticated technical alternatives to face-to-face communication.

The Feasibility of Potential Substitution

As was the case with the improved use of vehicles discussed in the previous section, little quantitative research on the potential for telecommunications to replace travel has been conducted in developing countries. Research has, however, been carried out in industrial countries on the extent to which business travel (and the use of the telephone or other media to conduct business) are substitutes in the functional sense that they produce the same result with the same user satisfaction.[18]

The primary research methods used have comprised the following:

a. Controlled experiments in which businessmen and civil servants carried out realistically simulated tasks (such as decisionmaking and bargaining) by audio and video communications media and (in the control group) face-to-face meetings
b. Detailed monitoring and evaluation of telecommunications field trials
c. Detailed surveys on the scale and content of communication activity and the extent to which respondents believe that business currently carried out face-to-face could be carried out effectively and acceptably over the telephone or by other means of telecommunications.

The research findings are complex. In the United Kingdom the overall results showed that where travel substitution was feasible, video was rarely necessary to achieve the required interchange, and audio (simple telephone service where only two people are involved, loudspeaking telephone, a telephone-conference link, or a studio-based audio conference system) was almost invariably cheaper than travel. As for effectiveness in completing various business tasks presently carried out in person, a U.K. Department of Transport study found that approximately 41 percent of the total activity could be effectively and acceptably carried out by audio telecommunication.[19] Another study in the United Kingdom concluded that, although a substantial portion of all conversations with people outside an individual's immediate work group took place by telephone (rather than face-to-face), a much greater portion of all remaining meetings involving travel could also have been carried out by telecommunications without loss of effectiveness.[20]

Telecommunications can also substitute for travel by linking employees who work at home with their employers or entrepreneurs who work out of their homes with their clients. Sometimes referred to as telecommuting, working at home was first "rediscovered" in the 1970s as a way to avoid time and energy wasted in weekly commuting; it has caught on relatively slowly, however, despite predictions that it would revolutionize the workplace. At the end of the 1980s, about 630,000 people in Great Britain were working at or out of their home (excluding domestics, including sales personnel), or about 2.5 percent of the total work force.[21] In the United States, the figure is estimated to be much higher, at about 25 million.[22] No data are available on the extent to which these workers depend on telecommunications.

Some descriptive studies exist on telecommuting arrangements.[23] In general, they have found that adoption of telecommuting has been slowed, on the one hand, by internal resistance to nontraditional forms of organization and management within large companies and, on the other hand, by external criticism from researchers and labor unions that telecommuting makes workers vulnerable to economic exploitation similar to that suffered by cottage workers in the early days of the industrial revolution. One area of significant growth has been in offshore data processing. Large insurance carriers, for example, increasingly send information such as claims via high-speed data links to be processed abroad in countries with lower wage scales. Telecommuting by independent workers, such as entrepreneurs, designers, salespersons, writers, software experts, and high-level professional consultants has also grown significantly, aided by widespread use of innovations such as facsimile machines and computer modems and supported by next-day package delivery services that enable finished products to be delivered in a timely manner.

Empirical Evidence

Having established that a partial substitutability of telecommunications and travel seems to be feasible when a full range of costs and behavioral factors are considered, the next question is how much travel is actually saved in practice? A partial answer is that only costs associated with the direct substitution of business trips by teleconferencing have been observed in most of the current field-trial studies.[24]

Teleconferencing is successfully used in the United States by organizations such as NASA, Bank of America, Ohio Bell, Exxon, Procter and Gamble, Texas Instruments, Ford Motor Company, Union Trust Bank, ARCO, Aetna, Boeing, Honeywell, and IBM, which report that the use of selected audio teleconference systems has replaced much travel.[25] For example, "almost all participants in the trial [Union Trust Company] substituted use of the system for at least 50 percent of their face-to-face meetings; more than one-third substituted for 80 percent or more."[26] Other examples include reports that IBM's teleconferencing system reduced travel costs by $414,000 in 1979, its first year of operation, and by $830,000 in 1980;[27] that a subsidiary of RCA Corporation held a video conference for 450 sales people around the United States costing $85,000, compared with $555,000 for a similar conference involving travel;[28] and that during a recent three-day period each of seven branches throughout the United States of a division of the 3M Company conferred for one hour using audio conference facilities with a panel of marketing, sales, and technical experts at the home office. The total cost of the audio conference was only 15 percent of what the travel bill alone would have been to bring everyone together.[29] Finally, a study of audio conferencing in a U.K. civil service department, Her Majesty's Stationery Office, also showed substantial travel savings.[30] As have several others, this study concluded that a significant degree of additional travel substitution over and above that achieved by the basic telephone system can, even in industrial countries, be—and already is—achieved with simple means such as the loudspeaking telephone.[31]

Overall, the findings of such studies indicate that the withdrawal of vision and separation by distance have no measurable effect on the outcome of normal conversations that exchange information and solve problems. However, in conflict situations, in bargaining, and in instances in which first contacts between strangers require interpersonal judgments and eye contact, the telephone is sometimes less effective than a face-to-face meeting.[32] One reviewer of such studies concluded that the potential for substitution between travel and teleconferencing (presumably in industrial countries) is between 20 and 25 percent.[33]

Somewhat less rigorous studies in the former Soviet Union also have provided evidence of travel substitution and resulting gains in work productivity. A study by the Moscow Electrotechnical Institute of Communication analyzed the extent to which intensified use of long-distance telephone communication (made possible by improvements in the quality and availability of telephone service) improved labor productivity. Two kinds of gains were identified: direct labor savings through the substitution of telephone calls for the preparation and sending of letters and telegrams and for personal visits, and indirect productivity improvements associated with more efficient management arrangements made possible by improved telecommunications—more operational control of production, smoother production flows, speedier marketing, and an improved supply of materials and machinery.[34] Both effects were estimated from the results of a survey of fifty-seven industrial enterprises and twenty-one construction organizations, which were then entered into a cost and productivity model.[35]

With regard to the substitutability of telecommunications, the Soviet surveys showed that, in the opinion of operational managers, some 24 percent of letters and 17 percent of telegrams could appropriately have been replaced by telephone service if it were of adequate quality. Although its role is not clear, price did not seem to be a very important factor compared with physical availability. Similarly, the responses indicated that on average some 4 percent of the volume of business travel undertaken was replaced by telephone calls (at a rate of about three calls per trip) once a reliable automatic trunk service was introduced and that a further 12 to 13 percent could be so substituted. The study concluded that the value of future increases in productivity and, hence, the reduction in production costs predicted to follow improved long-distance telephone communication would be 4.4 times greater than the annual cost of the telecommunications investments needed to bring about those improvements. A particularly interesting side result was that, given the basis of pricing for trunk telephone service in the former U.S.S.R., only 5 to 10 percent of these benefits would accrue as direct revenue to the telecommunications carrier itself—a result closely in accord with the low price elasticity and large consumer surplus found in many developing countries.

There is some limited evidence from Kenya about telephone-travel substitution. In 1979 telephone calling rates and the prevalence of other means of communication were surveyed for two regions in Kenya.[36] One, the Eldoret region, is farther from Nairobi and was the more industrial of the two regions examined. The other, the Nyeri region, although closer to Nairobi, also had a higher density of population and settlement. The survey showed that telephone subscribers in the Eldoret region tended to make about 80 percent more telephone calls per day than those in Nyeri and that they tended to rely much less on personal visits and vehicles to facilitate communication. The researchers concluded that some relative substitution of telephones for travel was taking place in the more sparsely populated and distant Eldoret region.[37]

More recently, in 1983, a survey of telephone users in four rural and semirural districts in Thailand found that 64 percent of office workers who used a telephone for their jobs would have made a personal visit to carry out their task if a phone had not been available and 11 percent would have sent a messenger. Only 17 percent would have written a letter or delayed the message. Although most of the calls made during the survey period were not local (about half were to other locations within the province and about one-quarter were outside the province, including 6 percent to Bangkok and 2 percent abroad), respondents emphasized that the distance involved did not make any difference and that they would have traveled to other provinces or even Bangkok to convey their message.[38]

The Question of Travel Generation

In addition to the possible substitution of telecommunications for travel, improved and extended telecommunications systems will likely also generate telecommunications traffic. Such a possibility is also important in the context of energy consumption, since improved telecommunications could actually generate new transport traffic as well, thereby to some extent offsetting the substitution effect. In the absence of reliable data on users' communication choices, opinion varies widely: some claim that there will be substantial reductions in travel, while some believe that improved telecommunications will, at least in industrial countries, stimulate a net increase in travel.[39] Moreover, care must be taken in evaluating such effects. Additional traffic is not necessarily undesirable; indeed, the benefits of the additional energy consumption may well exceed the costs. This will be more likely as the relation between energy prices and the economic costs of supply grows closer. The general concern is that in some instances energy prices may be less than the economic costs of supply.

Little empirical data explicitly identify transport traffic that has been induced by an overall increase in rapid communication. The use of teleconference systems in the United States has, in fact, produced the general view that there will be relatively little stimulation of traffic. For example, the use of teleconferencing by the NASA Apollo manned lunar landing project reduced travel expenditure 25 percent per staff member.[40] An independent interpretation of these data (not necessarily endorsed by NASA) concluded that this represented a substitution effect of up to about 35 percent and a generation effect in the region of 10 percent, resulting in a net reduction of 25 percent.[41]

Also, the mix between travel substitution and travel generation varies in different parts of the system, both among regions and among user groups; the effect may be different in rural than in urban areas. For example, in a study in rural Central Lincolnshire, England, in which 361 respondents kept diaries of their contacts for five days, researchers found that the major effect of the telephone was to bring individuals closer together. They contended, however, that the tele-

phone did not and could not acceptably replace the face-to-face contacts upon which most social relations were based. Hence, they
concluded that, at least in Great Britain, increased telecommunications was likely to stimulate the demand for social travel in rural
areas.[42]

From the sole view of saving energy, the travel-generation effect of
telecommunications investment should perhaps be minimized. Even
in the extreme case in which the introduction of telecommunications
services results in increased travel, however, the overall energy efficiency of communication will increase, since presumably a larger
share of total communication will be carried by telecommunications.
Hence, in most instances, the total economic efficiency gained from
the interaction of travel and telecommunications should be at least as
great as the gains calculated on a simple transport-substitution cost-
savings basis.

The Potential for Transport Substitution in Developing Countries

None of the fragmentary evidence on the interaction between telecommunications and transport can be regarded as conclusive. It is,
however, sufficient to suggest that a substantial degree of cross-
elasticity or substitutability is possible and that a shift in the balance
of investment between the two sectors in favor of telecommunications
could be beneficial in at least two ways.

First, in most developing countries, where telecommunications networks are highly inadequate, trips are made that could be replaced by
telecommunications by any standard of judgment. Deficiencies in
public telecommunications facilities are made up either by the use of
more costly modes of communication or by the construction of private, high-cost, special-purpose networks. Alternatively, the activity
that generated the need for special communications is abandoned altogether or carried out in a second-best way, which requires less communication. Hence, given the relatively undeveloped state of existing
telecommunications services in developing countries, improved service should decrease the overall costs of communications on a per
unit basis and increase efficiency.[43]

Second, in developing countries, where the total amount of travel
possible is sometimes directly constrained by absolute limitations on
the amount of petroleum, vehicles, and road maintenance equipment
that can be imported, increased opportunities for substituting telecommunications for travel could directly increase the absolute volume

of business that can be transacted as well as reduce the amount of energy consumed per unit of business activity.

Travel and the use of the telephone or other telecommunications services are clearly not perfect substitutes: most of the studies cited earlier scrupulously emphasized the limits of substitutability. However, where excess demand for telecommunications is chronic—as in most developing countries—travel is often used reluctantly as a substitute for telecommunications, with high costs in terms of energy, use of capital in the transport system, and time. This last point is worthy of particular emphasis, especially if the objective is to maximize the benefits derived from the communication sector, as distinct from merely minimizing the costs of attaining specific communication access or penetration targets. In such a case it will almost certainly be worth addressing the issue of attaining a more efficient use of scarce and costly human capital, as represented by the time wasted by persons with professional, technical, or managerial skills.

Given that telecommunications can and does substitute for travel in a wide variety of circumstances, what are the economic benefits? Few studies have explicitly calculated a rate of return on telecommunications investments aimed at travel substitution, primarily because causally related changes are difficult to identify. One exception is the ITU study on the People's Democratic Republic of Yemen cited earlier; in this case, the economic model focused tightly on the relationship between communications infrastructure and transport costs, with necessary assumptions held to a minimum because both the economy and the transport network were small.

Travel Substitution and Energy Saving

Although few studies have attempted to measure systematically the full benefits of telecommunications resulting from travel substitution, several attempts have been made to measure the effect of investment in telecommunications on energy consumption. These studies have been conducted primarily in industrial countries, for example the studies undertaken in Canada and the United Kingdom by Bell Canada and the former British Post Office.[44] Studies by these two entities used a similar conceptual basis, which distinguishes between direct and indirect energy use, between average and marginal rates of energy use, and between the energy dissipated by the system under consideration and the primary energy input (coal, oil, or nuclear-generated heat) used.[45]

The estimates discussed here are for direct energy consumption. They do not include energy embodied in goods or services consumed by the telecommunications system (for example, the energy consumed in refining copper for cables). They do, however, account for all of the primary energy input used to supply the system's direct energy needs. The estimates are of marginal energy consumption, that is, the increment of energy consumption associated with an increment of call traffic. They assume that system capacity is adjusted proportionately to such changes in traffic (so that the lumpiness in providing a new airline scheduling data network or a new microwave route is smoothed out) and that overhead energy consumption by administrative, maintenance, or similar functions in the telecommunications and transport corporations is fixed; it will not increase proportionately.

The Canadian and British studies yielded similar results.[46] The Canadian study, in projecting the comparative energy costs in 1985 for a three-hour meeting between people in Montreal and Toronto, showed that the energy used by an audio conference call was far less than one-hundredth of that which would have been used if the persons had met face-to-face after traveling by railway, automobile, or airplane.[47] Overall, the estimate of the savings that might be realized by replacing 20 percent of Canadian intercity business travel by telecommunications amounted to 3 percent of the total energy consumed by the transport sector in Canada. This was equivalent to 1.3 percent of national petroleum consumption, or about 0.8 percent of total energy use.

A British study of a substitution for travel between London and Glasgow (400 miles compared with 325 miles between Montreal and Toronto) presented its results in terms of savings per kilowatt-hour. It showed that for a three-hour meeting including four people, two of whom would have to travel from London to Glasgow if travel were undertaken, the relative efficiency of a studio-based audio telephone conference hookup compared with railway, automobile, and air travel was 1:225, 1:800, and 1:1,250, respectively.[48] A second study carried out in Great Britain compared the savings per kilowatt-hour of making a one-hour telephone call rather than traveling by automobile, railway, and air for several distances and estimated that the telephone was more cost-effective than rail and automobile by, respectively, ratios of at least 3:4 and 3:13 for 10 kilometers, 3:18 and 3:63 for 50 kilometers, and on up to 3:233 and 3:800 for 650 kilometers. The comparison at 650 kilometers for air travel was 3:1,241.[49]

Exercises have also estimated the potential for realizing petroleum savings by substituting telecommunications for interurban travel for business meetings. Lathey estimated that just under 3 percent of the total 1974 U.S. petroleum demand could have been saved by substituting interurban travel by car. The estimate was made using a U.K. estimate that 40 percent of business trips could be substituted by telecommunications and then assuming that the 40 percent estimate would hold for all interurban travel in the United States.[50] Similarly, Tyler, using an approach like Lathey's, estimated that 40 percent of business trips in the United Kingdom could be replaced by teleconferencing, and that if this were true about 0.3 percent of total U.K. national primary energy consumption would be saved.[51]

How far would the energy consumption equations be modified by including indirect energy consumption? In the absence of detailed energy input-output analyses of transactions between telecommunications common carriers, transport firms and agencies, and their suppliers, no definitive answer can be given. However, there are indications that including indirect energy consumption would shift the comparison further in favor of telecommunications. Studies of motor transport in the United States suggest that indirect energy consumption adds between 30 and 90 percent to the energy used directly.[52] A comparison of aggregate results from input-output studies suggests that the proportion would be considerably lower for telecommunications. A 1974 input-output study by INSEE in France estimated that a 100 percent increase in the price of all energy sources would increase the costs for telecommunications only 1.9 percent as against 12.7 percent for transport.[53]

Conclusion

This brief literature review has supported intuitive notions that there are strong linkages between the transport, energy, and telecommunications sectors. The extent of these linkages and the relative importance of the three sectors in any one country depend upon many geographical and economic factors. Industrial countries with relatively mature transport and telecommunications systems seem to have some potential for short-run changes in the relative use of transport and telecommunications as well as for a longer-term spatial reorganization of economic activity, which might save energy. However, in the absence of either major government restrictions on transport, petroleum, and so forth or further major price increases affecting transport, such changes are likely to be only marginal, at least for the next few years.

For developing countries, however, the picture may be different.

The changes in the relative costs of transport and telecommunications; the relative immaturity of both sectors; the widespread inefficiencies in the transport sector resulting from information deficiencies (unnecessary trips, empty return trips, badly timed trips); and the limited spatial dispersion of trade, commerce, and industry (also partly related to information deficiencies) all indicate that potential gains from increased coordination between the two sectors and from increased penetration of the telephone network into both urban and rural areas may be relatively large. The studies reviewed were limited, however, and mainly related to industrial countries. Although such studies showed that significant transport and energy savings are possible, only the ITU study on the People's Democratic Republic of Yemen proposes a ready means for determining the optimum level or mix of investments in the telecommunications and transport sectors, assuming that market signals can be relied upon. In many countries, where prices for telecommunications and transport services or major inputs such as energy often do not approximate their costs of supply, efficient solutions are exceptionally hard to determine.

Notes

1. International Energy Agency (1989) and World Bank (1989).
2. Portions of this chapter draw on a review of selected literature published on the substitution of transport and telecommunications by Michael Tyler and Emma Bird.
3. Early sources of information on technical and operational matters include Moon and others (1977); Lathey (1975); Schoppert and others (1960); and Stephens and others (1968).
4. Moon and others (1977).
5. *Wall Street Journal* (1983).
6. Lathey (1975). The rapid growth in the demand for mobile communication terminals in the United States, which in large measure reflects recognition of such potential efficiency gains, was acknowledged by a mid-1970s decision of the U.S. Federal Communications Commission to double the range of the radio frequency spectrum allocated for land-mobile uses. Savings in vehicle mileage of from 30 to 40 percent have been reported as typical, and the rate of return on investment is high.
7. Pye Telecommunications, Ltd. (1976).
8. Medinikov (1975). A model was developed to attempt to estimate the total economic benefits associated with these kinds of changes, and the Regional and Vocational College of Agriculture in Rostov reportedly carried out a program to measure the numerical parameters from the actual performance of the Rostov experiment.
9. Gorelik and Efimova (1977). Both of the U.S.S.R. studies relied heavily on judgmental assessments of gains rather than detailed measurements of physical quantities before and after the use of telecommunications was provided.
10. Schoppert and others (1960).
11. Gordon and Wood (1970).

12. General Motors Research Laboratories and Delco Radio Division (1962); United Nations (1977); and Stephens and others (1968).

13. Favout (1970). Before such systems are introduced in any one area, three area-specific questions would have to be addressed. (1) A behavioral question: will the availability of an enhanced information device actually lead drivers to rationalize better their choice of routes and parking places? (2) Financial or market questions: will the system generate sufficient cost savings (public and private) to justify the public and private costs involved, and can the concept be sold to the individuals who would buy in-vehicle display units and the public agencies that would provide infrastructure? (3) Broader questions of economics and energy analysis: is the direct effect of introducing an electronic route guidance system clearly to reduce energy consumption? It is likely, although not demonstrated in detail, that the energy cost of such a system would be trivial. However, the improved traffic flow resulting from electronic route guidance might encourage greater use of private automobiles, longer journeys, and increased traffic generally.

14. Kaul (1978).

15. Several short examples relating to the potential of telecommunications for fleet control in developing countries were outlined in chapter 1.

16. ITU (1988).

17. Pool (1979), p. 181.

18. Basic research was carried out at the Communications Studies Group, University College, London, and at the Johns Hopkins University, the Stockholm School of Economics, the London School of Economics, Bell Canada, and elsewhere. It was comprehensively reviewed in Short, Williams, and Christie (1976); Communications Studies Group (1975); and Johansen, Vallee, and Spangler (1979).

19. Tyler, Cartwright, and Collins (1977) and Tyler (1978).

20. Pye (1974). Several objections to this conclusion spring naturally to mind. For example, although travel substitution on a significant scale may be feasible, will the preference of individual business travelers allow it? Survey research in Canada and the United States addressed this question. The results tend to show that the answer is yes—most of those business travelers who travel more than the average would prefer to reduce the amount of travel, and they account for by far the greater part of the total volume of business travel. See Kollen and Garwood (1975) and Pye and Weintraub (1977).

21. Kinsman (1987).

22. Williams (1991).

23. National Research Council (1985) and Stern and Holti (1986).

24. Some scattered evidence exists of cost savings derived from telecommuting. For example, see Hughson and Goodman (1986).

25. Johansen, Hansell, and Green (1981) and Charles (1981).

26. Tomey (1974).

27. Satellite Communication Services (1981).

28. *Wall Street Journal* (September 4, 1980), as reported in Charles (1981) p. 300.

29. Telephony (1982).

30. Trevains (1978).

31. Williams and Young (1977).

32. For a summary of much of this research, see Reid (1977) pp. 386–414.

33. Charles (1981) p. 298.

34. The model and the aggregation of the results to give predictions of the economic effect of telecommunications investment at the national level are described in Gorelik, Efimova, and Kareseva (1975) and in Gorelik and Kareseva (1975).

35. The survey method relied heavily on the judgment of the respondents (industrial managers and specialists) about such matters as (a) the extent of substitutability of telephone calls for letters, telegrams, and visits; (b) the extent to which increased use of trunk

telephone systems had already saved costs through such substitution; (c) the extent of further substitution and cost savings expected to result from planned improvements in the quality and availability of trunk telephone service; and (d) the more subtle indirect gains in productivity. The problems with such judgmental surveys are well known, and in a planned economy where respondents may have an incentive to emphasize their need for improved facilities, such problems may be even more significant. There are other problems also. For example, it is not clear what benchmark of comparison is used in the published work to assess the productivity benefits of existing long-distance telephone service, although the benchmark seems to be a state of zero utilization.

36. Cleevely and Walsham (1980) pp. 21–22.

37. A World Bank project analysis in Pakistan in 1969 also indicated that substitution does indeed occur in a developing country and that even at the lower wage and salary levels, the additional staff time and fuel costs involved are often as much as three times the cost that would have been incurred by using the telephone service to do the same job.

38. Chu, Srivisal, and Supadhiloke (1985).

39. Respectively, Pierce (1962) and Hall (1969).

40. Fordyce (1974).

41. Tyler (1976).

42. Clark and Unwin (1981).

43. The development of low-cost portable audio conference terminals, such as the Videograph developed in France or the Electronic Blackboard developed in the United States, and the use of slowscan video to transmit images over regular voice telephone lines have been proposed as options to supplement audio conferencing in developing countries. Such subscriber apparatus can transmit handwriting, computer graphics, or simple visuals simultaneously with voice among several locations over ordinary telephone lines. Although this kind of system has been experimented with in developing countries, implementation has proven to be technically complex, and few systems are currently in operation despite nearly ten years of availability. See Goldschmidt, Tietjen, and Shaw (1987).

44. Respectively, Tyler, Cartwright, and Bush (1974) and Katsoulis (1974). See also Katsoulis (1976).

45. Since the parameters of the calculations are largely an engineering matter, the results for a developing country are not likely to be enormously different, except perhaps that given their relatively inadequate telecommunications networks and varying geographic and social conditions, the existing energy efficiency of the transport sector may be appreciably lower.

46. It is important to remember the limitations of the existing comparisons between transport and telecommunications as a basis for estimating overall energy savings. For example, the energy studies compared the energy consumption for a single teleconference with that involved in travel to an equivalent face-to-face meeting. Business trips, however, often combine several meetings, especially in cases of long-distance travel. A Bell Canada survey of intercity business travel showed that the average trip included 2.7 meetings, but this effect was less pronounced for the much more numerous short trips. See Kollen and Garwood (1975).

47. Tyler, Cartwright, and Bush (1974).

48. Pye, Tyler, and Cartwright (1974).

49. Tyler, Cartwright, and Bookless (1974).

50. Lathey (1975).

51. Tyler, Elton, and Cook 1977). Primary energy was defined as energy available to the final consumer (secondary energy) plus conservation losses and waste in the industries that supply energy.

52. Tyler, Cartwright, and Bookless (1974).

53. Tornato (1974).

Part III
Microeconomic Analysis of Benefits

Chapter 8

Price Change and Best-Alternative Estimates of Consumer Surplus

THE RELATION BETWEEN telecommunications investment and economic activity and development is highly complex. The survey of the literature and experience presented so far suggests that the benefits of a specific telecommunications project or investment program cannot readily be identified and measured by the aggregate international comparison of input-output tables, by the analysis of relations between GNP and telephone availability or usage, or, apparently, by analysis of national transport, energy, or locational objectives. This part of the book therefore focuses on the experience of microeconomic analysis in addressing the issue of telecommunications benefits.

Cost-benefit analysis can be used to help determine not only the amount of resources that should be devoted to telecommunications but also the best way to allocate those resources within the sector, that is, local versus long-distance facilities, urban versus rural, public call offices (PCOS) versus subscriber telephones, alternatives for new types of service, regional choices, and so on. Traditional economic theory prescribes that investment in every sector should be increased until all projects that yield rates of return in excess of the opportunity cost of capital have been financed. The capital rationing invariably encountered in developing countries requires that the rate of return produced by expanding the telecommunications sector should be compared with the returns offered by the best-alternative investment program that would be implemented if the funds were not spent on telecommunications; this in effect represents the real opportunity cost of capital. Because of artificial restrictions on the growth of telecom-

157

munications and massive unsatisfied demand, in most developing countries, the telecommunications sector offers scope for expansion according to this criterion, that is, rates of return to investment in telecommunications tend to be relatively high.[1]

Some argue, however (as noted in chapter 1), that these high financial rates of return may reflect benefits accruing only to higher-income groups of society and that expansion of the sector merely exacerbates income inequality. To examine the validity of such arguments normally requires supplementing the evidence on willingness to pay for telecommunications investments with more qualitative analysis of the nature and likely effect of telephone usage. Hence, two kinds of information should usually be reviewed to determine the effects of a particular project or program in a particular developing country. First, the estimated investment and associated operation and maintenance costs should be compared with the estimated benefits to the users of telephone and other services. These estimates should be based on their willingness to pay for access and calls and perhaps on their willingness to travel to attain services, to purchase or rent connections from intermediaries, or to pay more to communicate in the absence of a telephone or other telecommunications service. Second, information should be gathered on the characteristics of the persons and groups benefiting from the improved and expanded services and on how those beneficiaries use them; these data should then be qualitatively analyzed in light of that country's national development policy goals and programs. The first point is addressed in this and the following chapter; the second is the subject of chapters 10, 11, and 12.

Internal Rate of Return and Consumer Surplus

As noted above, identifying and measuring the economic benefits relative to the costs of a proposed telecommunications project or investment program typically require observation of how much subscribers and callers actually spend or are willing to spend to have access to and use telecommunications services. An assumption underlying this approach is that decisionmakers (individuals, households, firms, government agencies, and other organizations) are rational and understand their activities better than outside planners, so that the amount of money they are demonstrably willing to spend on telecommunications services is at least a minimum measure of the worth of those services to them.

The first step in such a benefit and cost calculation is to estimate the incremental revenues that will be generated by the additional telecommunications investment being proposed (a proxy for benefits) and then compare that amount with the investment, operating, and maintenance costs that will be incurred to generate them. Such a comparison, of course, must be made through time since the costs incurred and the revenues produced by the new investment are associated with both the present and the future. Hence, the principle of discounting plays a part; a given unit of revenue received today is of greater value than that same unit received five years from now; likewise a given unit of cost incurred today involves giving up more than if that same unit of cost were incurred in five years.

Given this, the economic criterion for the acceptability of a project is determined by estimating the present (discounted) value of the project's benefits net of its costs, where the benefits and the costs are defined in incremental terms compared with what the situation would be without the project. Risk and sensitivity analysis also must be incorporated in such predictions. In the telecommunications sector this analysis is usually presented in terms of an estimated internal rate of return (IRR) on the new investment.[2] The IRR is defined as the rate of interest (discount rate) that would be required to yield a net present value of zero.[3] Before such a return is calculated, however, at least two types of adjustments are necessary to ensure that the data on financial revenue and cost reflect the real costs and benefits of resources according to the national allocation of resources.[4]

First, adjustments must be made in the prices at which inputs (such as equipment purchases, foreign exchange, local currency, or labor) are valued, to compensate for any distortions in the price system. The intention is to reflect the true social cost or forgone benefits associated with each type of input used in terms of its scarcity and the opportunity cost of diverting it from other uses. This can be done by using shadow prices to value inputs.[5] Shadow prices are discussed in more detail in appendix D.

Second, items that are considered to be revenues and costs by the telecommunications operating organization, but that are pure transfer payments from one entity to another and do not represent flows of real goods and services in the economy, must be removed from the cost and revenue streams. For example, government taxes and customs duties are a cost to the operating entity but are a pure transfer and not a real cost according to the national use of resources.[6]

The IRR resulting from the above calculation, which essentially uses revenues from the operating entity as an estimate of the benefits that consumers derive from telecommunications services, usually underestimates the economic worth of those services because

a. Subscribers or PCO users may value the service more highly than the amount they are required to pay for it, that is, there might be consumer surplus that is not quantified
b. New telephone subscribers not only incur benefits for themselves but also increase the benefits of being connected to the system for current subscribers; that is, there are subscriber-related externalities
c. The willingness to pay a given price to make a telephone call reflects only a minimum estimate of the benefits incurred by the caller and does not reflect the benefits received by the callee or those who the caller or callee then contact, that is, there are call-related externalities.[7]

Points b and c are discussed in chapters 14 and 15. With regard to point a, documentation exists that in developing countries telecommunications subscribers or PCO users are often charged less than they are actually willing to pay for telecommunications services, or they in fact pay higher prices for the services than are recorded by the financial receipts of the telecommunications operating entity.[8] In other words, the benefits that consumers perceive to receive from the service exceed those measured by the revenue received by the telecommunications entity. This difference between what users are prepared to pay—the real measure of the benefits they perceive—and the revenues actually collected is known as "consumer surplus."[9]

The concept of consumer surplus has been relied upon extensively in attempts to supplement the IRR as an indication of the benefits of proposed telecommunications investment. Three methodologically different, but in practice interrelated, approaches have generally been taken at the level of a project or investment program to measure a portion of the consumer surplus:

a. Methods based on observing the consumption effects of price changes
b. Methods based on comparing the difference between the cost of carrying out a given activity using telecommunications and the cost of the best-alternative means of communicating
c. Methods based on estimating more completely the costs that telecommunications users actually incur when communicating.

For brevity, these are referred to as the price-change method, the best-alternative method, and the expenditure method.

Estimating Consumer Surplus by the Price-Change Method

Observing the response of users to a change in price can provide some information on consumer surplus at that time.[10] For instance, if call traffic is not congested, the price elasticity of demand of telephone call traffic might be estimated by observing the change in traffic associated with a change in the charge for calls. With this information, a portion of the demand curve for calls can be estimated, and an impression of the consumer surplus associated with calls can be obtained.

In developing countries, however, where demand is seldom fully met, observations of price and quantity rarely represent actual points on demand curves. Nonetheless, since what users actually pay for services reveals at least part of their valuation of the benefits received, even the empirical observation of points in the supply curve (although probably not on the demand function) can yield some insight into consumer surplus. The following two examples illustrate possibilities for partially estimating consumer surplus using the price-change method.

Example 13. Increase in Call Charges (Costa Rica 1976)

Consider figure 8-1, where DD' is the demand curve for telephone traffic. At price p_1 per metered pulse, calls are made that result in q_1 metered pulses a day. When the pulse price is increased to p_2, traffic drops to q_2 pulses a day. Since before the price increase telephone users were willing to pay at least p_2 for each of the first q_2 pulses, but were only asked to pay p_1, they were receiving a consumer surplus at least equal to $(p_2 - p_1)q_2$ for these calls. For the remaining $(q_1 - q_2)$ calls, they received a consumer surplus equal to the area of the triangle ABC below the demand curve DD'.

The sum of both consumer surpluses is shown in the shaded area in figure 8-1. If the observed price change is small, the straight line through points A (p_1q_1) and B (p_2q_2) would closely approximate the corresponding segment of DD' and the area of ABC below the curve would therefore be approximately equal to $(1 / 2) (p_2 - p_1) (q_1 - q_2)$. The total benefits received by users making q_1 calls were at least equal to what they paid (p_1q_1) plus these consumer surpluses.

Actual benefits were even higher, however, because the total consumer surplus also includes the area above p_2 and below DD' in figure 8-1. A better (but still conservative) estimate of consumer surplus can be obtained by adding the area of the unshaded triangle p_2Ba to the shaded area described above. It can thus be shown that total benefits equaled or exceeded $1 - (1/2\varepsilon)$ times the revenue (p_1q_1), where ε is the price elasticity of demand for pulses (a negative number) in the neighborhood of price p_1.[11]

Figure 8-1. *Estimating Consumer Surplus from the Effect of a Price Change*

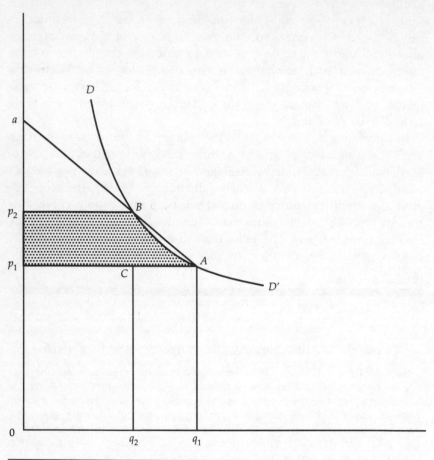

Such a method was used in 1975/76 to examine the demand for public telephone use in rural Costa Rica. An analysis of traffic data for ninety-two PCO telephones in eighty-two small villages before and after a 25 percent increase in call charges (from $p_1 = 12$ centimos per pulse to $p_2 = 15$ centimos) generally indicated that the average price elasticity of demand for pulses in the eighty-two villages was about -0.5. Hence, a village generating $q_1 = 100$ pulses a day would have reduced its traffic to $q_2 = 87.5$ pulses a day shortly after the call charges increased. At the initial price p_1, the consumer surplus on the first 87.5 pulses would have been at least 3 centimos per pulse or 262.5 centimos (2.63 colones) a day. The remaining 12.5 pulses would add a further 18.75 centimos of consumer surplus to make a total of 2.81 colones a day. Since at p_1, revenues were 12 centimos per pulse times 100 pulses a day, or 1,200 centimos (12.00 colones) a day, the measured revenues received by the telephone operating entity represented only about 81 percent of the quantified benefits perceived by the telephone users.

This, however, still understates the benefits. Using the better estimate of consumer surplus, initial telephone revenues generated by the Costa Rican village were found to represent at most only half the benefits perceived by the population. For further details, see appendix C.

Example 14. Changes in Real Tariffs because of Inflation (El Salvador 1977)

In an economy affected by price inflation, prices in real terms paid for telecommunications services decrease during the periods between increases in tariff levels. Looking back through time, this means that during periods of price inflation telecommunications users revealed a willingness to pay higher real prices than those being charged today. Looking forward, users will increasingly be charged less in real terms than they pay today, until tariffs are adjusted upward. These facts can be used to quantify part of the consumer surplus for telephone service, as in the following case relating to El Salvador's 1978–82 telephone development program.[12]

In 1977 El Salvador had long waiting lists of individuals and business firms wanting telephones. Telephone tariffs had remained virtually unchanged for more than ten years, during which time general price inflation was significant. Hence, telephone users had paid considerably higher prices in real terms (relative to other goods and services) than they were currently paying for telecommunications services. Given this, consumer surplus was partially estimated by tabulating the prices in real terms that existing consumers had been willing to pay when they acquired telephone service and by assuming that in the existing situation of excess demand new subscribers would exhibit characteristics similar to those of the average existing consumer.

With one minor exception, telecommunications tariffs had not changed since 1964. Between 1964 and 1977, however, domestic consumer prices inflated approximately 80 percent. As a result, subscribers using telephones in 1964 demonstrated in real terms that they were willing to pay at least 80 percent more for telephone service with lower quality and more limited access than current subscribers were paying. To estimate the prices in real terms that all existing telephone subscribers were willing to pay when they received service between 1964 and 1977, the average price paid in real terms and the quantity of new telephones supplied as well as the existing supply were taken to represent a point on the telecommunications supply curve at the end of each of the thirteen years. This point lay below the demand curves that actually existed at those times, since even at the higher real tariff, large excess demand (waiting lists and traffic congestion) was expressed during each past year. Given these tabulations of historical real prices and quantities, a weighted average price paid by present consumers when they joined the network was calculated. Given this estimated average price that present subscribers had in the past demonstrated a willingness to pay, the IRR on the future 1978–82 telecommunications investment program was estimated to be 36 percent.

This rate of return might still underestimate total benefits, however, since (a) the benefits to subscribers connected in past years were only measured by what they actually paid, with no allowance made for their consumer surplus at those times, (b) new subscribers to be added under the 1978–82 investment program would obtain a better quality of service (less congestion, improved transmission performance, fewer faults, and so forth), and (c) new subscribers would be gaining access to a larger number of other subscribers than was the case in earlier years. Even so, the estimated 36 percent rate of return was more than double the IRR of 16 percent that was estimated if only the projected financial returns to the telephone operating company were considered.[13]

Estimating Consumer Surplus by the Best-Alternative Method

Instead of observing the responses of telecommunications subscribers and other users to changes in price, a portion of the consumer surplus can be measured by estimating how much a user gains by using the telecommunications service rather than an alternative means of communication. Therefore, this section examines how much individuals or groups save by having a telephone, or, alternatively, how much they spend by not having ready access to a telephone.[14]

Five examples are reviewed. In one, a telephone demand function is derived and then tentatively verified by estimating the amount of money that at least some users spend communicating by using the next-best means in the absence of telephone service. In the second, the cost savings realized by using a telephone rather than the next-best medium are calculated for telephone users in two rural areas. In the third, consumer willingness to pay is estimated by examining alternative postal costs and priority access surcharges on long-distance calls. In the fourth, partial transport cost savings resulting from the installation of one telephone at each of two rural factories in Bangladesh are estimated. In the last example, the savings realized by using a telephone rather than bus transport are examined for thirty-six villages in India.

Example 15. Derivation of Demand Functions Based on Cost Savings for Communications (Pakistan 1968)

In this example a portion of the economic benefits associated with Pakistan's 1968–72 telecommunications investment program was measured. As a first step the demand curve for telephone services was estimated.[15]

To account for benefits associated with making local calls as well as those incurred solely by being connected to the network, the price variable was defined as a typical subscriber's annual expenditure on local telephone service; it comprised the annualized initial connection charges, one year's rental, and the charges for a typical 3,000 local calls. Under prevailing tariffs at the time, this amounted to Rs880. With regard to the quantity variable, at the beginning of the investment program the demand for telephone connections was estimated to be 196,000, of which 98,000 were in service, 65,000 were outstanding applications, and 33,000 were estimated as unreported demand. This, together with the annual price estimate (Rs880), fixed one point on the demand function. The demand curve itself was arbitrarily assumed to be a straight line. Based on available data from other countries and on qualitative considerations of the differences in demand composition between those countries and Pakistan, a price elasticity of demand of −0.25 at the identified point of Rs880 and a quantity of 196,000 connections demanded were assumed.

Since the shape and slope of the demand curve were largely based on judgment, an independently derived control was considered necessary to test the demand curve for overall validity. Hence, visits were made to several industrialists who had established factories on the outskirts of towns or in new industrial areas before 1968 and who were still waiting for telephone service. From interviews with these businessmen and information about employee salaries and transport costs, it was roughly estimated how much these industrialists were spending due directly to their lack of a telephone. The following is a composite case of what was found.

A business firm located about 4 miles from the center of a large town could be reached by car in about fifteen minutes. Junior executives at this establishment received a salary of about Rs2,000 a month, which, with 60 percent allowance for pension, leave, overhead, and so forth, and assuming 175 working hours a month, resulted in an average cost per executive-hour of about Rs19. Lacking a telephone, these executives often had to contact suppliers, buyers, merchants, transporters, government officials, and so forth in person. On average, they were out of their offices for such meetings about half a day every second day. It was estimated that, if they had telephones, these absences could have been cut to about half a day per week.[16] An estimate of the direct savings from having a telephone for an executive was therefore considered to be, first, the value of the time involved in traveling downtown, moving between meetings, and waiting for individual interviews due to the difficulty of prearranging and verifying times, and, second, the costs of transport for the journeys. Using a conservative estimate of two and one-half hours a week for lost time, at Rs19 an hour, the value of time saved by having a telephone to make local calls would have been about Rs2,500 a year per executive.

Transport costs, which were high partly because of import duties on vehicles and gasoline as well as heavy traffic congestion on downtown streets, might have been evaluated at anything from Rs200 to Rs500 a year depending on whether allowance was made only for the short-run marginal costs involved in these trips or for the full economic cost of transport, including depreciation of vehicles, and so forth. Hence, considering the time and transport cost saved, at least some telephones that were to be added under the 1968–72 telecommunications investment program could have been valued at about Rs2,700 to Rs3,000 a year.[17] This amount was much higher than the price at which demand would have been equal to current supply according to the demand curve that was initially estimated.

Since industrial subscribers were given priority, the foregoing information was relevant for the major uses to be made of the expanded telephone network. The data suggested that the demand curve initially specified probably represented a conservative estimate of demand. Hence, the demand curve could be used to calculate a somewhat conservative estimate of the benefits of Pakistan's 1968–72 telecommunications investment program. The IRR on the program was estimated at 19.1 percent. Adding the consumer surplus estimated by using the above demand curve approximately quadrupled this figure. With a shadow price of foreign exchange equal to twice the official price and with the customs duties on imports paid by the telecommunications entity removed, the estimated economic rate of return on Pakistan's 1968–72 telecommunications investment program exceeded 50 percent.

Example 16. Cost Savings for Communications in Two Rural Areas (Chile 1978)

In 1978 the costs and benefits of telephone service were studied for two rural areas in Chile.[18] One area included San Vicente de Tagua Tagua, a town with 239 subscribers and three PCO telephones in neighboring small villages. The area is located in a prosperous fruit-growing region and is well integrated into the modern sectors of the national economy. The other area included Cabildo, a town with sixty-four subscribers and two PCO telephones in neighboring small places. This area is in a relatively depressed, drought-prone, small-scale mining and cattle-grazing region.

To collect information on the perceived benefits of PCO long-distance telephones, 520 calls were immediately followed by a structured interview of the caller. From the information collected a portion of the cost incurred by each caller in making each call was estimated; this comprised the price paid for the call and estimates of the cost of round-trip transport and the time required to reach the telephone from the caller's place of residence or work.[19] The caller was also asked whether, for that particular call, he would have resorted to an alternative means of communication if a telephone had not been available.[20]

A portion of the consumer surplus associated with each call was then estimated as the monetary savings obtained by using the telephone rather than the least-cost alternative means of communication. No consumer surplus was computed in the few cases in which the alternative means would have cost less than the call actually placed.[21] Likewise, no consumer surplus was calculated when the alternative was not to communicate. Given the callers' reluctance to discuss income and the fact that in these rural areas a part of income is nonmonetary, seasonal, and difficult to estimate without detailed studies, no attempt was made to assign a monetary value to the estimated time savings obtained by using a telephone.[22]

Three alternative measures of benefits were calculated from the PCO telephone user

Table 8-1. *Internal Rates of Return from Investments in Rural* PCO *Telephones in Chile, 1978*
(percent)

Region	Financial	With tax and shadow price adjustment	With tax, shadow price, and partial consumer surplus adjustment
Cabildo	2.4	4.3	9.6
San Vicente	13.3	16.4	21.5

Source: Adapted from Nicolai and Wellenius (1979).

survey: an IRR, which measured the financial return on the investment in PCO telephones by the operating company; an IRR, after taxes were removed from both the cost and benefit streams, and foreign exchange and unskilled labor were shadow priced; and an estimate of the economic rate of return, where, in addition to the tax and shadow price of adjustments, the estimated consumer surplus (a comparison of the sum of call charges and transport costs with the cost of the alternative communication medium that would have been used) was added to the benefits stream. Table 8-1 summarizes the results of the three rate-of-return estimates for the sampled PCOs in the two regions.

For both village areas the estimated economic returns on the PCO telephones were greater than the financial returns on the investment to the telecommunications operating entity, and the returns on the PCO investment in the more economically prosperous region of San Vicente were the largest. This latter result is consistent with the findings of the cross-village correlation analysis in Costa Rica reviewed in example 12.

Example 17. Demand for Telephone Calls Based Partly on Costs of Postal Communication (India 1969)

This study addressed the substitutability between telephone calls and letters.[23] In 1969 in India the cost of sending a message by a local letter was estimated as a minimum of 45 paisa (postage was 15 paisa and typing, handling, and stationery were at least 30–35 paisa). It was assumed that important telephone conversations would generally substitute for at least two letters: an initial letter and a reply. It was further assumed that telephone users on average considered at least 5 percent of all calls made to be important enough to warrant paying as much as the cost of a letter and a reply; other calls could be replaced by one letter or other means of communication, or they would simply be eliminated. Given these assumptions, one point on the demand curve for local telephone calls was estimated to be at a quantity equaling 5 percent of all local calls and a price of 90 paisa.

A second point on the demand curve was derived under the assumption that the value of the most valuable local call was at least eight times the charge for a local call. This assumption was based on the fact that, on the long-distance manual trunk network, some callers were opting to be placed in a "lightning call" category under which the P&T telephone operator placed the call immediately rather than putting the caller at the end of the queue where he would wait up to four hours for a connection. The charge for this service was eight times the normal long-distance rate. Since the charge for local calls was only a fraction of that for trunk calls, the assumption that the value of the most valuable local call was also at least eight times the standard charge for a local call was considered to be conservative.

Finally, a third point on the demand curve for local calls was taken as the number of local calls made each year at the actual charge rate of 15 paisa. An S-shaped demand curve was then drawn through the three points. The curve as drawn generally implied that local calls per direct exchange line would fall by about 50 percent if the local call charge were increased threefold (demand was relatively inelastic).

Given this demand curve, the consumer surplus above the call charge rate of 15 paisa was estimated. Without consumer surplus, the IRR on India's P&T 1969–72 telecommunications investment program was estimated to be 6.5 percent. With the consumer surplus added to the revenue stream, the economic rate of return on the program was conservatively estimated to be about 25 percent.

Example 18. Cost Savings for Transport at Two Rural Factories (Bangladesh 1981)

In early 1980 a radiotelephone was installed at both the Nabaran Jute Mill (NJM) and at the Ghorasal Fertilizer Factory (GFF) outside Dhaka. The two telephones were linked directly to a central telephone exchange in Dhaka. Data on both telephone usage and cost savings were gathered through interviews with the general managers and several senior staff of both enterprises. Although the data are considered indicative, they were not confirmed through independent sample observation.

Based on these data, for one year, before and after differences in direct expenditures on gasoline and on the wages of managers communicating with Dhaka were estimated. Local gasoline prices were assumed to be roughly in line with world market prices. The initial data and the estimated expenditure savings are presented in table 8-2. These differences in before and after expenditure include only direct cost estimates related to gasoline and some transport time related to management wages. A more complete calculation would include savings in vehicle capital, other operating and maintenance costs, and other direct and indirect expenditures.

The capital cost of one very high-frequency radiotelephone installed in each factory plus a proportionate share of the base station cost of the receiver/transmitter at a central exchange in Dhaka were roughly Tk190,000.[24] It was not considered necessary to shadow price foreign exchange. Each factory reimbursed the Bangladesh Telephone and Telegraph Board for this cost when the telephone was installed. Installation cost was about Tk10,000, and annual maintenance cost beginning after one year of operation was assumed to be Tk20,000. These costs were also reimbursed by the factories. Based on available data, it was assumed that ninety-six outgoing telephone calls were

Table 8-2. Factory Expenditure Savings Data from Bangladesh, 1981

Item	NJM	GFF
Approximate distance from Dhaka (miles)	20	40
Average number of round trips to Dhaka a week		
Before installation of telephone	18	11
After installation of telephone	2	1.5
Assumed number of working weeks a year	48	48
Average round-trip gasoline cost (taka)	350	650
Average total round-trip time (hours)	4	7
Average number of staff making trip (including driver)	3	3
Average wage a month of management staff making trip (taka)	1,000	1,200
Average number of working hours a week	42	42
Gasoline expenditure savings (taka)	268,800	296,400
Reduced wage expenditure for transport communication for two management staff (taka)	36,557	39,197
Total expenditure difference (taka)	305,357	335,597

Note: NJM, Nabaran Jute Mill; GFF, Ghorasal Fertilizer Factory.
Source: World Bank data.

made each week at Tk0.75 a call (roughly reflecting the Board's operating cost), telephone operation costs to the factories (and the Board) averaged about Tk3,456 a year, and a similar number of telephone calls were received by each factory each week with the Tk0.75 call charge being paid by the caller. The additional capacity cost that the network as a whole incurred as a result of the two additional telephone connections was assumed to be trivial.

Given the above, the cost of telephone service to each factory during the first year of operation was approximately Tk203,456. The direct communication expenditure savings by NJM and GFF was Tk305,357 and Tk335,597 a year, respectively; in the first year direct communication expenditure savings exceeded costs by about 1.5 times at NJM and 1.6 times at GFF. During the second and each succeeding year, quantified direct savings exceeded costs (Tk23,456) by about thirteen times at NJM and fourteen times at GFF.

Example 19. Bus Transport Cost Savings for Telephone Users (India 1981)

A survey of 174 users of public telephones in thirty-six villages in six districts in Andhra Pradesh State in India found that 120 of the users regarded their most recent PCO telephone call to have been so important that they would have traveled to the destination of the call if the public telephone had not been available.[25] Hence, the cost that would have been incurred by each caller to travel to the call destination rather than use the telephone was estimated. Travel cost was taken as the fare for the cheapest means of travel—by bus. The value of the time lost by making the journey was conservatively estimated by using the prevailing wage rate for unskilled labor. Table 8-3 summarizes the results of the analysis.

For the 120 PCO telephone users who judged that their most recent call was urgent enough to warrant travel in the absence of a telephone, the estimated consumer surplus exceeded the cost of the telephone call between 2.5 and 5.5 times, depending on the distance involved.

Table 8-3. *Expenditure for Communication by Bus and by Telephone for Thirty-six Villages in Andhra Pradesh State, India, 1981*

Distance called (kilometers)	Average distance (kilometers)	Expenditure on telephone calls (Rs)	Cost of bus fare and bus terminal access (Rs)	Value of time lost through bus travel (Rs)	Total bus travel cost (Rs)	Consumer surplus for telephone call (Rs)
0–20	11.24	1.37	4.53	2.00	6.53	5.16
20–50	34.57	3.54	8.45	4.00	12.45	8.91
50–100	80.54	4.56	16.19	8.00	24.19	19.63
100+	149.00	5.44	27.69	8.00	35.69	30.25

Source: Adapted from Economics Study Cell, Posts and Telegraph Board, India (1981).

Notes

1. Examples of these high rates of return are presented in chapters 1 and 9, as well as in this chapter.

2. In the absence of good data, estimating such costs and revenues through time and calculating the resulting IRR can be highly sensitive to the assumptions made. For example, one analyst may predict rapidly growing demand and falling unit costs because of new technology and may assume that the telecommunications entity has the technical and managerial capacity to expand rapidly. Another analyst may predict less rapidly rising demand, constant unit costs, and a much lower implementation capacity of the telecommunications entity. In such an instance, the two analysts would reach different conclusions about the size of investment programs that could be justified and the extent of that justification, even though both followed the same analytical procedures; that is, the results of an IRR calculation to a certain extent depend on the assumptions used. Hence, the importance of sensitivity analysis.

3. Another way of stating this is that the IRR is the discount rate that equalizes the present values of the project's cost and revenue streams, that is, IRR $= r$ when

$$\sum_t (R_t - C_t)(1 + r)^{-t} = 0$$

where R_t is revenue and C_t is cost in time period t.

4. For a more general discussion of investment appraisal among alternative uses of funds and how choice depends on the decisionmaker's objectives and attitudes toward risk and time, see Littlechild (1979), chap. 7.

5. If a more detailed knowledge of the size and incidence of project benefits is available, social weights might also be used to value project output. See Squire and van der Tak (1975). In telecommunications projects, however, such information is rarely available. Furthermore, although the concept of social weighting attracted considerable interest in the 1970s, little progress has been made in its practical application to project analysis generally.

6. In some instances two additional problems must be dealt with when revenues are used as a minimum measure of benefits for telecommunications projects. First, the telecommunications price structure may not correspond to the real costs incurred in providing the added or improved service. Therefore, reliance on willingness to pay as the sole justification for investment may send misleading investment signals. Second, reliance on willingness to pay may not explicitly consider the achievement of other national goals, such as income redistribution, the generation of public savings, regional development, and so forth. These problems are considered in following chapters.

7. In some cases such externalities have "public-good" characteristics, that is, excluding noncallers or noncallees from reaping at least some external benefits (direct or indirect) of telephone service is not feasible. However, given the generally accepted definition of public goods—that "each individual's consumption of such a good leads to no subtraction from any other individual's consumption of that good" or that they "must be consumed in equal amounts by all" or be "a good for which the resource costs are not attributable to beneficiaries"—it would seem that telecommunications services that are sold to individuals (telephone rental, call charge, and so forth), whose consumption on the whole precludes others from that consumption, should be generally categorized as private goods, with which externalities are associated in some instances, rather than "public" goods. See Burkhead and Miner (1971), chap. 2; McKean (1968), pp. 43–47; and Arrow (1970). Nevertheless, some advocates of government subsidies for telecommunications services have focused exclusively on the public-good characteristics of telecommunications investment externalities and have based their subsidy arguments on a contention that telecommunications services are or should be treated as public goods. See, for example, Hudson and others (1979),

chap. 4, and Pierce (1979). Of course, on theoretical grounds the existence of externalities alone, whether or not they have public-good characteristics, can be enough under some circumstances to suggest subsidies, although in developing countries where demand for telecommunications services generally far exceeds supply, financial subsidies of general telephone service are very difficult to justify on economic grounds.

8. Several examples in this and the next chapter are illustrative.

9. The concept of consumer surplus is closely related to the concept of the downward-sloping demand curve. For example, a consumer (subscriber) would generally be willing to pay a higher price for only one call in a day than for the twentieth most important call of that day. Since telephone call charges are usually fixed, however, consumers pay the same amount for all calls. Therefore, the caller receives a "surplus" of benefit (satisfaction) for all calls for which he would have been willing to pay a price greater than that which he was actually charged. The same concept, of course, holds for connection fees and monthly rental charges.

10. Of interest to all telephone administrations is the question of how changes in the number of calls made or connections requested (Δq) relate to changes in price (Δp). To avoid possible confusion associated with different units of measurement and the absolute sizes of the variables, this responsiveness of quantity demanded to price is usually represented in terms of a price "elasticity" coefficient defined as

$$\varepsilon = (dq \ / \ dp) \cdot (p \ / \ q).$$

For a simple approximation, an arc price elasticity of demand can be calculated by dividing the percentage change in the number of calls made (or connections requested) by the percentage change in price, defined as

$$\varepsilon = (\Delta q \ / \ \Delta p) \cdot (p \ / \ q).$$

Given this estimation, if the unit call charge (or connection fee) is increased 50 percent and the number of calls made (or connections requested) then falls 25 percent, the price elasticity coefficient would be -0.5. Demand is said to be inelastic if the absolute value (that is, not taking into account the sign) of the coefficient is less than 1, elastic if the coefficient is greater than 1, and of unitary elasticity if it is equal to 1.

11. In figure 8-1 let the straight line AB be $p = a - bq$ (a and b are constants). Differentiating both sides gives $dp \ / \ dq = -b$. Substituting $dp \ / \ dq = (1 \ / \ \varepsilon)(p_1 \ / \ q_1)$ (from the definition of ε) yields $-b = (1 \ / \ \varepsilon)(p_1 \ / \ q_1)$. Substituting this expression of b in the equation of line AB then gives $a = p_1[1 - (1 \ / \ \varepsilon)]$. The consumer surplus CS at price p_i estimated from the area of the triangle $p_1 A a$ is $CS \geq \frac{1}{2} q_1 (a - p_1)$; substitution of the above expression for a yields $CS \geq p_1 q_1 \ / \ 2\varepsilon$. Revenues R are given by $R = p_1 q_1$. Hence the ratio of total benefits to revenues is $(CS + R) \ / \ R \geq - 1 \ / \ (2\varepsilon)$. This ratio equals 3.0 for $\varepsilon = -0.25$, 2.0 for $\varepsilon = -0.5$, and 1.5 for $\varepsilon = -1.0$. Under the (usual) assumption that the demand function is convex toward the origin, the tangent will lie wholly below DD', and benefits will continue to be underestimated by some margin; this is acceptable since the calculation of an economic rate of return is usually used in project analysis to test against low returns.

12. The consumer surplus exercise outlined for El Salvador in this example was originally undertaken with Mihkel Sergo.

13. A second estimate of the economic rate of return was obtained by assuming that new subscribers to be connected during 1978–82 would be willing to pay at least what was being charged in real terms to subscribers connected in 1977. This alternative estimate of the rate of return was between 31 and 33 percent, depending on whether the connection fee for new subscribers was taken to be that prevailing when they applied for service or that prevailing when they were actually connected.

14. In concept, this approach is only useful for estimating consumer surplus under fairly strict conditions. For example, if the user has no preference between a telephone call and

some other means of achieving a given result, for example, a personal visit, then the difference between the cost of the telephone call and that of the alternative can (assuming the alternative is more costly) indicate the size of the consumer surplus.

15. This example is based on a study undertaken by Christopher Willoughby. At the time, the official exchange rate for the Pakistan rupee was PRs1,00 = $0.21.

16. Letters were not believed to be a relevant alternative to the telephone and face-to-face meetings for local communication, that is, within a metropolitan area. They were primarily used to convey documents and confirmatory notes, which had to be in written form. In Pakistan, messages borne by a private messenger may be, in some cases, a poor but partially relevant substitute for telephone communications. However, messages were not believed to be an important substitute in the type of composite business dealings described in this example.

17. Actually, some telephones would be valued higher, since the figures used reflect something of a group average, not extreme values, and a single telephone would benefit more than one executive. The original demand curve assumptions were also checked using information on private transfers.

18. Nicolai and Wellenius (1979).

19. Like the previous exercise, this one involves a mix of the best-alternative method and the expenditure method of estimating consumer surplus. The exercise up to this point involves the "expenditure method" of benefit analysis.

20. More than 80 percent of calls were made by local residents or workers. Use of accepted alternative means thus involved an additional trip somewhere out of town, which could in many cases result in carrying out several other activities besides communicating. Hence, not surprisingly, the preferred alternative means was sometimes not the one with the lowest cost for the communication alone.

21. About 3 percent of the calls could have been replaced by letters, according to the callers, which would have cost less. The fact that the callers actually used the telephone, despite its being more costly, indicates that they attached a value to the qualitative differences and to the faster response times (these benefits were not quantified).

22. Since the alternative means generally involved considerable additional travel or response times, the time saved by using the telephone was on average quite high.

23. This example is based on a study undertaken by Christopher Willoughby. At the time, the official exchange rate for the Indian rupee was Re1 = 100 paisa = $0.13.

24. At the time the official exchange rate was Tk16 = $1.00.

25. Economics Study Cell, Posts and Telegraphs Board India (1981), pp. 48–50.

Chapter 9

Expenditure Method Estimates
of Consumer Surplus

THE EXPENDITURE METHOD of estimating a portion of the consumer surplus associated with telecommunications services attempts to measure more completely the costs that users actually incur in the process of communicating. In many situations it is possible to observe the amount of telecommunications services demanded at a given time by users who, for one reason or another, willingly incur additional costs while using the service.[1] For example, people traveling to a nearby public call office (PCO) use time and may incur transport costs in addition to paying the call charge.[2] Hence, if it is assumed that public telephone users in and around a rural village in a developing country are relatively homogeneous, with similar income and tastes, but that they must travel different distances and incur different costs to use the telephone, a rough demand curve can be estimated by comparing the number of visits to the telephone with telephone tariff charges plus costs associated with accessing the telephone.

It is also possible to estimate partial consumer surplus using the expenditure method without observing the costs of travel and travel time. In many countries, existing telephone installations are sold privately, either on the black market or legally, with the new user paying the old subscriber a substantial price in addition to the connection or transfer charges levied by the operating telephone company. Also, higher rental and purchase prices are sometimes paid for residences or offices with installed telephones than for those without. Finally, because call traffic is frequently congested during business hours, telephone users spend potentially valuable time in repeated attempts to make a call. Circumstances such as these provide further opportuni-

173

ties to estimate the value of telecommunications services over and above payments made to the telecommunications authority itself.

Estimating Consumer Surplus by the Expenditure Method

The following examples illustrate how indications of users' willingness to incur costs in excess of official telephone tariffs can be used to estimate at least part of their consumer surplus.

Example 20. Transport Cost and Time Spent to Reach a Telephone (Chile 1975)

In Chile, the costs and benefits of providing 200 towns and villages with a PCO telephone were estimated.[3] The 200 places were categorized into twenty-three groups according to locational characteristics and income level. A control village with similar income and locational characteristics, but which already had a PCO telephone, was assigned to each group.

Linear demand functions for telephone calls were then estimated separately for each of four income strata in each village without a telephone. One point on each demand curve was arbitrarily fixed, assuming that, were a telephone available, traffic in the town would be approximately equal to that currently observed in the associated control place if similar tariffs applied. Since observed calls were not homogeneous, an "equivalent price" was calculated as the weighted average price paid for calls in all of the control places, and the number of "equivalent calls" was calculated for each control place and for each income stratum by dividing actual revenues for calls by this equivalent price.

A second point on each of the 800 demand curves was estimated using interview information on the calls residents made on the nearest telephone outside the village. An "equivalent price" was worked out for each village and income stratum by adding average transport cost and estimated value of the time used to get to and from the telephone to the price calculated for the set of control places. The number of equivalent calls was calculated as the total expenditure incurred by callers from a given place and income stratum divided by the appropriate equivalent price per call. The estimated additional revenue and consumer surplus that would be generated by a public telephone were then estimated for each place by calculating the areas under the assumed linear demand curves, together with what were thought to be reasonable assumptions about the growth of call traffic through time.

Assuming an equipment life of fifteen years (twenty years for open wire installations), a discount rate of 12.6 percent, and a shadow price for foreign exchange equal to 1.3 times the official price, the present value of the total benefits expected to result from providing PCOs in the 200 villages averaged 4.8 times the present value of the forecast incremental revenues alone. For seventy-seven of the villages the present values of the estimated net benefit streams were positive. Hence, the immediate provision of telephones could be justified in terms of expected effect on the economy as a whole. In contrast, from the viewpoint of the telephone company, adequate financial returns would be obtained in only five of these seventy-seven places. For the remaining 123 places, investment was not recommended at the time; however, given forecast increases in traffic, it was estimated that twelve additional places would have a positive net present value of benefits if the PCO installations were postponed three or four years.

Example 21. Transport Cost and Time Spent to Reach a Rural Telephone (Costa Rica 1976)

During one week in 1970, everyone who used the only telephone in the village of Puerto Cortes in Costa Rica was interviewed. Information was assembled on the price paid for each call, the caller's income, travel time to the telephone, reason for the trip, transport cost, and hours worked during the week.[4] Assumptions were made about how the time of the callers should be valued and how the separate and joint costs of various means of transport should be allocated to the telephone call. The underlying premise of the exercise was that persons who lived closer to the PCO and therefore used it more than persons living farther away obtained a consumer surplus for at least some of their calls.

Summary results of the exercise are shown in table 9-1. Given the somewhat arbitrary nature of the assumptions involved in the analysis, two sets of assumptions were used to estimate transport cost and time value. As can be seen, the resulting estimates of consumer surplus are highly sensitive to those assumptions.

Table 9-1. *Two Estimates of Consumer Surplus for Telephone Calls Made in a Rural Costa Rican Village, 1976*
(colones)

Item	High estimate	Low estimate
1. Average call charge	4.41	4.41
2. Average transport cost per call[a]	19.02	2.54
3. Average travel time (hours)	0.59	0.59
4. Average monthly income of caller	422.90	422.90
5. Average number of hours worked during week	39.31	39.31
6. Opportunity cost of time[b]	1.51	0.38
7. Consumer surplus per call[c]	20.53	2.92
8. Ratio of consumer surplus per call to average call charge	4.65	0.66

a. Of the 404 calls used to tabulate the information in this table, only fifty-four of the callers were able to offer what was considered to be reliable information on transport costs. It was therefore assumed that, for the high estimate, transport costs for the missing observations were equal to the average for the persons responding and that, for the low estimate, nonrespondents incurred no transport costs.

b. For the high estimate, it was assumed that the value of time for all callers was equal to the average income for that time in the village, that is, $(6) = (3) \cdot (4) / [(5) \cdot 4.2]$. For the low estimate, it was assumed that many of the calls were made during periods when real resource costs in terms of time were minimal, and therefore the low estimate was arbitrarily set equal to one-fourth of the high estimate.

c. $(7) = (2) + (6)$.

Source: See appendix C.

Example 22. Rents and the Availability of a Telephone
(Egypt 1977)

If telephone connections in offices, houses, or apartments can be transferred when the property is sold or leased, the relation between the value of real estate property and the presence of a telephone line can provide information on the willingness of users to pay for telephone service.

In 1977, Egypt had an acute shortage of telephones, and local businesspersons and middle-income residents often obtained telephones by renting furnished offices or residences in which, legally, the telephone could be transferred as one of the furnishings. Rental advertisements in local newspapers commonly listed the availability of telephone service as one of the prominent attributes.

Differences in rent of from £E50 to £E150 a month[5] were observed for offices or residences that were similar except for the presence of a telephone.[6] Places with a telephone that was only one of several extensions on a shared line or places that were in areas known to suffer from heavily congested telephone traffic tended to command less of a premium than those with private lines or located in areas with less call traffic congestion. The larger and better located offices or apartments tended to carry the highest telephone premiums (£E150 and more), probably reflecting the high value that large businesses and higher-income residents attached to the telephone.

Given these findings, the consumer surplus for the monthly rental of a telephone was estimated using two assumptions. First, the demand curve for telephone rentals was downward sloping, reflecting a situation in which a relatively small proportion of the total population was willing to pay high monthly rentals for telephone service, whereas most of the population was willing to pay only much lower telephone rental charges. Hence, the demand curve was assumed to be convex to the origin of the price and quantity axes; mathematically, the demand curve was specified as a rectangular hyperbola with elasticity equal to -1. Since empirical estimates of the price elasticity of demand for telephone service are generally between 0 and -1, this was a conservative assumption.[7] Second, two points on that demand curve are 15,400 telephones (less than 3 percent of the total number of telephones in service) at a price of £E51.50 a month (the official monthly fee of £E1.50 for automatic message rate exchange service plus £E50 for the typical difference in rent) and 530,000 telephones at the official price of £E1.50.[8]

The bounded area under the demand curve above the official monthly price of £E1.50 and below a monthly price of £E51.50, and between the quantities 0 and 530,000 telephones was equal to £E2,807,669. This amounted to an average of £E5.30 per telephone a month or £E63.60 per telephone a year. Hence, for estimating benefits, the value of renting a telephone averaged £E63.60 a year above the average price actually paid to the telecommunications operating entity (£E18.00).

It was also assumed that this estimate of average consumer surplus would continue to be valid for users added to the system during the next four years. This assumption was considered reasonable since, despite rapid expansion of the system, a large proportion of the excess demand would not be satisfied—and hence most new users would be high-value users—for several years to come. Thus, for the following four years, the benefits arising from having a telephone were estimated to average at least four times the revenues paid to the telephone operating entity for telephone rentals.

Example 23. Value of Time Wasted in Call Attempts (Egypt 1977)

In addition to benefits derived from being connected to a telephone system, benefits also arise when calls are completed. In Egypt call charges paid by telephone users underestimated the users' costs of completing a call since the high proportion of failures to complete a call, time spent waiting for a dial tone during peak periods, and the necessity to repeat calls because of poor transmission quality and interruptions wasted a significant amount of otherwise productive employee time.[9]

The following assumptions were believed to reflect conservatively the situation in Egypt's two largest cities, Cairo and Alexandria, in 1976–77:

a. In 1976 the average urban caller earned an estimated £E6.00 a week.[10] Assuming the caller worked an average of forty-five hours a week, each working minute was worth approximately £E0.0022.

b. A minimum of one and one-half minutes per failed call was spent during business hours trying to complete a call that was unsuccessful because of technical problems or call traffic congestion.

c. Of calls made in business hours, 95 percent were related to business or government.

d. Unsuccessful calls resulting from technical faults or call traffic congestion during prime business hours constituted approximately 75 percent of all attempts to make local calls in Cairo and 55 percent in Alexandria.

e. An average of about 1,820,000 and 560,000 calls were completed, respectively, in the two cities during the five or six peak calling hours each working day.

f. During peak hours an unsuccessful call attempt rate of 25 percent would have been acceptable.

An estimate was made of the value of time wasted by government and business callers in Cairo and Alexandria when more than 25 percent of their attempted calls were unsuccessful. (No attempt was made to extend the calculation nationally.)

Using the above estimate of a portion of the costs that consumers willingly incurred in making a telephone call (official call charge plus value of time wasted) and the estimate from example 22 of the yearly value of having a telephone connection (average monthly telephone rental charge plus the estimate of consumer surplus due to rent differential), a stream of project benefits was estimated for Egypt's 1978–82 telecommunications development program. With program costs appropriately shadow priced, the internal rate of return in real terms on the proposed new investment was estimated to exceed 25 percent. This was about two and one-half times the rate of return that was calculated by using only revenues from tariffs as a proxy for benefits.

Example 24. Cost of Unauthorized Transfers (Myanmar 1978)

In many developing countries, the private transfer of telephone connections is either legally permissible or informally tolerated.[11] A good example is provided by the experience of Myanmar (formerly Burma), where during several years, numerous telephone subscribers revealed a willingness to pay a price in excess of that prescribed in the official tariff schedule to acquire a telephone.[12] In Yangon (formerly Rangoon), in particular, an unofficial private market existed in which telephone connections were bought and sold. In 1977 the national telephone organization began an intensive effort to trace unauthorized connections and, as a result, found about 520 cases in which telephones had been privately transferred at prices averaging K10,000.[13] All the unauthorized subscribers identified were fined K2,400 each but were allowed to keep their telephone after they paid the fine and the official connection charge (K450) and deposit (K300). When word spread that the unofficial connections were being legalized, about 470 additional unofficial subscribers in Yangon voluntarily turned themselves in, paid the fine, connection charge, and deposit (K3,150), and had their telephone status legalized.

Given this information and the fact that in 1977, 15,545 potential subscribers without telephones were registered on waiting lists, thus demonstrating a willingness to pay at least the official connection charge of K450, the consumer surplus associated with obtaining a telephone in Myanmar was estimated using the following two assumptions.

First, the demand curve for telephone connections was shaped so that a relatively small proportion of the total population was willing to pay high initial fees to obtain a telephone, whereas most of the population was willing to pay only relatively lower fees. Hence, the demand curve for telephone connections was assumed to be convex to the origin of the price and quantity axes and to have a price elasticity of −1.[14] Since empirical estimates for the price elasticity of demand for telephone service are generally between 0 and −1, this was a conservative assumption. Second, one point on that demand curve was 1,522 telephone connections (the number of recently legalized telephones in Yangon in fiscal 1977 plus the estimated number that were expected to be legalized in a forthcoming drive against unauthorized connections in large places other than Yangon) at a price of K12,850 (K10,000 private market price plus K2,400 fine plus K450 connection fee). This point, together with the slope of the curve implied in the assumed rectangular hyperbola with a price elasticity of −1, specified the demand curve.

Calculating the bounded area under this demand curve above the official connection fee of K450 and below K12,850 and between the quantities of 0 and 17,067 (1,522 plus the 15,545 waiting applications in Myanmar as of March 1977) and dividing that by 17,067 gave an average consumer surplus of K3,466 per telephone connection. Hence, to estimate benefits, the value of acquiring a telephone in Myanmar during 1977–78 and for the five-year period thereafter, during which a large excess demand was forecast to continue, was estimated to be the sum of the connection fee of K450 plus a one-time consumer surplus of K3,466.

Example 25. Value of Time Saved by Introducing Subscriber Trunk Dialing (India 1969)

In portions of India before the introduction of subscriber trunk dialing service, consumers wanting to make long-distance calls during business hours normally had to wait lengthy periods for their call to be completed. They could, however, move up in the queue by agreeing to pay more, from twice the ordinary charge for so-called "urgent" calls to eight times the ordinary charge for "lightning" calls. With the introduction of subscriber trunk dialing service, call waiting times were projected to disappear on some routes.

In an attempt to estimate the value of the time that would be saved by introducing this service, outgoing trunk traffic was analyzed on the principal routes serving the regional center of Madras. As shown in table 9-2, the correlation, irrespective of distance and hence of call charge, was high among three factors: the percentage of calls booked as urgent, the average delay in being connected for ordinary calls, and the average reduction in the delay for calls booked as urgent rather than ordinary.

Table 9-2. *Delay of Ordinary and Urgent Calls Booked, by Major Route in Madras, India, 1969*

Destination of call, by area	Percent of calls booked "urgent"	Average delay on ordinary calls (minutes)	Average difference in delay: urgent vs. ordinary (minutes)
Bombay	39.8	80	50
Calcutta	25.3	32	11
Coimbatore	35.0	70	31
Delhi	23.2	30	..
Hyderabad	25.9	41	20
Madurai	27.5	50	17
Salem	10.8	16	..
Tiruchi	28.4	46	28
Vellore	14.6	32	12
Vijayawada	11.5	13	..

.. Zero or negligible.
Source: Unpublished data from India P&T.

An analysis of the raw data indicated that telephone users were willing to pay a premium of about 25 percent of the base charge to save twenty minutes' delay in obtaining an ordinary call, 35 percent to save thirty minutes, 40 percent to save forty minutes, and 50 percent to save one hour. The average difference in delay between urgent calls and ordinary calls was about twenty-five minutes. Hence, with the introduction of subscriber trunk dialing, which was supposed to eliminate these delays, the time saved was assumed to be worth at least 30 percent of the base call charge to the average caller.

Example 26. Foreign Exchange Impact of Business Telecommunications (Kenya 1981 and 1987)

The Kenya Cashew Nuts Company was located between Mombasa and Malindi as part of a government program to decentralize industry to rural areas. It served as a collection and processing point for cashew nuts and marketed them through five agents located in centers throughout the world. Several of the eighteen grades of cashew nuts sold by the company occasionally experienced significant price movements on a day-to-day basis and through time. For example, one primary grade declined in price 67 percent in three months.

Telex was the accepted means of communicating with the company's five market agents and keeping abreast of global prices and contracted quantities for cashew nuts. In mid-1981 the nearest telex connection was located in Mombasa 35 miles distant; the company used this connection periodically with the agreement of the subscriber, which rents the telex from the Kenya Posts and Telecommunications Corporation. This arrangement resulted in delays for the company of up to two days in the receipt of, analysis of, and transmitted response to market information.

Given such information delays, together with the sometimes rapid price movement

and slight difference in price among markets, managers estimated that if they had access to two telex lines at the processing plant, which would allow them to reduce message turnaround time to less than half an hour, they could on average secure 5 to 10
percent better prices for the cashew nuts they market. Managers cited several instances
in which, because they lacked timely information or response, they sold low in a rising
market or missed a sale in a falling market. Given the company's current foreign exchange earnings of about $15 million a year, a 5 percent increase in the average sale
price of their cashew nuts would amount to an additional $750,000 a year of foreign
exchange earnings for the Kenya Cashew Nuts Company and for Kenya.

Management planned to open a branch marketing office in Mombasa, which would
have access to a shared telex line and would maintain some contact with the processing
plant through a magneto telephone exchange connection. The company estimated that
the Mombasa branch office would cost the equivalent of $1,000 a month to operate and
maintain. This was a small fraction of the increase in sales revenue that was estimated to
result from this somewhat improved but still cumbersome communications arrangement.

Additional microeconomic analysis of the role that telecommunications plays in
foreign exchange savings in Kenya was carried out by Communications Studies and
Planning International with the support of the ITU in 1987. Twenty companies, representing nearly 18 percent of Kenya's total export earnings, were selected from the agricultural, industrial, and service sectors and were interviewed in depth about how
telecommunications affects their exports and imports. The benefits were calculated
using improvements in infrastructure that were expected to be the result of a forthcoming investment by the World Bank of $3.26 million in Kenya's telecommunications sec-
tor. The twenty businesses interviewed were calculated to save over $2.0 million in
foreign exchange as a result of this project. Extrapolated for the export sector as a
whole, savings in foreign exchange were calculated at $11.6 million, yielding a benefit to
cost ratio for foreign exchange of 3.6:1.0 for a major telecommunications project.[15]

Example 27. Efficiency Gains by Business Firms (Kenya 1981)

A survey of how telephones were used by nine business firms in Nairobi examined
the hypothesis that some businesses could not exist in their present form without telephones, whereas others could, but only on a smaller scale, with lower productivity, and
with higher costs.[16] Three means were examined to improve a business firm's performance by improving communications services. These were a cost reduction effect
(lower unit costs through lower-priced inputs and increased management efficiency), a
selling price effect (higher selling prices through better market information), and a
business expansion effect (sales gains through better and wider area marketing).

Two of the nine firms surveyed were engaged in manufacturing and processing (East
Africa Industries and House of Manji), five provided services or transport (Alliance
Hotels, Industrial Distributors, Interfreight, Pan African Travel, and Standard Newspaper), and two were engaged in agriculture or horticulture (Kenya Horticulture Export Corporation and Kenya Nurseries). Within the above three categories the study
documented telecommunications benefits relating to business expansion, sales price
effects, purchase price effects, vehicle use, production stoppages, distribution costs,
labor time, and managerial time. Table 9-3 shows estimates of the average increment
in the percentage of profits to revenues for each form of benefit that the nine firms
would experience if access to the national telecommunications system were extended
and if the quality and reliability of the service were improved.

Overall, the study found that the single most important potential benefit to the

Table 9-3. *Estimated Average Increment for Nine Firms in the Percentage of Profits to Revenues, by Benefit Category, in Nairobi, Kenya, 1981*

Benefit category	Estimated average percent increase in profits to revenues
Business expansion	2.8
Sales price effects	0.4
Purchase price effects	0.2
Inventory costs	0.0
Vehicle use	0.1
Production stoppages	0.0
Distribution costs	0.3
Labor time	0.9
Managerial time	0.3
Mean for the nine firms	5.1

Source: Adapted from Communications Studies and Planning International (1981).

nine firms would be an increase in sales made possible by better access to markets for buying and selling. It was estimated that the major part of the efficiency benefits gained from expanded sales would come from cost reductions obtained through improved capacity utilization rather than from greater economies of scale; that is, diminished machine downtime, better allocation of managerial time, greater regularity of sales and purchases, and similar effects.

Table 9-4 shows the estimated gains or benefits to the nine firms compared with

Table 9-4. *Estimated Benefits and Expansion Costs, by Firm, in Nairobi, Kenya, 1981*

Firm	Estimated benefits to firm (K Sh)	Estimated telecommu- nications costs (K Sh)	Benefit/ cost ratio
East Africa Industries	11,930,000	125,460	95.1
House of Manji	4,417,900	14,760	299.3
Alliance Hotels	870,000	13,480	64.5
Industrial Distributors	2,380,000	17,080	139.9
Interfreight	5,565,000	57,900	96.1
Pan African Travel	275,400	2,320	118.7
Standard Newspaper	6,035,600	31,060	194.3
Kenya Horticultural Export	2,627,500	31,460	83.5
Kenya Nurseries Corporation	27,400	2,560	10.7
Total or average	34,128,080	296,080	115.3

Source: Adapted from Communications Studies and Planning International (1981).

what the study authors acknowledged was a very rough indicative estimate (likely to be on the low side) of the average cost per telephone line and per call if telecommunications services were available throughout Kenya. On the basis of an analysis of the interview data the study concluded that telephone calls made by employees of these nine Kenyan firms were on average worth over ten times the cost; that is, a large consumer surplus was associated with expanded telephone service for businesses in Nairobi.

Example 28. The Role of Telecommunications in Improving the Reliability of the Power System (Brazil 1980)

In 1980 an analysis attempted to determine the effects of an improved telecommunications network on the operation and maintenance of the COELBA power system in the Brazilian state of Bahia.[17] It was hypothesized that an improved telecommunications network would improve the operation and control as well as maintenance of the existing COELBA power system and that, as a result of such improvements, the reliability of the supply of electricity in the state would improve. Specifically, it was contended that improvements in operation and control would allow the COELBA system to operate more efficiently with fewer interruptions in supply (outages) resulting from overloaded circuits, burned-out transformers, and so forth. The principal economic benefits of improved operation and control were measured by the decreases in outage costs resulting from reductions in outage frequency. With regard to maintenance, it was assumed that improvements in telecommunications would allow failures within the COELBA system to be identified and corrected more quickly. The major economic benefits of such improvements were measured by the projected decreases in outage costs resulting from shorter and less-frequent outages.

The estimation process involved the following: (a) the effects of outages on important industries in Bahia were determined from a survey of seventeen industrial firms around Salvador, the central city; (b) the effects of outages on residential consumers in Bahia, and estimates of consumers' willingness to pay to avoid such outages, were determined from a survey of residential consumers; this survey covered 182 consumers in five economic regions in Bahia; (c) data on value added according to industrial sector for 1970–75 were obtained from the state government and served as a basis for estimating industrial value added over the lifetime of the telecommunications project; (d) data on the number of residential electricity consumers in the COELBA system in 1980 and estimates of the growth rate of such consumers were obtained from COELBA; (e) information on the effect of the telecommunications system on the quality of electricity supply within the COELBA system, that is, estimates of resulting reductions in the frequency and duration of outages, were obtained from interviews with officials from COELBA; (f) estimates of project capital and operating costs were obtained from the various telecommunications project feasibility studies; and (g) data on the number of street lights used in Bahia were supplied by COELBA.

Table 9-5 presents the final estimates for the frequency and duration of power outages in the state of Bahia with and without the proposed telecommunications system (point e, above). Given these assumed estimates and the derived estimates of the costs of power outages to both industrial and residential users in the state (points a through d, above) as well as the project costs, the internal rate of return on the investment in the proposed telecommunications project was estimated to be 98 percent.

Table 9-5. *Yearly Average Frequency and Duration of Outages Affecting Industrial Consumers, with and without the New Telecommunications System, in Bahia, Brazil, 1981–2000*

Time of day	Outages without new telecommunications (hours)		Assumed outages with new telecommunications (hours)	
	Frequency	Duration	Frequency	Duration
0800–1800	96.00	0.80	90.00	0.60
1800–2400	58.92	0.80	54.00	0.60
0000–0800	36.00	0.85	30.00	0.64

Source: COELBA and industrial survey.

Example 29. Gasoline Consumption Resulting from Inadequate Telephone Access (Tanzania 1982)

As of January 31, 1982, there were 17,718 connected telephone lines in Dar es Salaam and 17,816 people registered as waiting for telephones. Some of the effects on gasoline consumption of such an imbalance in telephone supply and demand were indicated through interviews with the management of three local business organizations in April 1982. The three organizations were the Sanitary Appliance and Hardware Company, the Board of External Trade (a government agency), and H. J. Stanley and Sons, Ltd. In reviewing communications needs and patterns, managers of the three organizations discussed ways in which inadequate telephone access and quality of service forced them to rely on motor vehicle transport for the simple function of soliciting and exchanging information.

The Sanitary Appliance and Hardware Company sold and distributed a variety of appliances and hardware and, beginning in 1978, manufactured locks, hinges, and other small hardware. The main retail and wholesale store in the center of Dar es Salaam had a telephone; the factory about 6 miles distant was enrolled on the waiting list for a telephone in January 1979. Frequent contacts between the main store and the factory, and between the store, the factory, and other places in Dar es Salaam, were necessary to check stocks, coordinate deliveries, order supplies, arrange transport, contact banks, customs officials, and buyers, and report problems and occasional emergencies at the factory.

The Board of External Trade was a government marketing information and coordination agency that served exporters and buyers. It had two telephone lines to contact ministries, banks, customs officials, exporters, buyers, and so forth. The Board attempted to maintain contact with about 200 exporters in the Dar es Salaam area and with another 200 suppliers, transporters, and exporters throughout the country. Although Board staff attempted to complete thirty to forty local contacts each working day, they were able to contact an average of only fourteen exporters by telephone. Assuming thirty daily contacts, an average of sixteen exporters either drove a vehicle to the Board office or a Board staff member traveled to them (an average of 10 miles one way in the Dar es Salaam area). Assuming a fifty-fifty split and that some of the trips were multipurpose and might therefore more properly be counted as four net Board-related trips rather than eight, the Board was associated with about twelve local vehicle trips approximately 20 miles long each working day.

H. J. Stanley and Sons, Ltd., was a multifunction business firm, which among other activities extracted and marketed salt, operated a coastal vessel, and dealt in coconuts and coconut processing. The Stanley group offices had three telephone lines with several extensions; office space was rented to a fourth firm, which also used the Stanley telephones. Many of the business contacts of the Stanley group complained that they could not complete calls into the Stanley offices since the Stanley telephone lines, when not out of order, were busy most of the time. In reviewing their daily business contacts and total miles driven, the two senior partners at H. J. Stanley and Sons calculated that they could each reduce their personal vehicle travel by an average of 225 miles a week if they and their salt and coconut facilities, some port facilities, and selected suppliers and wholesalers had access to reliable telephone service. The Stanleys also noted that if telephone service in the Dar es Salaam area became so congested as to be totally unusable during business hours, not only would they have to increase their own personal driving, but they would also at a minimum have to add two senior purchasing clerks at a salary of 4,800 Tanzanian shillings a month ($8,861 a year). Each clerk would on average drive at least 35 miles a day to contact suppliers, transporters, banks, distributors, and so forth.

Table 9-6 summarizes the value of the direct petroleum consumption that managers of these three establishments estimated could be saved given improved and expanded telephone service in the greater Dar es Salaam area. Only net mileage savings were tabulated, and the Tanzanian shilling was shadow priced at T Sh13 per $1 rather than the official rate of T Sh9.28. One year was assumed to consist of forty-eight working weeks. The price of gasoline at retail petrol stations—$2.54 per imperial gallon—was assumed to reflect the foreign exchange cost of gasoline to Tanzania, and an average

Table 9-6. *Estimated Foreign Exchange Expenditures Related to Petroleum Consumed by Three Business Organizations because of Inadequate Telephone Service, in Tanzania, 1982*

Item	Assumed net round trips saved per day	Miles	Working days per week	Gasoline expenditures per year (dollars)
Sanitary Appliance and Hardware				
Owner/manager	3.5	12[a]	5.5	—
Assistant manager	2.0	12[a]	5.0	—
Total				3,421
Board of External Trade	12.0	20[a]	5.0	11,697
H. J. Stanley and Sons, Ltd.				
Partner	—	41[b]	5.5	—
Partner	—	41[b]	5.5	—
Total	—	—	—	4,386
Two senior clerks	—	70[b]	5.0	3,412

— Not available.
a. Miles per trip.
b. Miles per day.
Source: World Bank data.

vehicle mileage figure of 12.5 miles per gallon was used to reflect the variety of motor transport used, including small cars, four-wheel-drive vehicles, and trucks.

The managers of all three organizations, however, emphasized that they considered the savings in gasoline consumption to be quite unimportant compared with the savings in scarce managerial time. The two H. J. Stanley and Sons partners estimated that they spent at least two hours each day in motor vehicles, which would be unnecessary if adequate telephone service were available (a total of 22 hours of management time lost each week). The Sanitary Hardware and Appliance Company manager and assistant manager estimated that they spent at least 12.8 and 6.7 hours, respectively, a week of unproductive time in motor vehicles, which they could spend on managerial tasks if better telephone service were available. The owner/manager of the company contended that, given current local demand for hardware and several new products on which he would like to work, the company could grow at least 25 percent a year more rapidly if he could trade time wasted driving for time spent on management and product development.

Example 30. Reduction of Wasteful Agricultural Travel and Transport (Uganda 1982)

The Busoga Growers' Cooperative Union was Uganda's largest producer of cotton and the fourth largest producer of coffee. It comprised about 200 farmer cooperatives, 7 cotton ginneries, 2 coffee factories, and a small dairy farm. Telecommunications facilities were limited to three telephone lines at headquarters and one line each at three of the other places.

During the annual four-month cotton marketing season a typical ginnery's manager traveled about 100 kilometers to the union's headquarters three times a week to follow up on orders for production imports, make arrangements for transport, enquire about likely pay dates, and take care of other routine administrative matters, which could all have been handled by telephone if such service had been available. During the remaining eight months of the year, the manager traveled for these purposes about once a week. Also during the cotton marketing season, the secretary of a typical cooperative rode a bicycle or hitched a ride three or four times a week to the ginnery, an average of some 13 kilometers away, for reasons similar to those requiring travel between the ginnery and headquarters. He also went twice a week to the nearest bank (an average of 65 kilometers away) to collect or enquire about payments for past deliveries of cotton. During the remaining eight months of the year, the secretary traveled about one-third as much. Each trip required a full day of the manager's or secretary's time. Assuming that the value of one day's time equaled one-twentieth of monthly salaries plus allowances, and using average costs of public transport for the distances traveled, these trips cost the union the equivalent of $0.12 million a year for all units and cooperatives. This understated the true cost, since it did not include losses related to unit and cooperative performance resulting from repeated absence of managers and slow administrative communication.

The union had a fleet of twenty-nine 8-ton lorries and hired fifty-five lorries from private transporters, mainly to carry the cooperatives' produce to the union's units (for example, cotton to the ginneries). During the four-month marketing season, cotton was carried three times a week between the typical cooperative and ginnery. About 20 percent of the time the lorry returned empty, however, because a lack of timely communication prevented effective coordination with the cooperative's management.

Thus, for each ginnery, about twelve empty return trips were made during each cotton marketing season. For all 200 cotton growers' cooperatives, these empty trips cost the equivalent of $0.13 million each season, considering only vehicle depreciation, maintenance, and fuel. Moreover, 150 of these cooperatives also produced coffee, which was collected three times a week during nine months of the year; about 10 percent of the coffee-related lorry trips were ineffective for the same reason, at a total cost of about $0.22 million a season.

Management contended that with adequate telephone service, these excess travel and lorry movements could be mostly eliminated. Furthermore, management argued that both administrative communication and fleet movements could be managed with greater flexibility; empty lorries could be diverted to pick up produce at alternative cooperatives nearby.

The benefits of providing all cooperatives with good-quality telephone service were estimated to be the sum of the excess costs incurred in travel and freight transport as calculated above, or about $0.5 million a year. Extension of the existing high-frequency public radio call service to the 200 cooperatives would meet the low call traffic requirements of administrative and transport coordination; the total cost of such investment could be recovered by reducing wasteful travel and freight in about one year. A facility giving higher-quality and more reliable service (for example, very high-frequency multiaccess radio) would cost about four times as much; such costs would still be more than offset by the direct savings in three to four years, which is much less than the life of properly maintained equipment.

Example 31. Value of Improved Telecommunications Infrastructure at the Microeconomic Level, Extrapolated to the National Level (the Philippines 1984 and Costa Rica 1986)

In 1984, in-depth interviews were conducted with the managers of 250 businesses in Northern Luzon and Northern Mindanao. These interviews asked managers how limited telephone service affected the performance of their company and sought to ascertain the possible impact of expanding and enhancing that service.[18] The survey results were used to estimate aggregate benefits to the economy of a planned telecommunications project. At the time, the government of the Philippines was beginning two major telecommunications programs, the Rural Telecommunications Development Project and the National Telephone Program (phase 1). The study took into account the specific expansion being planned for each geographic area.

The data were based on the assessment of benefits from the changes planned, not on actual results. Benefits as estimated by the businesses interviewed were examined according to three categories. Direct benefits included cost reductions that were the direct result of improved telecommunications services in the course of normal operations (such as reduced travel and use of messengers or more efficient use of management time). Consequential benefits represented the cost of opportunities lost due to poor-quality telecommunications; they could be realized only if normal operations changed somehow. Business expansion benefits accrued to firms that could expand more rapidly if telecommunications were improved; they were calculated according to improved economies of scale in production. Costs were calculated based on the capital costs of improving the network as planned by the two government projects.

The results were examined by sector for each region and are summarized in table 9-7. Some businesses were not likely to benefit from improved telecommunications

Table 9-7. *Benefit to Cost Ratio Following Installation of Telephone Service, by Sector and Region, in the Philippines, 1984*

Sector and region	Number in sample	Benefit to cost ratio	Proportion of sample with net benefits
Agriculture	90	44.6	45
Farming, forestry, fishing	40	18.0	28
Agricultural manufacturing	23	69.0	57
Agricultural trading	27	40.8	52
Health	47	33.4	45
Public and private health care delivery	42	35.5	38
Distribution of health care products	5	26.7	100
Other businesses	94	13.1	64
Manufacturing	54	10.5	57
Construction	2	39.4	100
Wholesale and retail trade	20	7.4	80
Transportation	6	31.6	83
Financial services	5	21.1	60
Other services	7	13.6	43
Northern Luzon			
Agriculture	47	35.7	47
Health	27	6.4	41
Other business	63	12.1	75
Northern Mindanao			
Agriculture	44	56.3	39
Health	20	63.1	50
Other business	51	12.1	53

Source: Communications Studies and Planning International (1986).

(those with a benefit to cost ratio less than 1). For the purposes of the study, it was assumed that these firms would probably not choose to subscribe, and therefore all benefit to cost ratios included only those businesses showing positive benefits.

Overall, the study found that the economic benefits of the planned expansion clearly exceeded the costs; the average business subscriber would obtain yearly benefits equal to thirteen times the annualized cost per line of the project in Northern Luzon, where the investment was targeted toward the countryside, and equal to twenty times the cost in Northern Mindanao, where the investment was targeted toward towns. Although the benefits to primary agricultural producers were moderate, the highest cost to benefit ratios were for agricultural industries and traders (rice and grain mills and commodity exporters). The study also found a direct relation between increased business size as measured by number of employees (ranging between one to four employees and twenty to forty-nine employees) and increased telephone-related benefits (ranging between $120 and $13,700 a year for these same categories). Since telephone service was estimated to have a total economic cost of $200 a year, using telephones was not viable for very small businesses.

Aggregating the microeconomic data involved two steps: first, extrapolation from the firms studied to obtain regional results and, second, conversion of benefits initially expressed as financial profits into benefits expressed as consumer and producer welfare costs incurred by the economy for the total costs of constructing and operating the network. On the basis of lines per business, the benefit to cost ratios were 13.5 and 20.1 for Northern Luzon and Northern Mindanao, respectively. Looking at the telecommunications system and including only business line benefits, the respective benefit to cost ratios were 1.0 and 1.2. This means that the benefits accruing to business users were sufficient to pay back the entire investment cost of the system; the benefits to the other 90 percent of subscribers would be added to this, making the total benefit to cost ratio even more favorable.

A similar study was carried out by the same researchers in Costa Rica in 1986. This study compared economic benefits that were estimated by nonsubscribers, by recent subscribers, and by persons or businesses that had subscribed for five years. Nonsubscribers estimated their potential gains to be $480 a year, which was very close to the gains estimated by recent subscribers, $500 a year. This gives some support to the accuracy of nonsubscribers' estimated benefits. However, long-standing subscribers realized benefits of $720 a year, which suggests that long-term changes in the way a business is run are required in order to take full advantage of the potential benefits and that these changes take time to learn and implement. Based, in broad terms, on the results of all three studies, a new telecommunications project was expected to yield benefits to its business customers of approximately 5 percent of the total revenues of all establishments they served.[19]

Example 32. Social and Economic Benefits of Rural Public Call Offices (Vanuatu 1988)

Although urban areas and large businesses in rural areas are adequately served by telecommunications services in Vanuatu, the majority of the rural population does not have easy access to telephone service. About sixty rural telephones (forty-four of which are public) serve about 80 percent of the total population of 130,000. In the expectation that expansion will occur, a study was done to determine the extent and optimal spatial layout that would yield the highest benefits to the rural users of public telephones for a given cost.[20] Using econometric methods, the study directly measured the losses resulting from the lack of adequate telecommunications.

Data were collected on round-trip distance traveled (either on foot or by vehicle) to all existing public phones in rural areas. Demand for domestic calls per 100 inhabitants a year was then expressed as the cost incurred by the individual callers, based on the minimum subsistence wage and the cost of paid transportation by vehicle. The valuation of travel distance was determined to be V30 per kilometer, which was, in turn, used to calculate consumer surplus.

The number of domestic calls (combined domestic and business) made annually per 100 inhabitants in rural areas was inversely proportional to the distance of the caller from a telephone and declined roughly from 290 calls a year at 0–1 kilometer to 101 calls a year at 2 kilometers and to 40 calls a year at 5 kilometers. At distances between 5 and 20 kilometers, calling rates were clustered around 40 calls a year and then fell further to about 16 calls a year at distances over 20 kilometers. Government calls fell much more sharply, from 176 calls a year at 0–1 kilometer to about 15 calls a year at 2 kilometers. Two spikes at 9 and 18 kilometers represented an unusual frequency of calls in locations where a government office did not have its own phone.

A spatial model was then developed to determine the benefit of extending rural telecommunications by reducing the distance rural dwellers had to travel to reach a telephone. The country was divided into ninety-six zones, and the model was used to compute the call rate per 100 inhabitants, volume of telephone traffic, distance to the nearest telephone, zone in which the nearest telephone was located, and consumer surplus for each zone and for the entire study area in aggregate. Expanding the number of telephone locations from forty-four to sixty-four was calculated to yield an incremental benefit of V2.6 million a year (56,000 kilometers of travel saved a year); if 100 zones received a phone, the incremental benefit would be V11.4 million a year (125,000 kilometers saved a year). The average annual benefit realized per location from the addition of ten locations (increasing the number of telephones from forty-four to fifty-four) was V94,100 and from the second addition of ten locations (from fifty-four to sixty-four) was V78,200. Overall, the economic benefits gained from the telephone network increased as the number of locations increased, but at a decreasing rate; most of the benefit was obtained in the first 100 locations, since as it increased, the traffic tended to include more lesser-valued calls. One limitation of the study is that benefits were not evaluated net of costs, since relevant information was not available on the costs of providing more service.

Conclusions from Microeconomic Analysis

The microeconomic or project-specific analysis of telephone benefits examined in this and the previous chapter revolves around the assumption that total benefits of telephone access and use can be perceived and assessed best by those who actually have or are demanding access to and use of a telephone. As such, the analysis does not usually consider benefits external to the telephone subscriber or caller, the indirect or secondary benefits, or the distributional effects of the benefits.

Within this limited framework, the analysis focuses on case-by-case attempts to quantify the willingness of consumers to pay for telephone service not only by observing the official tariffs that are paid but also by attempting to derive at least minimum estimates of benefits that consumers actually accrue, over and above tariffs, in other words, consumer surplus. Specific examples involving three procedures to estimate consumer surplus were reviewed. To varying degrees, some of the examples encompassed more than one estimation technique or method of analysis; considerable subjectivity was involved in placing some examples in the expenditure method group rather than the best-alternative method group and vice versa. More fundamentally, however, the benefit estimation exercises themselves generally suffered major limitations.

The fundamental limitation of the exercises reviewed, and in fact of much project analysis, is that adequate data are not readily available to assure accurate results.[21] In the absence of adequate inputs, somewhat arbitrary assumptions must often be made. Given a sufficient number of assumptions in any one exercise, a rate of return or other numerical benefit total can be obtained. In some instances, however, the results depend more on the assumptions used in the analysis than on the data reviewed. Generally, in the examples reviewed in this and the previous chapter, dependence on two types of assumptions was critical: assumptions about the analytical framework or concept and assumptions about the interpretation of the collected data.

With regard to the framework, examples of critical assumptions in the studies cited included "other things remained equal," "the observations were homogeneous," and "the demand curve had a particular shape and elasticity." In abstracting a field situation, it is often assumed that variables not directly included in the analysis remain unchanged. In many instances, this assumption may not be entirely realistic. For instance, consider the determination of price elasticities of demand for telephone calls in example 13. If, as is common, a secular growth trend underlies short-term variations in telephone traffic, elasticities determined from traffic measurements before and after a tariff increase could be underestimated and lead to an overestimate of consumer surplus. Also, as countries become more able to adapt to changes in relative prices, the possibility of substituting other expenditures for telecommunications usage grows; that is, in the long run, demand becomes more elastic. However, a consumer's reaction to a price change after one month might be more severe than it would be after three months. Hence, an estimate of price elasticity of demand made one month after a price change could result in a much lower estimate of consumer surplus than one made after three or four months. Example 14 illustrates the problem in another context. Clearly, many relevant factors besides the real level of telephone tariffs changed between 1964 and 1977 in El Salvador, and the econometric problem of disentangling the effect of this variable from the host of other forces at work was probably insurmountable. In a growing and increasingly modern economy and telephone network, neglecting these factors probably resulted in an underestimation of consumer surplus.[22]

Assumptions that communication events are homogeneous are also rarely valid. Telephone calls vary in duration, distance spanned, purpose, and transmission quality, and these differences produce different values for the persons making and receiving the calls. Likewise, users differ in their occupation, education, income, and other factors, and these differences affect the way in which they value calls and on their calling patterns. Although these issues may be partially addressed (by averaging a number of heterogeneous calls to a more aggregate "equivalent call," as in example 20), they usually could benefit from more definitive analyses.[23]

A related problem is that of establishing analogies among towns or villages (example 20), residences or offices (example 22), and the distribution of future subscribers (example 24). There is the obvious difficulty of matching things that are different by definition. In example 20 the very possibility of finding meaningful control places can be questioned on grounds that the villages with telephones probably were already significantly different (had closer economic ties with cities, as well as higher average income and educational levels) when the decision was made to provide them (and not others) with telephones. Thereafter they probably evolved differently because they had telephones.[24]

A third critical assumption about the analytical framework, which directly affected the size of benefits calculated in several of the examples, relates to the shape and elasticity of the demand curves. In all cases reviewed, this choice was of necessity made on a somewhat arbitrary basis. In examples 13, 15, and 20, the demand curve was assumed to be linear, an assumption unlikely to be realistic particularly for large changes in quantity. In examples 22 and 24, the demand function was specified as a rectangular hyperbola, whereas in example 17 it was taken to be an S-shaped curve. Although such shapes may be more consistent than straight lines with what might be expected given the usual pattern of income distribution in a developing country, and they usually yield a more conservative estimate of consumer surplus, they are still choices with little empirical support.

Four different price elasticity estimates were used in the examples: -0.5 for call charges in example 13, -1.0 for rental charges in example 22 and for connection fees in example 24, -0.25 for a composite of telephone service charges in example 15, and roughly a -0.17 for local telephone calls in example 17. Except for example 13, none of these elasticities had much empirical basis in the particular situations in which they were used. Finally, using an implied single demand function to estimate price, quantity, and benefit values through time, as was done in several of the examples, involves critical assumptions about other things remaining equal and does not explicitly consider the fact that demand for telecommunications services in developing countries has grown, and will probably continue to grow, at a fast pace even when tariffs are increased in real terms.

Besides the above problems relating to the formulation of the general framework for the analysis, a general class of problems is associated with actually interpreting or extrapolating the data collected. Examples observed include placing a value on time or transport and sorting out multiple trip objectives. With regard to trip objectives, when individuals journey to a town or village center to use a telephone, they frequently perform other tasks in addition to making the telephone call. The decision on whether and how to communicate is not independent of other actions. Hence, the travel-related data used in examples 15, 16, 18, 19, 20, 21, and 32 are by definition affected by the assumptions about time allocation that were used.

Furthermore, placing a monetary value on the total cost of travel (examples 15, 16, 18, 19, 20, 21, and 32), on the value of time (examples 15, 18, 19, 21, and 23), and on postal communications (example 17) is a complex undertaking. Even if time saved could be valued at the users' average or marginal income (often this would be too high), measuring income at a sufficiently disaggregate level is in many instances not possible. As shown in the studies from which examples 16, 21, and 32 were drawn, this is particularly so in rural areas of developing countries, where (a) a large part of income is received in kind rather than in money and hence is not included in the basic sources of economic information (census or national accounts), (b) income is subject to large seasonal variations and thus difficult to estimate at any particular time, and (c) as occurs in some countries, rural inhabitants are far more reluctant than their urban counterparts to give accurate information on income during interviews. Example 21, which provides two divergent but equally plausible estimates of the value of time and the cost of transport, illustrates the problem well. Also, given the many variables and accompanying assumptions involved in example 20, it is likely that the authors of that exercise could have rationalized at least half a dozen equally plausible outcomes.

Other problems relating to data interpretation were apparent in the examples that were based on interpretations or forecasts made by interviewers or managers with potential vested interests. Examples 26, 27, 28, 29, 30, and 31 rely heavily on interpretive analysis or unverifiable claims about future expertise or telecommunications-related savings. As such, the conclusions reached in these examples are perhaps more dependent on assumptions than are most. In example 31, some confirmation of data was obtained in the Costa Rican study, where projected estimates were compared with actual first-year benefits.

A final point remains to be noted. The calculations in the examples, to the extent they are valid, probably yield only partial measures of consumer surplus. To some extent, the systematic biasing of calculations toward low or conservative estimates is what makes these simple methods useful. Since cost-benefit comparisons are generally used to identify and eliminate low-return projects, an underestimation of benefits, if understood, helps to ensure that overprovisioning of services does not occur. It can be argued, of course, that underprovisioning is an equally serious problem; in most developing countries it is certainly the most prevalent of the two.[25]

Although their limitations are formidable, the studies described serve two useful functions: they assist in examining project justification, and they assist in making decisions about project composition. With regard to project justification, such limited analytical exercises at the project or program level have often provided, at low cost, approximate but reasonably credible quantifications of some portion of consumer surplus. This has been especially useful in examining potential projects that are only marginally acceptable when judged by purely financial criteria. Such exercises also have attracted the attention of decisionmakers, have furthered an understanding of how users value telecommunications services, and ultimately have highlighted opportunities for additional investment.

With regard to project composition, estimates of price elasticities have helped shape tariff reforms, and the more detailed analyses of telephone usage in rural areas of developing countries have on occasion served to define such programs better. In some instances they have also assisted in selecting the particular rural locations for early investment. Such analysis may, in fact, be more useful for determining priorities within the telecommunications sector than for assessing the merits of the total program.

There is much scope for improvement and innovation in microeconomic analyses to assist investment decisionmaking. Some of the experience outlined in the transport economics literature might be useful in dealing with problems of valuation of time and allocation of cost in multipurpose trips. Computer-assisted traffic and user analysis could provide operating entities with better estimates of demand, usage by type of subscriber, value by function, and so forth. Operations research approaches to inventory management might be applied to the analysis of telecommunications benefits in specific sectors or for large entities (food storage and distribution). And, as has been done in the power sector (example 28), the responses of users to major outages of telecommunications service might yield information on the benefits derived from these services.

Notes

1. The expenditure approach has been extensively used in other sectors to evaluate recreational benefits, following the pioneering work of Marion Clawson. A typical example is Kavanagh and Smith (1969).

2. Several approaches to the evaluation of time have been developed in the literature of transport economics. See, for example, Beesley (1965).

3. Baeza, Bunster, and Schenone (1975).

4. The data for this exercise were collected by Instituto Costarricense de Electricidad and the University of Costa Rica. See appendix C for details.

5. One Egyptian pound (£E) was equivalent to $1.43.

6. This exercise used a small sample identified initially from newspaper advertisements. Similarity of accommodation was judged on the basis of advertised description and the opinion of real estate brokers.

7. For estimates of the price elasticity of demand for telephone connections and calls, see Littlechild (1979), pp. 35–37; Alleman (1977), pp. 67–82; Meyer and others (1980) app. C; Taylor (1980), chaps. 3 and 4; and Taylor (1983).

8. The difference of £E50 a month in rent was considered representative for modest one- or two-bedroom second- or third-floor walk-up apartments or offices with three-party lines occupied by visiting and middle- or upper-middle-income Egyptian businesspersons.

9. For estimates of the value of time wasted in the United Kingdom in 1974 and earlier because of network congestion, see Short (1976).

10. This estimate was derived by adjusting the 1968 national average weekly wage upward by changes in the cost of living. This was considered to be a conservative adjustment, since it was thought that wages generally grew more rapidly than the cost of living over the period, that wages in large urban areas were higher than the national average, and that wages of telephone users were on average higher than the average urban wage.

11. For example, in Chile, Costa Rica, India, Thailand, and Venezuela, among others, telephone connections were bought and sold in the middle or late 1970s through advertisements in local newspapers. In Brazil's state of Bahia in the 1980s, the telephone company facilitated the private transfer of telephone connections by acting as a broker between buyers and sellers and by publishing the current average price of such transfers by exchange areas. In addition, the company collected fees when telephones were moved from one location to another.

12. This example was originally cited in Saunders (1979).

13. In 1977 the exchange rate for the Myanmar (Burmese) kyat was approximately K1.00 = $0.149.

14. As in example 22, the demand curve was assumed to be a rectangular hyperbola.

15. ITU (1988b).

16. Communications Studies and Planning International (1981).

17. This exercise was initially undertaken by Mark W. Gellerson for Robert F. Gellerman at the Inter-American Development Bank, Washington, D.C. The methodology for valuing savings brought about by less frequent or less lengthy power outages is outlined in Munasinghe (1979) and Munasinghe and Gellerson (1979).

18. This study was carried out by Michael Tyler and Charles Jonscher, with the support of the ITU. See Communications Studies and Planning International (1986); Booz Allen and Hamilton (1984). A similar, complementary study was carried out in Costa Rica in 1986. See Booz Allen and Hamilton (1986).

19. Communications Studies and Planning International (1981) and Jonscher (1987).

20. ITU (1988c).

21. To obtain data through surveys, interviews, observation, or other methods normally requires a large amount of field work. In many instances, however, the results are not worth the effort required to prepare and evaluate a telecommunications project that has a high internal financial rate of return.

22. In particular, the technical quality of service increased during the period, as did the number of subscribers that would be called by anyone with access to a telephone.

23. Also, communications media, which under given conditions might be acceptable substitutes of one another, are not necessarily valued equally; for instance, it was found in examples 16 and 26 that a high value was often attached to the speed of the communication and response that are obtained by using the telephone. This was also found in several of the surveys reviewed in chapters 10 through 12.

24. There is also the problem of grouping places on the basis of judgment—this, besides yielding rather arbitrary association, is not a practical proposition when the number of entities to be analyzed is large.

25. Also, it is difficult to compare different program components (long distance, urban, and rural) or alternative programs with different compositions, since the extent to which benefits have been underestimated is unknown and probably varies from case to case.

Part IV

Telephone Access and Use

Chapter 10

Characteristics of
Telephone Subscribers

AS DISCUSSED IN PREVIOUS CHAPTERS, unmet demand, high economic returns on investment, a wide diversity of effects related to development, and declining unit costs suggest that telecommunications
should be accorded a relatively high priority for new investment in developing countries. In fact, were economic efficiency the only criterion, such evidence would typically be sufficient to justify a rapid
expansion of the sector. Despite such evidence, however, planning authorities in developing countries often assign other needs a higher
priority and thus postpone investment in the telecommunications sector. Such decisions are often explained by the view that, although
telecommunications investments are profitable in a financial sense,
only a relatively narrow and privileged sector of the community benefits directly from them. Such opinions are, however, usually based on
intuition rather than substantive analysis.

Little is known about the extent to which lower-income groups in
developing countries actually need to communicate rapidly beyond
their immediate environment.[1] Also, the extent to which they currently have effective access to communications facilities and understand how to use them is not well documented. Access is not only a
question of the physical location of facilities relative to that of population but also a matter of service quality, pricing, and social, cultural,
and educational constraints on use.

In this and the following two chapters, evidence primarily from developing countries is examined on the distributional aspects of telephone benefits and access to service. The following questions are

199

addressed: What is known about the characteristics of business, residential, and public telephone users? What are telephones used for? What is known about telephone use by different employment, educational, or income groups? And, are there consistent differences in reasons for use, or use patterns, by type or location of user? Although a full analysis of the distribution of telecommunications access and use in developing countries would be complex, even if reliable data were available, several limited studies have been reported. A review of these studies gives some useful indications.

Many of the data examined, particularly on telephone use, were derived primarily from surveys of subscribers or other users. As such, they are subject to the limitations of interpretation and potential bias normally associated with survey techniques.

Business and Residential Telephones

For tariff purposes some telecommunications entities divide telephone subscribers and applicants into residential and business (non-residential) categories, and the latter sometimes into various other groupings (government, commerce, and professional). Although user categories are labeled differently by different entities, across the array of countries, the proportion of residential telephones clearly tends to increase with telephone density, that is, with the size of the telephone system relative to that of the population. This is predictable since in the early stages of system development, business users (including government) demand a high priority and are willing to pay a high price in the expectation that reliable and rapid communications will reduce costs and increase efficiency. Such is the case in many countries in Asia and Africa, which typically have about one telephone main line per 100 inhabitants, and where 45 to 65 percent of existing lines are used by business or government subscribers. Furthermore, many of the telephones officially classified as residential are, in fact, provided to meet needs related to business or government.

At higher levels of national telephone penetration, business demand tends to be more adequately met, and residential users receive a larger share of expanded telephone services, when the demand for telephones tends to increase with education and income. For example, in South America, where most countries average between three and ten telephone lines per 100 persons, the proportion of residential telephones is generally in the range of 60 to 80 percent. This level is equal to that in countries like the Sweden, Switzerland, and the

United States, which have more than fifty telephone lines per 100 persons, and where residential subscribers also account for 70 percent or more of all telephones.

Published data on the relative proportions of business telephones or telephone lines do not necessarily reflect the primary use to which the telephones are put.[2] In many of the developing countries that differentiate between types of subscribers, business subscribers pay a higher monthly rental fee than residential subscribers. One result is that when a building is used for both business and residential purposes (small shops, farms, and so forth), subscribers and potential subscribers have an incentive to list their telephones as residential even though they are used mostly for business.[3] A few countries have at least partly overcome this problem by classifying telephones located in shared accommodations as business irrespective of where they are specifically placed; in other countries the practice is ambiguous.[4]

The general tendency observed among countries—that the proportion of residential telephones tends to increase as national telephone penetration and density rise—holds within countries as well. In Malaysia the proportion of residential telephone lines increased from 24 percent in 1956 to 40 percent in 1964, to 47 percent in 1978, and to 71 percent in 1988 as density increased from 0.6 telephone lines per 100 inhabitants to 0.9, 2.0, and 7.2, respectively.[5] Table 10-1 shows that the proportion of residential telephones also increased in the two largest towns in Nepal, Kathmandu and Biratnagar, as the number of telephones increased from 1,947 in 1968 to 5,620 in 1977.[6]

The proportion of nonresidential telephone lines also tends to be larger in towns and villages than in bigger centers, and in cities with more than one exchange the proportion varies greatly with location. For example, in 1977 the proportion of lines connected to government and business subscribers in Kathmandu (Nepal's capital and largest city) was 43 percent, whereas in Biratnagar (the second largest city) it was 58 percent, and in the rest of the country it averaged 71 percent. In Ethiopia in 1982 the proportion of subscribers registered in the government, public service, business, and commercial categories was 28 percent in Addis Ababa, the capital city (see table 10-2). This contrasted with 39 percent in the nation as a whole and 69 percent in a sample of four small manual-exchange towns. In Uruguay, in 1980 the proportion of lines officially registered as nonresidential was 20 percent in Montevideo (the capital and largest city) and about 33 percent in the rest of the country. In Thailand in 1980, the proportion of nonresidential telephone lines in Bangkok varied between about 10 and 80 percent depending on the location of the exchange

Table 10-1. *Estimated Location of Telephones, by Type of Establishment, in Nepal, 1968 and 1977*

	1968			1977		
Type of establishment	Kathmandu	Biratnagar	Ten other towns	Kathmandu	Biratnagar	Ten other towns
Percent of telephones located in primarily residential quarters	43	33	0	57	42	29
Percent of telephones in government sector	15	18	0	11	5	12
Percent of telephones in business, commerce, industrial, service, and mining sectors	42	49	0	32	53	59
Total telephones	1,643	304	0	4,737	883	1,671
Percent of national total	84	16	0	65	12	23

Note: Kathmandu is the capital and largest city; Biratnagar is the second largest city.
Source: Adapted from Integrated Development Systems (1980).

Table 10-2. *Telephone Subscribers, by Type of User, in Ethiopia: Addis Ababa and Four Small Towns, 1982*

Type of subscriber	Ethiopia		Addis Ababa		Four small towns[a]	
	Number	Percent	Number	Percent	Number	Percent
Business and commerce	17,542	23	4,630	11	51	27
Public service, government, and parastatals	12,203	16	7,225	17	81	42
Registered as residential	45,000	59	30,416	70	60	31
Other	1,525	2	995	2	0	0
Total subscribers	76,270	100	43,266	100	192	100

a. See table 10-7 for a finer breakdown of subscribers in the four small towns.
Source: Ethiopian Telecommunications Authority.

area, averaging 61 percent; in the provinces, where there is rarely more than one exchange per city, the proportion varied only in the range of 60 to 80 percent.[7]

Finally, a similar relation holds in Burkina Faso and Mali. As shown in table 10-3, in Burkina Faso in 1980, 19 percent of the telephones in three larger provincial towns were in the hands of individual subscribers without a readily identifiable occupational listing compared with only 9 percent in six smaller towns. In Mali in 1979 the proportion of telephones in the hands of individual subscribers fell from 60 percent in Bamako, the capital city, to 11 percent in the very smallest places with some telephone service. The low proportion of only 9 percent of telephones classified as residential in small towns in Burkina Faso corresponds closely to the proportion of telephones classified as residential in twelve small exchange areas in various parts of Bangladesh. There, only 8 percent of the telephones were identified as residential (see table 10-6).

Users of Business Telephones in Urban Areas

On a national basis, the input-output analysis reviewed in chapter 5 suggested that the most frequent users of business telephones in developing as well as industrial countries are persons employed in trade, services, government administration, and to some extent transport. In other words, the tertiary sector of an economy (rather than the primary and secondary sectors) generally purchases the largest quantity of telecommunications services. This finding also tends to be re-

Table 10-3. *Number of Telephone Subscribers, by Functional Category, in Selected Towns in Burkina Faso, 1980, and Mali, 1979*

| | Burkina Faso | | | |
| | Three larger towns with 200-line automatic exchanges | | Six smaller towns with smaller manual exchanges | |
Subscriber	Total	Percent	Total	Percent
Public service activities				
Government administration[a]	120	n.a.	69	n.a.
Education[b]	15	n.a.	6	n.a.
Health[c]	18	n.a.	9	n.a.
Law and order[d]	12	n.a.	10	n.a.
Churches and missions	5	n.a.	8	n.a.
Total	170	37	102	67
Economic activities				
Commercial banks	23	n.a.	6	n.a.
Shops, hotels, garages, and gas stations	77	n.a.	17	n.a.
Other business establishments, factories, and managers[e]	91	n.a.	5	n.a.
Transport[f]	7	n.a.	3	n.a.
Total	198	43	31	20
Other				
Individual subscribers[g]	86	n.a.	14	n.a.
Public call office	3	n.a.	6	n.a.
Total	89	20	20	13
Total telephones	457	100	153	100

n.a. Not applicable.

a. Central and local government offices and officials; quasi-government organizations and enterprises, such as electric power company and posts and telecommunications (offices and officials); and international and aid organizations. Government offices include special programs such as forestation, agricultural extension, meteorological service, and so forth.

b. Secondary and technical schools, colleges, adult training programs, and teachers.

c. Hospitals, clinics, dispensaries, physicians and medical workers, and drug stores.

flected in data showing the mix of business telephone subscribers by economic category and by the call traffic they generate.

Information available for Thailand in 1980, and partly summarized in table 10-4, shows that business and government subscribers in Bangkok, which comprised 61 percent of all telephone lines in service, accounted for 83 percent of all calls made and a similar proportion of telephone revenues. Eighty-nine percent of business and government lines were connected to subscribers in the tertiary sectors, mainly wholesale and retail trade, finance, transport, communi-

	Mali		
Bamako, the capital city	Three towns with automatic exchanges	Six towns with small manual exchanges	Sample of rural services in Sikasso region
Total Percent	Total Percent	Total Percent	Total Percent
212 n.a.	66 n.a.	84 n.a.	47 n.a.
57 n.a.	17 n.a.	10 n.a.	4 n.a.
77 n.a.	28 n.a.	26 n.a.	13 n.a.
78 n.a.	22 n.a.	28 n.a.	14 n.a.
16 n.a.	11 n.a.	1 n.a.	0 n.a.
440 17	144 28	149 36	78 55
44 n.a.	8 n.a.	3 n.a.	1 n.a.
430 n.a.	74 n.a.	86 n.a.	18 n.a.
46 n.a.	12 n.a.	4 n.a.	16 n.a.
44 n.a.	14 n.a.	12 n.a.	5 n.a.
564 22	108 21	105 25	40 29
1,531 n.a.	266 n.a.	155 n.a.	15 n.a.
10 n.a.	4 n.a.	7 n.a.	8 n.a.
1,541 61	270 51	162 39	23 16
2,545 100	522 100	416 100	141 100

d. Police and courts of justice (offices and officials).

e. Could include some shops.

f. Railway and truck terminals and agencies.

g. Telephones listed under individual names (could include some shops and businesses).

Source: Data compiled by Gerald Buttex, Hurman Ruud, and Robert J. Saunders with assistance from the respective Office of Posts and Telecommunications, Burkina Faso and Mali.

cation, utilities, and other services.[8] Tertiary sector subscribers accounted for about 92 percent of all revenues generated from business lines.

With regard to changes through time, it is generally accepted that as an economy develops, the tertiary sector becomes an increasingly important source of output and employment. This relative growth through time of the tertiary sector and its heavy reliance on communications have incited efforts to define a fourth sector, the information sector. As outlined in chapter 5, research into the size and effect

Table 10-4. *Distribution of Telephone Subscribers, by Type of Economic Activity, in Thailand, 1980*
(percent)

Subscriber	Bangkok metropolitan area	Provinces	Thailand
Government (excluding state enterprises)	5.2	5.1	5.2
Services			
Wholesale and retail trade	24.5	34.2	26.9
Banking, insurance, and real estate	3.8	2.7	3.5
Transport and communications	2.4	1.4	2.2
Electricity and water	0.2	0.5	0.3
Other services	18.1	14.6	17.1
Total services	49.0	53.4	50.0
Manufacturing	5.4	4.2	5.5
Construction	1.1	0.6	1.0
Agriculture, fishery, forestry, mining, quarrying, and others	0.2	0.9	0.3
Residences	39.1	35.8	38.0
Total	100.0	100.0	100.0

Source: Adapted from Telephone Organization of Thailand (1980). Based on a stratified random sample of 1,993 lines in nine Bangkok exchanges and 4,669 lines in twenty-four provincial exchanges. Classifications were made using information available at the local exchanges supplemented by telephone and personal interviews.

of the information sector has been undertaken in several industrial countries. In conjunction with such work, it has been observed that within each sector of an economy the level of demand for telecommunications services is generally a positive function of the proportion of employment that works on information-related tasks rather than on physical production itself.[9] Some evidence from developing countries also supports such a conclusion.

For example, data on local telephone calls and employment were collected and analyzed in a 1971 survey of 2,000 business establishments in the metropolitan area of Santiago, Chile, of which 1,100 had telephones.[10] The results showed that for each firm, when the total number of persons employed was correlated with the total number of local telephone calls made during a typical day, three statistically significant relations were found when the firms were categorized as being manufacturing industries, wholesale and retail trade, or white-collar office companies. For all three categories, the number of daily calls increased with the number of employees, and, for any given size of firm, the white-collar office companies made more telephone calls than commercial establishments, and commercial establishments made more calls than manufacturing industries. Furthermore, as the number employed by the three types of firms increased, the number of calls made by the office firms increased faster than those of the other two.

The number of local telephone calls made by the firms surveyed were then compared with the number of white-collar workers and owners.[11] The result was that no statistically significant difference in telephone calls per white-collar person was found among the three categories of firms. Further, the observed relation between total number of telephone calls and the number of white-collar workers and owners was virtually identical to that previously observed when telephone calls and total employees were compared for only white-collar office firms. Such a result is consistent with the conclusion that business telephones are most important to persons employed in the tertiary sector or as administrators and managers in other sectors, that is, persons working in the information sector.

Users of Business Telephones in Towns and Villages

Several exercises provide information on the characteristics of telephone subscribers in towns or villages of developing countries: two each from India and Egypt, and one each from Burkina Faso, Mali, Bangladesh, Ethiopia, Nepal, Kenya, and Syria are discussed here. Table 10-5 summarizes the occupations and income levels of a sample of forty subscribers in ten small-town exchange areas in five states in India in 1978.[12] Of those subscribers sampled, 92.5 percent were self-employed, and most were businesspeople with incomes that were higher than those of the majority of the population in those mostly rural areas. In a 1981 study from India in which sixty telephone subscribers in twelve exchange areas in the Andhra Pradesh State were sampled, 67 percent of the subscribers were engaged in business, whereas 20 percent were engaged in agriculture.[13] Thirty-five percent of the subscribers earned annual incomes of less than Rs15,000 annually, whereas 40 percent had annual incomes of more than Rs25,000, which, even by urban standards in India, would be considered relatively high. Of the sampled subscribers, 72 percent had passed through high school, and 20 percent had taken at least some college courses.

Table 10-5. *Distribution of Telephone Subscribers, by Primary Occupation and Income, in Ten Small Rural Exchange Areas of India, 1977–78*
(percentage in category)

Annual income (rupees)	Business	Agriculture	Professional	Private	Official	Total
4,800–6,000	2.5	2.5	0.0	2.5	2.5	10.0
6,000–7,200	7.5	0.0	0.0	2.5	0.0	10.0
7,200–10,000	12.5	0.0	2.5	0.0	0.0	15.0
10,000+	52.2	7.5	5.0	0.0	0.0	65.0
Total	75.0	10.0	7.5	5.0	2.5	100.0

Note: Individual percentages in each cross-tabulated cell do not merit much attention due to the small size of the total sample.
Source: Adapted from National Council of Applied Economic Research, India (1978).

Likewise, in a pilot study of nine rural villages in Egypt, the more educated villagers had access to and used the available telecommunications and postal services for business, family, and emergency purposes more than lesser educated villagers. Also in the Egyptian study the relation was positive between the proportion of the population in nonagrarian occupations and telephone penetration in the villages.[14]

A follow-up study surveyed 2,000 individuals in 143 Egyptian villages, representing a national sample. The results confirmed those obtained earlier. Farmers in Egypt generally marketed their produce and purchased their inputs through government cooperatives, which tended to accentuate the shift of phone ownership from farmers to government administrators. The study found that approximately 63 percent of village omdahs (mayors or headmen) had telephones, as did 5 percent of individuals in commerce or professions but only 0.2 percent of farmers. Although each nonagrarian user averaged about eight calls a month, the population as a whole averaged 0.08 calls, or slightly less than one call a year. This means that one-quarter of 1 percent of the population made 25 percent of all calls.[15]

Table 10-3, referred to earlier, showed that in Burkina Faso in the six smaller towns with manual telephone exchanges, at least 82 percent of all telephones were located in tertiary sector offices or establishments or were used primarily by persons involved in tertiary sector activities. The corresponding figure was 59 percent for the three somewhat larger towns with 200-line automatic exchanges. This was primarily because government telephones were less important. Table 10-3 also showed that a similar relation holds for Mali. Table 10-6,

Table 10-6. *Telephone Subscribers, by User Category, in Twelve Small Exchanges in Four Divisions (Regions) and the Dhaka Multiexchange Area of Bangladesh, December 1980*

Subscriber	Twelve small exchanges		Dhaka multiexchange area	
	Number	Percent	Number	Percent
Public service				
Government administration[a]	562	30	8,392	18
Quasi-government and other				
Schools and colleges	61			
Doctors and nurses	46			
Hospitals and dispensaries	12			
Clubs and associations	6			
Total quasi-government and other	125	6	4,924[b]	11
Total public service	687	36	13,316	29
Commerce, business, and industry				
Shops	130			
Business offices	115			
Banks	100			
Industrial factories	77			
Lawyers	33			
Agricultural farms	30			
Hotels and restaurants	16			
Cinema and theater	10			
Business office and shop telephones listed under owners' names	540			
Total commerce, business, and industry	1,051	56	—	—
Residential telephones	155	8	—	—
Total business and residential	1,206	64	33,067	71
Total telephones	1,893	100	46,383	100

— Not available.

a. Includes some government residential telephones.

b. Includes eighty-seven working connections for the railway board and connections for electric power board, water board, and so forth.

Source: Telephone and Telegraph Board, Bangladesh.

which summarizes a survey by the Telegraph and Telephone Board in Bangladesh, shows that for twelve small exchange areas (three in each of the four regions of Bangladesh), approximately 86 percent of the telephones were located in or used by tertiary sector establishments or agencies. Finally, in four small towns in Ethiopia, with manual-exchange service in 1982, almost all of the telephone subscribers not registered as residential were providing tertiary sector services (table 10-7).

Table 10-7. *Telephone Subscribers by User Category, in Four Small Manual-Exchange Towns from the Eastern, Western, Northwestern, and Central Regions of Ethiopia, 1982*

Subscriber	Number	Percent
Business and commerce	51	27
Commercial bank offices	4	
Contractors, shoe factories, tailors	4	
Groceries, bakeries, hotels, bars	17	
Pharmacies, gas stations, other retail shops	24	
Distribution and warehousing	2	
Public service and government	81	42
Municipal government administration	4	
Central government administration	19	
Police	6	
Public utilities and public construction	15	
Agriculture and forestry administration	12	
Schools	10	
Health service clinics	3	
Social, political, and special-interest associations	12	
Registered as residential	60	31
Total telephone subscribers	192	100

Note: Of the four towns, Bedele had fifty-six subscribers, Butagira had fifty, Gursum had thirty-six, and Worota had fifty.
Source: Ethiopian Telecommunications Authority.

The Nepal study referred to earlier examined the specific use of subscriber telephones by different economic sectors and residential users in the capital city of Kathmandu and the small town of Pokhara. Table 10-8 shows that the three largest users—tourist facilities, financial institutions, and government—were clearly involved in tertiary sector activities.[16] Of the registered nonresidential parties on the waiting list (waiters) in those two towns in early 1980, almost 65 percent were from the five clearly tertiary sectors of tourism, financial institutions, government, services, and trade. In a sample of fifteen residential waiters, fourteen were employed in the services or trade sectors or were schoolteachers.

Another example of telephone calling rates is available from a sample of subscribers in several small towns and market centers in two regions in western and south-central Kenya. Table 10-9 shows that unlike the two somewhat larger principal towns in Nepal, calling rates in these more rural places were highest for the transport and storage

Table 10-8. *Estimated Average Monthly Calls per Telephone, by Sector, in Two Towns in Nepal, 1979–80*

Subscriber	Kathmandu	Pokhara
Tourism sector[a]	1,043	219
Financial institutions	688	—
Government	659	—
Mining and quarrying	658	—
Agricultural agencies and corporations	601	—
Construction sector	424	—
Services[b]	422	—
Wholesale and retail trade	365	73
Transport and communications	271	—
Residential	196	72
Electricity	160	—
Total average	335	102[c]
Average excluding residential	496	122
Number of observations in twelve-month stratified random sample	147	20

— No subscribers.

a. Refers only to hotels, tour groups and guides, airlines, and so forth.

b. Refers to hospitals, social organizations, international agencies, diplomatic missions, and so forth.

c. Factors leading to the lower average in Pokhara include the fact that Pokhara is served by one small manual exchange connecting about 125 subscribers. Call traffic is also congested.

Source: Adapted from Integrated Development Systems (1980).

sector. The two regions sampled contain mostly smaller towns and market centers, and they are located a considerable distance from Nairobi. Hence, these towns relied heavily on vehicle transport to move supplies and goods. Financial institutions had the second highest calling rates in both the Kenyan and Nepalese studies.

Finally, in a study initiated in Syria, information was collected on the location of telephones by type of establishment in eighteen small localities.[17] The communities had an average of 116 telephones, with the smallest having 33 and the largest 242. Telephone waiting lists existed in some of the communities. Table 10-10 shows that, just as in the other studies reviewed, businesses operating in the tertiary sector, even in the microcosm of these small Syrian communities, had the greatest access to telephones. The top five location-occupation activities, which can all clearly be categorized as tertiary, accounted for more than 80 percent of total business and government telephones.[18]

Table 10-9. *Telephone Calling Rates for Telephone Subscribers, by Economic Category, in Small Towns and Rural Centers in Two Regions of Kenya, 1980*
(number of calls per day per respondent)

Subscriber	Telephone calls made	Telephone calls received	Total
Transport and storage	38.1	43.0	81.1
Banking and finance	22.6	23.6	46.2
Soft industry, such as food processing, textiles, paper products	17.6	27.3	44.9
Public utilities	20.1	22.8	42.9
Public administration and security	16.5	21.9	38.4
Telecommunications, broadcasting, and printing	12.0	17.5	29.5
Mining, metal chemical, and machinery industries	14.3	14.5	28.8
Wholesale trade	9.0	19.2	28.2
Education, health, and other services	10.4	12.7	23.1
Construction	10.6	11.0	21.6
Retail trade	10.0	11.1	21.1
Plantation agriculture	5.4	14.9	20.3
Other agriculture	4.7	5.3	10.0
Mean	12.9	16.5	n.a.

n.a. Not applicable.
Source: Adapted from Cleevely and Walsham (1980).

Users of Residential Telephones

A study of urban residential telephones in Chile found that income was the most important variable influencing both the demand for telephones and local and long-distance calling rates.[19] About 80 percent of the sampled families in the upper decile of the income distribution demanded a telephone, whereas only about 10 percent in the fifth decile and 2 percent in the lowest decile requested telephone service. What is particularly interesting is that local calling rates increased sharply with income (from four calls a day at the lowest incomes to fourteen calls a day at the highest) even though there was no charge for local calls. Such a result suggests that urban residential telephone use was influenced not only by purchasing power but also by the extent to which different segments of the population participate in community life. In other words, income (which was also found

Table 10-10. *Distribution of Telephones, by Functional Location, in Eighteen Small Locales in Syria, 1976*

Location of telephone	Total number of telephones	Percent of telephones
Government	372	17.8
Business		
Shops	333	16.0
Traders	118	5.7
Doctors and pharmacies	49	2.4
Auto service facilities	42	2.0
Unions and clubs	42	2.0
Public telephones	39	1.9
Restaurants and cafes	29	1.4
General offices	27	1.3
Small industries	26	1.2
Cinemas	13	0.6
Education and religious groups	12	0.6
Societies	9	0.4
Hotels	8	0.4
Farms	7	0.3
Other businesses	4	0.2
Total business and government	1,130	54.2
Residential location	954	45.8
Total	2,084	100.0

Source: Data compiled by Syed Sathar and Yuji Kubo, with assistance from the Syrian Telecommunications Establishment.

to be closely correlated with education and occupation) may be related to urban residential telephone use not only as a constraint on total expenditure but also as a factor in, and indicator of, the intensity, complexity, and range of the family's communication pattern.[20]

Two complementary findings in the Chilean study are also of interest. First, at a given income level, the aggregate demand for residential telephones was significantly higher among urban households that were already intensively using other means of communication (mainly travel, post, and telegraph) than among those that were not. Second, families with residential telephones averaged two to four times as many communications with kin, friends, and special-interest groups as did families with like incomes but without telephones.

Somewhat similar results were found in an examination of residential telephone subscribers in Guatemala City. In July 1981 a sample of 2,360 residential subscribers was classified into low-, middle-, and high-income categories on the basis of a field survey in which the quality and value of their housing were judged by external visual inspection. On this basis, it was estimated that about 13 percent of current residential telephone subscribers in Guatemala City lived in housing that would be typical of relatively low-income inhabitants, 49 percent middle-income, and 38 percent high-income. As in the Chilean example, on average, the telephone calling rates for local, long-distance, and international service also increased with the proxy for income level. Occupants of the relatively high-income residences generated an average telephone bill almost three times that of middle-income homes and four times that of lower-income subscribers.[21]

Nevertheless, even though income (and education) is a factor in urban residential telephone demand and use, at least in some of the relatively high-income Latin American countries, telephones seem to be demanded by and reasonably well dispersed among the middle- and lower-income groups. For example, in Montevideo, Uruguay, in 1980 only 7 percent of residential telephone lines (and about 6 percent of outstanding applicants) were in areas of high income, 15 percent in upper middle income, 56 percent in lower middle income, and 22 percent in low income. In Medellín, Colombia, about 70 percent of the planned 1980–90 investment in residential telephones was proposed to meet the demand of middle- and low-income families.[22]

Evidence is also available on the characteristics of residential telephone subscribers in lower-income nonmetropolitan communities. The Nepalese study, noted earlier, provided information on the educational level and occupations of a sample of residential subscribers in the capital city of Kathmandu.[23] It showed that 87 percent of the residential subscribers had a general education and 13 percent a professional education (engineering, medicine, law, and so forth). Of the total, 23 percent could be categorized as "only literate," with the remaining 77 percent having attained higher levels of academic certification. The conclusion is that residential telephone ownership goes hand in hand with higher education. Table 10-11 shows that businesspersons and government officials dominate residential telephone access in a sample of residential subscribers in Kathmandu.

In Sri Lanka, most residential respondents in a survey of new subscribers in Colombo and Kandy were relatively well off. Even so, except for the 20 percent of respondents in the highest income bracket (about Rs5,000 a month) who spent a little over 3 percent of their income on telephone usage, most subscribers (in income brackets ranging from Rs2,000 to Rs5,000 a month) were willing to spend as much as 5.5 to 7.6 percent of their income to use the telephone.

Table 10-11. *Occupations of a Sample of Seventy-eight Residential Subscribers in Kathmandu, Nepal, 1980*

Occupation	Percent
Businesspersons	21.4
Government officers	18.6
Politicians	14.3
Engineers and doctors	10.0
Teachers	7.1
Social workers	4.3
Lawyers	2.9
Not mentioned	21.4
Total	100.0

Source: Adapted from Integrated Development Systems (1980).

Given that the connection fee was Rs10,000, even well-off users had to make a substantial expenditure to pay for their new phones. Business subscribers, by contrast, spent an average of only 0.38 percent of sales on their telephone usage.[24]

A general survey of four countries in Southeast Asia (Malaysia, Singapore, Sri Lanka, and Thailand) used demographic information and found that where residential telephones were available, all income groups above a certain threshold (occurring in the middle class) were subscribers. In urban areas, as many as one-third of all residential phones belonged to the "poorer" half of all households, but this same type of household had only one-fifth of telephones in rural areas.[25]

With regard to residential telephone users in even smaller, more rural communities, the survey of eighteen small localities in Syria, referred to earlier, also provided limited information on the occupations of residential subscribers. From the data presented in table 10-12 it can be inferred that in these Syrian villages the better-educated members of the communities were also well represented among the residential telephone subscribers. Furthermore, given that more than 70 percent of the subscribers were farmers, drivers, teachers, or other public officials and employees, most of the residential subscribers were probably from the middle-income group.

Finally, the respondents in a survey of selected residential telephone subscribers in twelve small towns in Costa Rica were employed as outlined in table 10-13. Of these residential subscribers, 29 percent did not complete primary school, 19 percent completed only primary school, 24 percent were educated at the secondary level, and 28 percent were educated at the postsecondary level. The conclusion is that

Table 10-12. *Distribution of Residential Telephones, by Occupation of Subscriber, in Eighteen Small Locales in Syria, 1976*

Occupation of residential subscriber	Total number of telephones	Percent of telephones
Public officials and employees	292	30.6
Farmers	214	22.4
Teachers	93	9.7
Drivers	92	9.6
General workers	73	7.7
Merchants	43	4.6
Owners of shops	40	4.2
Doctors	28	2.9
Owners of small industries	22	2.3
Lawyers	16	1.7
Contractors	13	1.4
Cafe owners	11	1.2
Pharmacy owners	11	1.2
Engineers	6	0.6
Total	954	100.0

Source: Data compiled by Syed Sathar and Yuji Kubo, with assistance from the Syrian Telecommunications Establishment.

Table 10-13. *Employment Categories for 163 Residential Telephone Subscribers in Twelve Small Towns in Costa Rica, 1980*

Employment category	Percent
Not working	5
Government	
Professional	24
Nonprofessional	13
Total	37
Private	
Bars, restaurants, and cantinas	6
Retail stores and shops	14
Small industries and nonprofessional services	10
Professional services	4
Transport and communications	6
Total	40
Agriculture and livestock	18
Total	100

Source: Adapted from Kilgour (1982), p. 209.

the subscribers roughly represented a cross-section of the rural middle class.[26]

Notes

1. The distributional issue in telecommunications, of course, has many other dimensions, of which distribution between different income groups is only one. The distribution between urban and rural areas, and between the more modern sector and the "informal sector" in urban areas, are also important, as is the overall balance of provision among regions, between subscribers and PCO users, between local and long-distance facilities, among different types of services, and so forth.

2. Such inaccuracies are probably partly responsible for some of the questionable results found in the analysis of statistics within and among countries done by Hardy (1980).

3. This is also sometimes the case in industrial countries. Mayer, in presenting data on the frequency of residential calls in the United States, explains some of the subscribers with abnormally high calling rates by noting that some "are undoubtedly businesses conducted from the home, and they pay a residential rather than a business rate." Mayer (1977), pp. 246–61.

4. One East Asian country abandoned the dichotomy of business and residential classification primarily because of the problems with subscriber relations, which the telecommunications entity was having with small shop owners in the capital city. These small business owners lived in the back of their shops and usually requested that their telephones be installed on the residential side of the wall separating their living quarters from their shop. It was estimated that during business hours the telephones were used 90 percent of the time for business purposes, but the shop owners insisted that the lower residential tariffs should be charged because the telephone was located in their residence.

5. American Telephone and Telegraph Corporation (1980, 1981, 1982, 1989, 1990) and World Bank project documents.

6. In the Nepal survey, "residential" was to a certain extent defined as "ownership of dwellings."

7. Telephone Organization of Thailand (1980).

8. In mid-1980 when these data were collected, telephone service was limited in rural areas in Thailand. Aggregate national data thus essentially reflect the situation in Bangkok and several smaller urban places.

9. For one example, see Pierson (1979).

10. Wellenius (1971). Original analysis carried out by Richard Meunier.

11. Owners were included in this category to accommodate small shops or establishments with only one or two employees and in which the owner may have been the only person serving in a primarily white-collar function.

12. National Council of Applied Economic Research, India (1978). Two exchange areas were surveyed in each of the states of Haryana, Punjab, Madhya Pradesh, Rajasthan, and Uttar Pradesh.

13. Economics Study Cell, Posts and Telegraphs Board, India (1981).

14. Kamal, Dessouki, and Pool (1980).

15. Pool and Steven (1983).

16. In 1979, 95 percent of the lines in Nepal's beginning telex system (with only fifty-six lines in operation at that time) were connected to entities in the tertiary sector, especially tourism (23 percent), commerce (23 percent), and transport and communication (16 percent).

17. Initial data for this exercise were compiled by Yuji Kubo and Syed Sathar with the co-operation and assistance of the Syrian Telecommunications Establishment.

18. Part of the reason for this, of course, may be that tertiary sector activities tend to be clustered near the spatial center of communities, where telephone access is usually available first. Hence, farms, small industries, and so forth, although they may desire service, are often located outside existing local service areas.

19. Wellenius (1978), chap. 9 and Wellenius (1969b).

20. Also some of the higher-income, higher-calling-rate residential telephone users could have been professionals working at least part-time from their residences, as is common in developing countries.

21. World Bank data.

22. World Bank data.

23. Integrated Development Systems (1980).

24. Kojina, Hoken, and Saito (1984).

25. Engvall (1986).

26. Kilgour (1982), pp. 209-20. Of further interest is the finding that 58 percent of the professionals surveyed considered themselves the primary user of the home phone; this compares with 52 percent of nonprofessionals and 41 percent of farmers.

Chapter 11

Use of Business and Residential Telephones

IN VIEW OF THE EVIDENCE in the previous chapter it is not surprising that telephones in developing countries are used—as distinct from the category of users—overwhelmingly for purposes related to government and business. For example, in 1980 in Bangkok, Thailand, although 39 percent of the lines were residential, an estimated 85 percent of all calls related to business and government activities. Estimates of telephone use in large urban areas in Egypt, El Salvador, and India showed that more than 90 percent of all calls made during the peak business hours were for purposes related to business.

Business Telephone Use

Even where most lines are residential, business use tends to dominate. In Montevideo, Uruguay, where almost 80 percent of the telephone lines are residential, survey data suggested that about 60 percent of all telephone calls and 70 percent of telephone revenues related to business functions. In Guatemala City, partly because tariffs paid by residences are lower than those paid by business and government subscribers, 67 percent of all subscribers were categorized as residential. As in the above instances, however, business and government subscribers are the most intensive users of telephones. From a sample taken in July 1981 of 4,400 lines registered in Guatemala City, business and government lines averaged 3.2 times as many outgoing international calls, 2.3 times as many long-distance calls, and 1.4

219

times as many local calls as did residential lines. Overall, subscribers officially categorized as being business and government accounted for more than 60 percent of all telephone company revenues.[1]

The 1971 survey of 1,100 business telephones in Santiago, Chile, showed that 88 percent of those telephones were used most frequently for work-related communication.[2] Of these business telephones, 10 percent were, however, used most frequently for social purposes, that is, to communicate with members of the family, kin, and friends. The remaining 2 percent were used for unspecified purposes, which apparently included household-related calls (to order goods to be delivered at home and so forth). Studies in several other Chilean cities showed the proportion of business telephones used most frequently for social purposes to vary between 9 and 33 percent.[3]

In the 1978 study of ten small-town exchange areas in five states of India, referred to in the previous chapter, 72 percent of the most recent calls made by the (mostly business) subscribers in the sample were to conduct business, 18 percent were to contact family members, and 10 percent were to reach friends or undertake public service, health, or other purposes.[4] Of particular interest is that callers perceived 25 percent of their most recent calls to be of an urgent nature and a further 50 percent to be calls that could not be delayed until the following day.

In the 1981 study of twelve small-town exchange areas in the Andhra Pradesh State of India, the purpose of calls varied greatly depending on whether the call was local or long distance.[5] Of the sampled subscribers, 70 percent frequently used the telephone for long-distance communications, and, as table 11-1 shows, calls relating

Table 11-1. *Proportion of Subscribers Making Local and Long-Distance Calls in Twelve Villages in Andhra Pradesh State, India, 1981*
(percent)

Purpose of call	Local calls	Long-distance calls
Agricultural business	18	40
Nonagricultural business	23	23
Both agricultural and nonagricultural business	13	22
Professional needs and contacting government offices	12	13
Emergencies and social calls	34	2

Source: Adapted from Economics Study Cell, Posts and Telegraphs Board, India, (1981).

to some form of agricultural, business, government, or professional activity involved roughly 98 percent of all long-distance calls, although they involved only 66 percent of local calls. With regard to the urgency of the most recent call made, 73 percent of the subscribers sampled stated that the last long-distance call made was of immediate necessity concerning their work and business; 27 percent felt that the call could have waited. Finally, two-thirds of the subscribers sampled made most of their long-distance calls during off-peak periods. This was probably partly because off-peak tariffs were lower and partly because long-distance and large urban local networks tended to be highly congested during business hours, greatly reducing the chance of getting through.

In the Nepalese study, a sampled group of trade and tourism establishments in both the capital of Kathmandu and the smaller town of Pokhara also responded that their most frequent reason for calling was for so-called productive purposes. As table 11-2 shows, these calls were even more important among persons who had to use someone else's telephone, and emergency calls constituted a significant 6 to 14 percent of all calls made or received.[6] Finally, business applicants waiting for telephone service to become available were asked to rank the reasons why they needed their own telephone. On average they ranked business promotion as number 1, saving transport and postal

Table 11-2. *Most Frequent Reasons for Using Primarily Business Telephones, for Two Types of Establishments in Kathmandu and Pokhara and for Waiters in Kathmandu, Nepal, 1980*
(percent)

Purpose of call	Wholesale and retail trade establishments		Tourism establishments: hotels, tour guides, airlines, etc.		Registered waiters in wholesale and retail trade, Kathmandu[a]
	Kathmandu	Pokhara	Kathmandu	Pokhara	
Social	29	34	32	26	6
Productive	50	45	50	60	81
Emergency	8	14	7	7	6
Unspecified	13	7	11	7	7

Note: Calls include both incoming and outgoing.

a. Waiters do not have telephone connections but use the telephones of friends or other private sources.

Source: Adapted from Integrated Development Systems (1980).

costs as number 2, and saving time as number 3. Public relations were ranked fourth and "social prestige" last.[7]

Somewhat similar results emerged from the sample of (mainly business) telephone subscribers in towns and market centers in the two regions in west and south-central Kenya. Of a total of 493 sampled telephone subscribers, 81 percent used their telephone only or mainly for business. Of the remainder, 11 percent used it for "some business," whereas 8 percent used it only for social purposes.[8] Table 11-3 gives a more detailed breakdown of the results. When 574 subscribers, applicants on waiting lists, and public coin box telephone users were asked to list and describe their most important telephone contacts, the dominant reasons noted were placing and receiving orders, giving or receiving information (which presumably takes place in almost all calls), and taking care of administrative matters.[9] Together these three highly interrelated categories accounted for 69 percent of all contacts.

Table 11-4 shows the results of a very similar survey undertaken in 1982 in three medium-size towns in Ethiopia. There, when 996 telephone subscribers and public call office users were given the summary list of reasons for making telephone contact that emerged from the

Table 11-3. *Frequency of Reasons Given for Telephone Contact by a Sample of Telephone Users in Two Regions of Kenya, 1980*

Reasons for telephone contact	Number of times mentioned	Average times mentioned per respondent	Times mentioned per 100 respondents
Placing or receiving orders	803	1.4	33
Giving or receiving information	653	1.1	27
Handling administration	210	0.4	9
Ordering spares or repairs	129	0.2	5
Handling payments or finances	121	0.2	5
Arranging transport	109	0.2	4
Arranging meetings	98	0.2	4
Soliciting business	54	0.1	2
Conducting urgent business	13	0.0	1
Negotiating	8	0.0	0
Other	255	0.4	10
Total	2,453	n.a.	100

n.a. Not applicable.

Note: Users comprised subscribers, registered applicants, and public coin box telephone users.

Source: Adapted from Cleevely and Walsham (1980).

Table 11-4. *Reasons Given for Most Frequent Telephone Contact by a Sample of Telephone Users from Three Medium-size Towns in Ethiopia, 1982*

Reasons for telephone contact	Number of times mentioned	Average times mentioned per respondent	Times mentioned per 100 respondents
Conducting urgent business	717	0.72	22
Soliciting business	337	0.34	10
Placing or receiving orders	286	0.29	9
Handling administration	242	0.24	7
Giving or receiving information	237	0.24	7
Handling payments or finances	224	0.22	7
Arranging meetings	163	0.16	5
Ordering spares or repairs	138	0.14	4
Arranging transport	119	0.12	4
Negotiating	53	0.05	2
Other	736	0.74	23
Total	3,252	n.a.	100

n.a. Not applicable.

Note: Users comprised both subscribers and public coin box and call office users. Of the three towns, Nazareth had 1,556 subscribers, Awassa had 547, and Shashemenē had 500. In total 996 users were sampled: 586 from Nazareth, 222 from Awassa, and 188 from Shashemenē.

Source: Ethiopian Telecommunications Authority.

Kenyan study, they most frequently identified urgent business (which in translation was generally interpreted to mean urgent business or social matters) and soliciting business as the two principal reasons for telephone contact. These were then followed by the three reasons most frequently mentioned in the Kenyan survey: placing or receiving orders, handling administration, and giving or receiving information.

There are sometimes, however, perceived problems in the strong relation between business telephones and business use. In Papua New Guinea, for example, a 1977 study found that business owners and managers felt that business and government telephones were used by employees too frequently for personal purposes, to the extent that employers would sometimes fit locks on telephone dials to deter outgoing calls.[10] Such an action, of course, also denied the employer some benefits related to communication.

Finally, the Kenyan survey threw some light on the nature of telephone use by the agricultural sector and reinforced the intuitive conjecture that although agriculture is generally a less-intensive telephone

Table 11-5. Percentage of Important Telephone Contacts between Sectors, for a Sample of Telephone Subscribers, Waiters, and Coin Box Users in Two Regions of Kenya, 1980

Sector making call	Sector receiving call												Residential customers	Percent of contact outside own economic category	Number of other economic categories contacted
	1	2	3	4	5	6	7	8	9	10	11	12			
1. Retail trade	14	5		13	39		8				8		7	85.7	6
2. Plantation and other agriculture	9	20	6	17	7		9		5		16	8		80.4	8
3. Transport and storage		5	26	12	15		15				4	5	16	74.4	7
4. Textiles, paper, and food processing	5	7	10	30	9		12		5		4	8	7	69.6	9
5. Wholesale trade				30	32		18						5	68.2	3
6. Utilities						33	6	9			6	15	21	66.7	5
7. Mining, metals, and chemicals	6	6	8		13		44				5	5	9	56.3	7
8. Construction	4	4		5	6		10	47			4	12	5	52.6	7
9. Banking and finance	6	6					6		48		6	6	25	52.1	5
10. Telecommunications, printing, and broadcast	4	4		13						50	9	7	7	50.0	7
11. Education and health services	12						5				50	12	10	19.8	4
12. Public administration and security						4					11	59	10	40.6	3

Note: This table should be read by rows. Values given are the percentage of time the calling sector mentioned the receiving sector as being a major recipient of calls. Each respondent could mention up to five sectors that were frequently called. Values of less than or equal to 3 percent were omitted to clarify the pattern.

Source: Adapted from Cleevely and Walsham (1980).

user, it needs, because of the diversity of its inputs and outputs, to communicate with a wider variety of other economic activities than do most other sectors. Table 10-9 showed that in the two regions sampled in Kenya the average calling rates for plantation and other agriculture were considerably below those for all other sectors. Such a finding is consistent with the input-output results observed in chapter 5, in which the agricultural sector purchased fewer communication services than did most other sectors.

A low average calling rate, however, does not necessarily mean that the agricultural sector has little need to communicate rapidly with other sectors. In fact, of all the sectors surveyed in these two regions of Kenya, the agricultural sector initiated the greatest variety of telephone contacts with other sectors. From table 11-5 it can be seen that about 80 percent of all calls originating in the agricultural sector were directed to other sectors, with the most important being to food processing and textiles, education and health services, retail trade, chemicals and metal fabrication and manufacture, government, wholesale trade, and transport and storage. In fact, if one were looking for a sector whose communication linkages are mostly productive, it is noteworthy that agriculture was the only sector that had insignificant contact with final consumers in the residential sector.[11]

Also of note from table 11-5 is that, other than agriculture, retail trade and transport and storage (both tertiary activities) were the most outward-communicating sectors, whereas at the other end of the spectrum both the public administration and security and the education and health services sectors (also tertiary activities) communicated with themselves and with each other well over 60 percent of the time.

Residential Telephone Use

Data available on the uses of residential telephones suggest that in general the largest proportion of calls made on such telephones pertain to social or family uses. Table 11-6 shows that for samples of residential subscribers from six countries, calls classified as social accounted for about 70 percent in three of them and around 50 percent in the remaining three. Of the developing countries shown, three had social calls in the 50 percent range, whereas one (Chile) approached 70 percent. The lower proportion of social calls in Papua New Guinea could be related to the fact that a few business subscribers might have been included in the sample and that in Papua New

Table 11-6. Reasons for Using Residential Telephones in Several Countries
(percentage of calls)

Use	Papua New Guinea[a]	Chile (urban)[b]	England[c]	United States (New York)[d]	Thailand[e]		Nepal, Kathmandu and Pokhara[f]	Nepal, Kathmandu, residential telephone applicants[g]
					Bangkok	Provinces		
Social	50	69	74	67	51	40	52	34
Friends	26	39	—	36	26	18	—	—
Family	24	30	—	31	25	22	—	—
Business	31	20	17	16	42	12	24	45
Goods and services acquisition for home	—	7	5	4	7	46	—	—
Emergency services	17	3	4	—	—	—	14	8
Other	2	1	—	13	—	2	10	13

— Not available.

a. Papua New Guinea data exclude expatriates and may include some business subscribers. In the survey, the subscribers sampled were asked what they used their telephones for; there were on average 1.4 answers per respondent. The figures presented are percentages scaled down to sum 100.

b. Figures represent the proportion of sampled telephone lines reported to be used most frequently for the given purpose. Figure given for "friends" includes communication with kin.

c. These results are for old customers in the British Post Office's 1971 residential research sample of current supply. Figures represent the proportion of telephone lines reported to be used most frequently for the given purpose.

d. These results were observed when 90,000 subscribers lost their telephone service for twenty-three days because of a fire in the switching center. A sample of those residential subscribers was asked what types of calls they most missed being able to make. "Other" includes 10 percent given as "medical."

e. These figures are based on a survey of 108 residential telephone subscribers and applicants in Bangkok metropolitan area and 100 in provincial cities and towns. "Family" includes communication with kin. Applicants used borrowed and public telephones and comprised 20 percent of the sample surveyed. Figures represent the proportion of sampled telephone lines reported to be used most frequently for the given purpose.

f. These include calls both made and received.

g. Waiters do not have telephones but use those of friends and other private subscribers.

Source: Evans and Bryan (1977); Wellenius (1978), chap. 9; British Post Office personal communication; Srivisal and Tamura (1980); Integrated Development Systems (1980); Wurtzel and Turner (1977), pp. 246–61.

Guinea (as in Nepal and Thailand) telephone density was so low that many friends and relatives of telephone subscribers did not have telephones.[12]

Nevertheless, residential telephones in all six countries were also used to an important extent for economic activities related to production and distribution. In the Chilean survey, 20 percent of the urban residential telephones were used most often to further work-related purposes, 7 percent to access goods and services for the household, and 3 percent to obtain services under emergency conditions (mainly to contact urgent medical assistance, firefighters, or police). In Papua New Guinea and overall in Thailand, approximately 31 and 28 percent, respectively, of the residential subscribers surveyed indicated business-related communication (excluding calls to suppliers of goods and services for the home) as the primary reason for using the home telephone, whereas in Nepal a high figure of 45 percent of the calls that persons waiting to receive a residential telephone made using the telephone of a neighbor or friend were reportedly related to business.[13] Similarly, in urban areas of Sri Lanka, residential phones were used for business as well as for private purposes about 40 percent of the time among recent subscribers surveyed in Colombo and Kandy.[14]

In provincial Thailand, each subcategory of business-related calls made from the home (calls to one's office, customers, and government services) had a considerably lower incidence than in Bangkok: only 12 percent of the families surveyed gave these as the most frequent uses of their residential telephone as against 42 percent in the metropolis; a far greater proportion of residential telephones in the provinces (46 percent) were mostly used to contact shops for home supplies (presumably a more viable and common practice in small places than in a large metropolis), bringing the total of residential telephones used primarily for purposes directly related to economic activity to 58 percent, even higher than in the Bangkok metropolitan area (49 percent).

The study in Sri Lanka examined the frequency with which non-subscribers were able to borrow subscribers' phones and to leave messages, an important aspect of telephone use in areas where density is extremely low. Although only one-third of businesses surveyed were willing to offer their telephones to borrowers, two-thirds of residential subscribers said that they would allow neighbors to use their telephone. Both groups were somewhat less willing to deliver messages left for nonsubscribers—one-quarter of businesses and two-fifths of residential subscribers said they would do this. The researchers concluded that in Sri Lanka, many telephone subscribers considered their phones to be public property and appeared to use their phones very efficiently. In contrast, Pool and Stevens found that many telephones in wealthy households in Egyptian villages were underused. Social custom seems to play a larger role than economic factors in determining the access that outsiders have to home phones.[15]

A survey of residential telephone subscribers in twelve small towns in Costa Rica did not ask the explicit purpose of telephone use but did ask if having a telephone in the home had changed how the respondents did their work. Of the 93 percent of the sample who were working, 79 percent said they had noticed a change, and 51 percent of these indicated that they used their home phone primarily for work purposes. When these respondents were asked how work had changed because of the telephone, 41 percent mentioned saving time, 16 percent getting information, 14 percent obtaining products, and 13 percent traveling less.[16]

When the small-town residential subscribers in the Costa Rican sample were asked whether the installation of the telephone had caused a change in the town, 93 percent replied affirmatively. Of those, 44 percent stated that the telephone allowed more contact with family or made contact easier, 19 percent noted it reduced the need to travel or made travel easier and saved time or money, 13 percent noted the telephone induced more progress and development in the town, 9 percent stated it brought more information and news so the town was less isolated, 7 percent noted it allowed problems to be solved more rapidly, and 5 percent stated the telephone allowed more rapid help in emergencies.[17]

One additional inference about how individuals use residential telephones for business can be made from the occupations of residential subscribers in the eighteen small towns in Syria, which were presented in table 10-12. Although, from the information available, it is not possible to calculate the extent to which residential telephones were used for business-related purposes, most persons in the occupations listed in table 10-12 (farmers, drivers, merchants, shop and industry owners, doctors, lawyers, and so forth) would certainly have occasion to use their telephones for reasons directly related to their employment.

Of course, the most frequent use of a residential telephone is not always regarded as the most important. For example, a 1971 survey of newly connected residential customers in Great Britain showed that they considered emergency uses to be the most important reason to have a telephone, whereas calls to family, kin, and friends were the most frequent use.[18] The families surveyed decided to apply for a home telephone both because of the potential to make occasional but highly important calls and because of the ability to make more common calls of lower individual value.

A similar pattern was found elsewhere. In a 1973 survey of 1,500 households in and around Tokyo, an overwhelming 72 percent categorized the residential telephone as "very useful" in case of accidents.[19] The second highest category to receive the "very useful" designation was work or profession, with 53 percent. Also, the residential subscribers sampled in the two countries with the lowest per capita income (Papua New Guinea and Nepal) tended to rank emergency use as much more prevalent than did residential subscribers in the other countries discussed in this section. In particular, applicants for residential (as well as business) telephones in Papua New Guinea stated a concern for security, protection, and the possibility of an emergency as the main reason for wanting to have a telephone, even though, in fact, such calls were likely to be relatively infrequent; 41 percent of the existing subscribers nominated emergency and safety as major reasons for having a telephone, whereas only 25 percent gave these as reasons for actually using it.[20]

Telephones and Other Means of Communication

The survey of residential telephone subscribers in Thailand also indicated the way in which subscribers perceived the importance of telephones relative to other forms of communication media. As table 11-7 shows, the telephone was overwhelmingly the preferred means of dealing with emergencies and setting up appointments, as well as communicating with kin and friends. The table also shows which other means of communication would have been preferred if the subscriber had not had a telephone, although over 40 percent still preferred a telephone, albeit a public one. In Bangkok only 2 percent said they would not communicate at all.[21]

There were also notable differences between Bangkok and the provinces. For example, in the provinces the telephone was not preferred for setting up appointments as often as it was in Bangkok (although in both cases it was the means preferred most often); meetings were indicated somewhat more frequently in the provinces, possibly reflecting shorter travel distances and sometimes poorer-quality service. The public telephone was perceived to be more important in the provinces than in the metropolis as an alternative to a residential telephone, accompanied also by a slightly greater preference for letters and tele-

Table 11-7. *Preferences for Alternative Communication Media of 208 Residential Telephone Subscribers and Applicants in Thailand, 1980*

Item	Alternative media	Percent of subscribers sampled	
		Bangkok	Provinces
Preferred media to report an	Telephone	86	86
emergency	Other people	8	4
	Other (fire alarm and so forth)	5	7
	Unknown	1	3
	Total	100	100
Preferred media to set up an	Telephone	83	77
appointment	Meeting	10	15
	Letter	2	6
	Messenger	1	2
	Other	—	—
	Unknown	4	—
	Total	100	100
Preferred media to	Telephone	45	42
communicate with kin and	Visit	34	31
friends			
	Letter	3	11
	Other	1	1
	Unknown	17	15
	Total	100	100
Preferred media to gather news	Newspaper	41	38
	Television	31	27
	Telephone	13	15
	Neighbor	1	5
	Other	9	10
	Unknown	5	5
	Total	100	100
Other communication means	Public telephone	43	49
preferred if the subscriber	Travel to meet party	16	16
did not have a telephone	Letter	14	19
	Telegram	7	11
	Other	17[a]	5
	Unknown	3	—
	Total	100	100

— Not available.

a. Neighbor's telephone, 6 percent; government radio, 5 percent; messenger, 2 percent; telex, 2 percent; and no communication, 2 percent.

Source: Adapted from Srivisal and Tamura (1980).

grams and less preference for other means (including neighbors' telephones, which are in many instances not available in small communities with low telephone densities).

In the Kenyan study of telephone use in two rural regions, both business and residential subscribers were asked what means of communication they used to contact other economic sectors; they were allowed to list up to five methods. The results shown in table 11-8 reveal that the telephone and postal service accounted for almost 80 percent of all economic contacts, with the telephone being the main or only means (among those listed) in more than half of the contacts mentioned. The survey of residential telephone subscribers in the twelve Costa Rican towns noted earlier yielded a similar result. When the sample respondents were asked which means of communication they used most frequently to communicate with San José or any other city in the course of their work, 94 percent stated the telephone, with the other 6 percent noting telegraph, radio, messenger, or other.[22]

Finally, the survey of 474 males in nine small Egyptian villages showed that where access to telecommunications services was minimal and the long-distance telephone service was unreliable and congested, persons who communicated outside the village by means other than physical travel felt that telephones were important.[23] Nevertheless, even those who used one or more of the means of communication listed in table 11-9 did so infrequently; the primary reason for this was that the persons they wished to contact could not be easily reached by existing postal or telecommunications facilities. Partly to compensate for this, village administrators used the telephone primar-

Table 11-8. *Means of Communication Used for Economic Contact by a Sample of Telephone Subscribers in Two Regions of Kenya, 1980*

Method	Percent of contacts using method[a]
Only phone	16.6
Mainly phone	34.7
Phone and letter	17.7
Mainly letter	10.5
Only letter	0.1
Personal visit and other	20.4

a. Respondents were allowed to list up to five economic contacts.
Source: Adapted from Cleevely and Walsham (1980).

Table 11-9. *Means of Communication Used by 474 Male Respondents during Twelve Months in Nine Villages in Egypt*

Medium of communication	Percent who had used listed communication medium	Median messages sent or received by users during year
Sent letters	16.0	5.0
Received letters	17.0	5.0
Talked on telephone	10.0	3.0
Sent telegrams	4.0	2.0
Received telegrams	3.0	2.5

Source: Adapted from Kamal, Dessouki, and Pool (1980).

ily for leaving dictated messages. Similar results were obtained in the follow-up study, which covered 143 villages. Personal travel was the villagers' principal means of conveying information, and if adequate phone service had been available, 60 percent of this travel would have been unnecessary. The message system was haphazard at best, since messages were easily misunderstood in translation, and their delivery was uncertain and frequently delayed.[24]

Notes

1. World Bank data.
2. Wellenius (1971). The original analysis was made by Richard Meunier.
3. Jabif (1971).
4. National Council of Applied Economic Research, India (1978).
5. Economics Study Cell, Posts and Telegraphs Board, India.
6. Integrated Development Systems (1980).
7. When residential subscribers on the waiting list were asked the same question they ranked time savings as first, cost savings as second, and business promotion as third. As with the business subscribers on the waiting list, residential subscribers ranked public relations fourth and social prestige last.
8. Cleevely and Walsham (1980).
9. Each respondent was allowed to mention up to five contacts. Respondents were asked to "please list the activities and locations of the most important businesses or farms or government departments or individuals that you or this establishment deal with—and what is the main reason for your contact for each of them?" The 574 respondents in the sample gave 2,453 replies.
10. Evans and Bryan (1977).
11. Limited evidence of the importance of telecommunications to agriculture was also found in the survey of nine rural villages in Egypt (discussed briefly in chapter 10). In several of the villages examined, the cultivation of commercial crops (rather than subsistence and traditional crops) was associated with fewer messages being carried among villages, on

the one hand, and much greater use of the telegraph and postal system, on the other. See Kamal, Dessouki, and Pool (1980).

12. In Papua New Guinea, almost 60 percent of the subscribers sampled stated that few or none of their friends or relatives in their town were telephone subscribers. The telephone density in 1977 in Papua New Guinea was 1.4, whereas in 1980, in Nepal it was about 0.12 and in Thailand, 1.1.

13. These figures are not strictly comparable and for Papua New Guinea do not necessarily indicate the relative frequency of each type of call. The Papua New Guinea data refer to multiple responses, at an average of 1.4 responses per subscriber. When the Papua New Guinea responses were prorated to sum to 100 percent, the business-related calls comprised 31 percent of the total. Business-related calls in Thailand were comprised of calls to shops, one's customers, one's workers, and government entities, in that order of importance.

14. Kojina, Hoken, and Saito (1984).

15. Kojina, Hoken, and Saito (1984) and Pool and Steven (1983).

16. Kilgour (1982), pp. 195–96.

17. Kilgour (1982) pp. 197–99.

18. British Post Office, Statistics and Business Research Department, personal communication.

19. Research Institute of Telecommunication and Economics (1974).

20. Of course, the emergency factor may be for receiving as well as for making calls. Although only 10 percent of the respondents in the Egyptian rural village surveyed reported using a telephone during the previous year (table 11-8), 30 percent stated that they would use a telephone for emergency purposes, and 9 percent claimed that at one time or another they had actually used a telephone to call a physician in an emergency. Kamal, Dessouki, and Pool (1980).

21. The figure of 2 percent who would not communicate seems to be too low since it implies an almost completely inelastic demand for communication, which has not been observed in other surveys.

22. Kilgour (1982) p. 201.

23. Kamal, Dessouki, and Pool (1980). Of the nine villages, three had small telephone exchanges (thirty to eighty-eight telephones), five had a few scattered direct lines, and one had no telephone access. Most of the telephones were in private hands. Telephone calls were generally viewed as time-consuming and difficult to make because call traffic was severely congested. Five of the villages had post offices, and three had telegraph offices.

24. Pool and Steven (1983).

Chapter 12

Use of
Public Call Office Telephones

IN MANY DEVELOPING COUNTRIES the population living in towns, rural villages, and low-income urban areas typically gain their first access to
rapid two-way communication through public coin box telephones or
public call offices (PCOs). Such service may be provided for reasons of
facilitating general economic development, social equity, or national
unity, even though the facilities sometimes do not generate enough
revenue to cover their full financial costs, especially during the first
few years of operation. Nevertheless, such public facilities are being
provided with increasing frequency in the developing world, and in
fact this provision has been encouraged by institutions such as the
World Bank and the Inter-American Development Bank.[1]

In reviewing the survey data and study results presented in this
chapter, the generally poor quality of PCO service in developing countries must be kept in mind. In many instances circuits are highly congested, telephone operators in urban areas give incoming PCO calls low
priority, and transmission quality is poor, partly because of inadequate
maintenance. Poor-quality service results in long waits for calls to be
completed and ultimately in the cancellation of many calls. It also de-
ters individuals from attempting to make any calls except those
judged to be of high priority. Table 12-1 shows survey results for the
town of Zway, Ethiopia, which reflect a typical situation. During a
five-day sample period in February 1982, of the PCO calls urgent
enough to be attempted, two-thirds were completed and one-third
were canceled. Moreover, the interurban calling rate was higher for
PCOs than for most subscriber telephones.

234

Table 12-1. Interurban Telephone Calls, by Type of Subscriber, for the Manual-Exchange Service Town of Zway, Ethiopia, 1982

| | | Interurban calls[a] | | | | | | |
| | | Outgoing calls | | | | Incoming calls | | |
Subscriber	Number of subscribers	Calls established	Calls canceled	Total registered call attempts	Attempts per subscriber per day	Calls received	Calls per subscriber per day	Total calls per subscriber per day
Government administration, public service, and political organizations	27	190	72	262	1.95	261	1.93	3.88
Business and commerce[b]	18	33	8	41	0.46	35	0.39	0.85
Registered as residential	32	18	6	24	0.17	16	0.11	0.28
Public call office	1	78	39	117	23.40	6	1.20	24.60
Total	78	319	125	444	n.a.	318	n.a.	n.a.

n.a. Not applicable.

a. Calls were recorded for a period of five days for about twelve hours a day.

b. Eight small hotels, nine retail stores, and one large-scale farm.

Source: Ethiopian Telecommunications Authority.

Characteristics of PCO Telephone Users

Table 12-2 presents selected characteristics of rural or small-town users in five developing countries.

In the 1978 Indian study of ten rural PCOs in five states, most PCO users outside large urban areas were individuals with relatively high levels of education, were either employers or self-employed rather than employees, and had incomes that were higher than those of most rural dwellers. Furthermore, a large proportion of the users were businesspersons, professionals, or civil servants, even though an average of 70 percent of the population in the areas served by PCOs were engaged in farming.

A feature of the 1981 Indian study, based on interviews of 174 PCO users in thirty-six villages from six districts in Andhra Pradesh State, is that users who gave agriculture as their principal occupation made up a relatively large 41 percent of those interviewed. Business and services (the definitions of which are somewhat unclear) made up 54 percent of the PCO users, which corresponds roughly to the 1978 figure of 65 percent. The 1981 group was slightly less well educated than the 1978 group, whereas the income of the two samples appears to be distributed in roughly the same ratios. The interesting comparison with regard to income, however, is with the 1981 Indian survey of rural telephone subscribers discussed in chapter 10.[2] Although only 35 percent of the subscribers had incomes of less than Rs15,000 a year, almost 76 percent of the PCO users fell in that lower-income category.

The study on Papua New Guinea presents a somewhat more restricted picture of the educational level of PCO users than the two Indian surveys. In Papua New Guinea, where 79 percent of the population had at most a primary education, this group accounted for only 5 percent of rural PCO callers and used PCOs infrequently. In contrast, persons with secondary, technical, vocational, or university education constituted only 3 percent of the population but comprised 75 percent of PCO users and tended to call frequently.[3]

Somewhat different results, however, were observed in Costa Rica. In a study of public telephones located in a cross-section of rural villages, PCO usage was fairly evenly distributed among economic and social groups.[4] The caller's occupation predictably tended to reflect the nature of the local economies (agricultural workers placed more than 80 percent of calls in an important agricultural area, businesspersons were the dominant callers in a river port, and so forth); users were not predominantly employers or government officials; no one small group dominated usage (about 80 percent of calls were placed by

Table 12-2. *Characteristics of Persons Using Rural PCO Telephones in Five Countries*
(percentage falling in category)

Characteristic	India (1978)	India (1981)	Costa Rica (1975)	Papua New Guinea (1976)	Mexico[a] (1978)	Peru[b] (1985)
Employment status						
Self-employed or employer	77.5	—	36.0	—	—	—
Wage earner-employee	22.5	—	60.5	—	—	—
Principal occupation						
Business	65.0	33.0	23.7	—	22.0	24.6
Professional and technician	7.5	4.0	23.2	—	22.0	23.8
Agriculture	7.5	41.0	32.7	—	15.0	9.7
Government service	12.5	—	—	—	20.0	—
Administrative personnel	—	—	8.9	—	—	—
Services	—	21.0	—	—	—	33.3
Other	7.5	1.0	11.5	—	21.0	8.6
Income[c]						
0–¢300 a month	—	—	45.5	—	—	—
¢300–¢600 a month	—	—	22.0	—	—	—
¢600+ a month	—	—	10.6	—	—	—
No response	—	—	21.9	—	—	—
0–Rs4,800 a year	20.0	—	—	—	—	—
Rs4,800–Rs6,000 a year	7.5	—	—	—	—	—
Rs6,000–Rs10,000 a year	32.5	—	—	—	—	—
Rs10,000+ a year	40.0	—	—	—	—	—
0–Rs5,000 a year	—	20.1	—	—	—	—
Rs5,000–Rs10,000 a year	—	41.4	—	—	—	—
Rs10,000–Rs15,000 a year	—	14.4	—	—	—	—
Rs15,000+ a year	—	24.1	—	—	—	—
$262.53 average	—	—	—	—	—	100.0
Educational level						
Primary or less	15.0	26.0	—	5.0	—	16.2
Middle	22.5	16.0	—	—	—	—
Matriculates (high school)	25.0	37.0	—	—	—	13.0
Secondary or technical	—	—	—	50.0	—	35.7
Attend college	—	—	—	—	—	17.6
Graduate	30.0	—	—	—	—	—
Postgraduate	2.5	—	—	—	—	—
College or university (degree)	—	21.0	—	25.0	—	17.6
Still in school	—	—	—	16.0	—	—
Other	5.0	—	—	4.0	—	—

— Not available.

Note: The category definitions are not completely comparable in all instances.

a. Results of a survey of all PCO users for one month in the rural village of Metepec (state of Hidalgo), Mexico. Survey carried out by Centro de Investigaciones para el Desarrollo Rural in June/July 1980.

b. Results of a two-year survey of seven small villages in rural Peru carried out by Florida State University with ENTEL, the Peruvian telecommunications authority. See Mayo and others (1987).

c. The mean income level of income earners in the sample villages in Costa Rica was ¢372 a month.

Source: See table 12-4, except for the Mexican and Peruvian studies.

users who called on average only two times a week or less, whereas heavy users, placing five or more calls a week, generated only 11 percent of the traffic); and the mean income of callers was close to that of the rural population at large.

In a study of seven small villages in rural Peru, the typical user was male, thirty-three to thirty-five years of age, born outside the project zone but a resident for fifteen to seventeen years, well educated, a frequent traveler outside the project area, and either employed in a professional or technical position or the owner or manager of a business. Business owners and managers increased their share of total use of the telephone system from 14 to 34 percent over the two-year period of the study. One factor in this increase was that businesses learned to use the system more efficiently over time, by using calls as well as telegrams and letters to arrange appointments for future calls. Persons classified as employees or workers averaged 17.4 percent of total use, varying without a clear trend over the period, and use by housewives fell from 20.3 to 9.6 percent with no easily detectable explanation. Overall, the income of users was, on average, 14 percent higher than that of nonusers.[5]

With regard to the distance traveled to use public telephones, table 12-3 shows that in Papua New Guinea, Costa Rica, and Mexico, public telephones were used primarily by persons who lived or worked nearby. The average rural PCO user surveyed in Costa Rica traveled only about 0.3 kilometer to reach a PCO, although 5 percent of the callers traveled more than 2 kilometers, and in fact this latter group averaged 8.5 kilometers of travel per call.[6] There was also some indication that the distance traveled was not independent of the purpose of the call. Individuals making emergency calls, for example, traveled on average about 60 percent farther than the average caller.

In Papua New Guinea, where telephone penetration was much lower than in Costa Rica, 81 percent of PCO users reported going to the telephone area specifically to place a call, and on average they traveled 6.9 kilometers.[7] Seven percent of the users traveled more than 24 kilometers, and, in extreme cases, a couple of days' walk followed by some motorized transport were involved. In some areas it was not uncommon to make half-day boat trips just to make a telephone call. The results of the survey in the Mexican village are roughly consistent with those from Papua New Guinea except that in the Mexican survey only 33 percent of the persons interviewed traveled to the village just to make a telephone call.

In the 1978 Indian study, the average rural PCO served thirteen neighboring villages, the number varying between four and twenty-six

Table 12-3. *Distance Traveled to a* PCO *to Make a Telephone Call in Three Countries*
(percentage falling in category)

Distance traveled	Costa Rica (1975)	Papua New Guinea (1976)[a]	Mexico (1978)[b]
0–500 meters[c]	79	—	—
0–2 kilometers[d]	95	—	—
0–5 miles[e]	—	77	—
6–10 miles	—	8	—
11–15 miles	—	2	—
15+ miles	—	7	—
Within the village	—	—	68
Village to 5 kilometers	—	—	9
5–10 kilometers	—	—	11
10+ kilometers	—	—	12

— Not available.

a. The Papua New Guinea results are for the 81 percent of the 916 respondents who made a special trip to make a call.

b. Results of a survey of all PCO users for one month in the rural village of Metepec (state of Hidalgo), Mexico. Survey carried out by Centro de Investigaciones para el Desarrollo Rural in June/July 1980.

c. From a sample of three villages (see appendix C).

d. From a sample of eleven villages (see appendix C).

e. Seventy-five percent of the natives of Papua New Guinea traveled less than 5 miles, compared with 90 percent of the expatriates.

Source: Except for the Mexican study, see table 12-4.

depending on the district concerned. Use was concentrated mainly in villages that had their own PCO, however, although occasionally users traveled much farther; the PCOs covered between 2.5 and 13.0 kilometers. The average distance traveled to make a call was tabulated for five rural PCOs; the range was between 0.2 kilometer for two of the PCOs and 3.8 kilometers for one that served an area with a radius of 11.5 kilometers and contained seventeen villages. The average distance traveled to place a call for all five PCOs was 1.1 kilometers.[8]

Finally, a study in Chile showed that 72 percent of all calls made from the rural PCOs surveyed were made by people who lived in the same village, 23 percent were made by users coming from other villages (not necessarily for this sole purpose), and 3 percent were made by persons traveling between other places (mainly vehicles stopping by).[9]

Less information is available on urban PCO users. One 1983 study in Senegal examined PCO users in both rural and urban environments.[10] Interviews were solicited from 381 telephone callers and 305 telegram senders at eighteen sites throughout all eight regions of the country. The average user of public telecommunications facilities was a male between the age of thirty-one and fifty-six years, who lived within 2 kilometers of a post office and had slightly more than a primary school education (six to eight years), which is well above the national average. Only 15 percent of users were female, and only 16 percent were engaged in agriculture or cattle raising. User occupations differed markedly between the most urban PCO in a residential suburb of the capital, Dakar-Pikine, and a very rural PCO in Podor (see figure 12-1). The educational level was highest for persons using urban PCOs, averaging almost nine years, but, surprisingly, was equally high in several of the most rural locations. It was hypothesized that government officials in these areas lacked their own telephones and were forced to rely on the PCO; in larger villages and towns, these officials had their own telephones, and educational averages for PCOs were thus lower. In seven of the eighteen sites, no housewives (the vast majority of women who used public telephones) used the PCO. These sites, which occurred in both rural and urban areas, tended to have the highest number of agricultural users. However, unlike all other sites, 30 percent of the Dakar-Pikine PCO users were housewives. This high figure was attributed to the important commercial role played by women in the markets of the urban capital.

Purposes of Public Telephone Calls

Some information is also available on the purposes of telephone calls made at town or village PCOs in developing countries. The summary data presented in table 12-4 show considerable variation in the survey results among countries.[11]

Calls made to participate in activities directly related to the generation or management of economic activity ranged from a high of more than 75 percent in the 1978 Indian survey to a low of between 25 and 30 percent in Kenya and Papua New Guinea, with an average among the nine surveys of about 50 percent.[12] Mirroring this, calls to kin and friends ranged from roughly 20 percent in the 1978 Indian survey to nearly 60 percent in Costa Rica and Papua New Guinea, and over 70 percent in Kenya.

The Kenyan results, however, reflected a tendency of business-

Figure 12-1. *Occupations of PCO Telephone Users in Urban and a Rural Areas of Senegal, 1983*

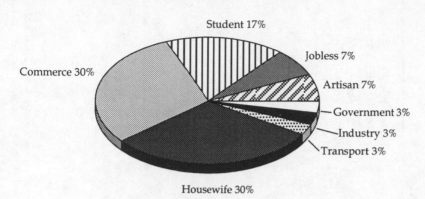

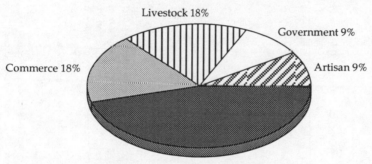

Source: Christopher W. Nordlinger, "Users of Public Telecommunications Facilities and Their Benefits in a Developing Country: A Case Study of Senegal," in ITU, *Information, Telecommunications, and Development* (Geneva, 1986).

persons, who presumably value their time relatively highly, to use modes of communication other than coin box telephones. This was primarily because the quality of service was poor; to use a coin box telephone individuals generally had to queue for up to half an hour before booking the call, only to lose it when the operator called back and found the next person using the telephone. The problem was also

Table 12-4. Reasons for Using Rural PCO Telephones in Several Countries
(percent of calls)

Use	India (1978) General use	India (1978) Most recent call	India (1981)	Chile (1978)	Korea (1979)	Costa Rica (1975)	Papua New Guinea (1976) General use	Papua New Guinea (1976) Most recent call	Remote Canada (1974)	Mexico (1978)	Rural Kenya (1980)
Call family or friends	5	23	32	28	43[a]	57[b]	75	60	35	50	76
Participate in economic activity	95	77	64	71	51[c]	38	25	31	41	50[d]	24
Business related	—	72[e]	47	55	30	28	20[e]	—	—	31[e]	—
Government administration	—	—	7	—	16	9	—	—	—	7	—
Goods and services acquisition for the home (except emergencies)	—	—	—	11	—	—	—	—	—	—	—
Emergency services	—	5	10	5	5	1	5	—	5[b]	12[b]	—
Other	—	—	4[f]	1	6	5	—	9[g]	19[h]	—	—
Total	100	100	100	100	100	100	100	100	100	100	100

— Not available.

Note: The data presented in this table are not necessarily strictly comparable since the wording of survey questions, sampling procedures, and characteristics of the target populations varied somewhat among the countries listed. Overall, however, the results roughly indicate the actual situations.

a. "Personal" calls, probably including calls to access goods and services for the home.

b. Includes calls related to both an emergency and health or a discussion of health problems.

c. In addition to the categories listed, 14 percent of calls were made to collect farming information, 10 percent to conduct other business, 6 percent to conduct military affairs, and 3 percent to report national security and police matters.

d. The Mexican data were from a survey of forty-six rural village PCOs in five states.

e. Includes calls to access goods and services for the home as well as to reach family and friends.

f. Health purposes.

g. Includes calls for emergency services.

h. Includes 11.9 percent for weather information, which is important because transport is by small airplane since the region does not have road or rail access.

Source: Appendix C. Figures in this table refer to the three-village survey; National Council of Applied Economic Research, India (1978); Economics Study Cell, Posts and Telegraphs Board, India (1981); Nicolai and Wellenius (1979); Chan-Kil (1979); Evans and Bryan (1977); Hudson and others (1979); Secretaría de Comunicaciones y Transportes Mexico (1979); and Cleevely and Walsham (1980).

aggravated because local calls were untimed, and therefore once a connection was made users tended to talk at considerable length.

In the Senegalese survey, the majority (53 percent) of calls made at the Dakar-Pikine PCO, the most urban of all surveyed, were made for financial reasons, while 20 percent were personal, but not urgent, and 27 percent were personal and urgent. No calls were considered to be related to government, perhaps because the respondents considered calls made to discuss government affairs with friends and relatives in government offices to be personal, based on their relationship to the person called.[13]

In Papua New Guinea a high proportion of the calls initiated on public access telephones were received on business or government telephones. Seventy-five percent of the respondents indicated that a business or government telephone was the destination of the PCO call they had just made, and 41 percent of the PCO calls made to family or kin were made to business or government telephones. This partly reflects the fact that in many instances friends or relatives in government were the caller's only contact who had access to a telephone. Such contacts thus delivered messages and information to other friends, relatives, or business acquaintances.

Finally, as seen in table 12-4, in several of the countries surveyed, a relatively high 5 percent of the rural PCO calls were made expressly to obtain emergency services. Such calls would presumably have a high value to the caller. The emergency call figure of 10 percent in the 1981 Indian survey may include urgent calls for purposes other than obtaining emergency services. Not surprisingly, the income of persons making emergency calls followed a different pattern from that of persons making business or social calls. Of all emergency calls in the Indian survey, 61 percent were made by callers with annual incomes less than Rs10,000 and 89 percent by callers with annual incomes less than Rs15,000.

These fragmentary results document the fact that a relatively high proportion of town or rural PCO calls relate directly to productive economic activity or to the provision of emergency services and advice. Given the probable high value of some of these activities, such calls probably account for most of the benefits derived from rural PCOs. Although the other major category of calls—those to kin and friends—presumably also affects the allocation of both human and physical resources, the most important effect of the calls is probably how they influence family and social change as well as the quality of life of rural dwellers.

Importance of PCO Calls and Alternatives for Communication

Two complementary questions relate to the purposes for which PCO telephones are used: how do PCO users perceive the importance or urgency of their calls, and how would they communicate in the absence of a telephone? Table 12-5 presents information on the perceived urgency of calls made from rural PCOs in India and Papua New Guinea. In all three surveys more than 75 percent of calls were for communication that was considered necessary immediately or on the same day. Hence, at least for these cases, the speed of the communication was highly important to public telephone users.

Table 12-6 presents information from the surveys in Papua New Guinea, Costa Rica, Chile, and India (1981) on how communication patterns would change if the PCO at which the call was made were not available. In India and Chile, most of the PCO calls would have been replaced by either traveling to use the next nearest telephone or traveling to meet the other party. The postal service, telegraph, and messengers would rarely have been chosen, apparently because they are slow and do not offer an immediate response.[14] Moreover, as was shown in table 12-4, more than 70 percent of all PCO calls in the Chilean survey and 64 percent in the 1981 Indian survey had to do with economic activities. In Costa Rica and Papua New Guinea, however, where 40 percent or less of calls were related to direct economic activity, there may have been a larger potential for substituting letters

Table 12-5. *Perceived Urgency of* PCO *Telephone Calls in India and Papua New Guinea*
(percentage falling in category)

| | India | | Papua New Guinea (1976) |
| | | | |
Urgency	(1978)[a]	(1981)[b]	
Immediate (emergency)	27.5	—	16
Same day	60.0	77.6	62
Within a few days	12.5	22.4	16
During the next week or so	—	—	6

— Not available.

a. Data are for the last call made.

b. Data are for the last call made. The two alternative responses were immediate urgency or could have waited.

Source: Same as table 12-4.

Table 12-6. *Rural* PCO *Calls That Would Be Substituted by Alternative Means of Communication if Telephones Were Not Available in Five Countries*
(percent)

Alternative	Papua New Guinea (1976)	Costa Rica (1975)	Chile (1978)	India (1981)	Peru (1985)
Travel to nearest other telephone	—	3	22	11 }	5
Travel to meet other party	30	—	21	69 }	
Letter	24	35	3 }	10	11
Telegram	18	18	3 }		34
Messenger	10	—	2	—	32
Other	—	26[a]	—	7[b]	7[c]
Would not communicate	18	18	48	3	11

— Not available.

a. Includes 22 percent that would be replaced by obtaining information from radio broadcasts. See appendix C.

b. Seven percent stated a combination of the listed alternatives.

c. Radio is also an option in Peru.

Source: See table 12-4 and Mayo and others (1987).

or telegrams for calls.[15] In Costa Rica, where information of value to farmers (wholesale market prices, weather forecasts, and so forth) is regularly broadcast by radio, the radio was considered to be an important alternative means of communication.

In Peru, the choice of an alternative to telephone service changed significantly over time for users of the system; when telephone service was first introduced, 36 percent would have sent letters, but this declined to 11 percent two years later. In contrast, the number who would have sent a telegram increased from 14 to 34 percent, and the number who would have done nothing increased from 0 to 11 percent. This phenomenon was attributed to the increasing familiarity with all kinds of telecommunications that customers gained after the introduction of telephone service. The reason for the growth in the number of persons choosing to do nothing is more obscure; either their calls were so urgent that they felt nothing else would have been an adequate substitute, or their calls were not important enough for them to seek an alternative.[16]

Table 12-6 shows that, except for India, in the absence of a PCO a substantial number of communications would not have occurred. The proportion of calls for which substitute communication would not have been undertaken was 18 percent for both Costa Rica and Papua

New Guinea and almost 50 percent for Chile. This suggests the possibility that the rural telephone generates communication and that although improved rural accessibility to telephone service results in some substitution among means of communication, it also increases the total volume of individual communications.[17]

In the Senegalese study, the use of telegrams and telephones appeared to have limited substitutability. Table 12-7 shows that only 6 percent of telegraph users would have tried telephoning as an alternative. The explanation lies in the difference between destinations. Two-thirds of all telegrams were sent for personal reasons, mostly to convey congratulations, condolences, and birth announcements, and telegrams were the chief means of sending messages abroad or to locations without telephone service. Personal visits and telephone calls were generally too expensive or simply not possible under these circumstances, so that posting a letter was often the only real option to sending a telegram.[18]

Constraints on Access to PCOs

PCO or coin box facilities in developing countries should be located where they are most accessible to the local population and where they are reasonably secure from vandalism. Hence, following the above examination of who uses public telephones and for what purposes, it is useful to examine briefly the factors that most limit or constrain the access and use of such facilities. For purposes of this discussion, constraints are grouped according to whether they relate to primarily physical, economic, or social and educational factors.[19]

Table 12-7. *Alternative Means of Communication Chosen by Telephone and Telegraph Users in Senegal, 1983*
(percent)

Alternative	Telephone users	Telegraph users
Do nothing	6	3
Use subscriber telephone	7	6
Go to another post office for the same communication	28	40
Post a letter	26	40
Send someone to the destination	8	2
Go to the destination in person	25	9

Source: ITU (1986).

Physical Constraints

The major constraint on public telephone use in developing countries is clearly physical: the location of the telephone relative to that of the population. The uneven distribution of telephone service between and within urban and rural areas has been noted throughout this book, which also documented that, in general, the farther a person is from a telephone, the less frequently he or she will use it, if at all. Related to this, public telephones tend to be used less if the quality of voice transmission is variable or poor, and if the waiting lines or waiting times are long for booked long-distance calls.[20]

Other physical constraints include the placement of the telephone so that it is not readily or continuously available. In many rural towns in developing countries, the PCO is located in a post office, cafe, store, or other such place that is open only during daytime business hours. This obviously reduces some of the potential emergency call benefits of the telephone. In at least one South Asian and one North African country, rural PCOs were used several hours each afternoon to deliver and receive the day's "telegram messages," thus precluding their regular use for telephone calls. Also, in several areas in Korea, many rural PCOs were placed in the residences of village chiefs, who were often out of town. This substantially limited the public's ability to use the PCOs, and uncertainty about whether the telephone was available discouraged persons from neighboring villages from placing calls.[21]

Finally, an additional physical barrier to public telephone use is the difficulty of contacting individuals without subscriber telephones. Communicating with someone located at a place with only public telephones requires a means to relay the message or summon the party to the telephone. Such messenger facilities are often inadequate. In the Korean villages surveyed, for example, incoming calls to rural PCOs were held while the party was sought and brought to the telephone, and this process was considered to be too slow. It was finally recommended that a public address system be installed to summon people to the telephone. In Costa Rica and several other countries, small boys would hang around the rural PCO either to run and summon or to take a message to the person being called. In areas of some countries, however, the telephone message is simply passed on by word of mouth.

Economic Constraints

A second constraint is the cost of making the call. Part of the cost is incurred in getting to a telephone (opportunity cost of time and direct expenditure) and, as such, is indirectly related to the physical access of the population to telephones.[22] The other cost, of course, is the call charge, and since most calls made at rural or town PCOs are billed as long distance, the charges are in many instances significant for the average rural dweller.

It can be argued, however, that the call charge is probably a factor holding down social calls to family, friends, and kin more than business or emergency calls, since, at least for subscriber telephones, the demand for business calls is generally thought to be more price inelastic than the demand for residential calls.[23] Evidence supporting such a contention for calls made at PCOs in developing countries is, however, limited.

Social and Educational Constraints

Social and educational barriers to the use of public telephones are widespread in developing countries. Among the more prominent barriers are insufficient experience with telephones and lack of adequate information about how to use them with confidence. In Papua New Guinea, for example, individuals needed to be taught what the telephone could do for them, where it was, and how to use it; this seemed to be important even in urban areas. Predictably, users tended to be mainly younger and better educated individuals, who had been exposed to ways of life in which the telephone plays an important role, and persons with white-collar occupations, where telephones are normally used as part of work.

In general, women tend to use PCOs significantly less often than men in most developing countries. In Egypt, no women were interviewed; in Senegal, the balance between male and female users was 85 to 15 percent; and in Peru, it was 68 to 32 percent. This reflects sociocultural restrictions as well as the lower average level of education and employment common among women in most countries.

In Papua New Guinea, having to ask someone for assistance or permission to use a telephone inhibited use. For example, many of the radiotelephones serving remote rural areas were kept in the residences of government officials, schoolteachers, missionaries, or planters. Although everyone had a right to use these telephones or to send and receive messages over them, very few villagers did so; villagers would sometimes travel half a day to use a public telephone, although a radiotelephone station was located within 3 kilometers of their village. Apparently this was independent of the personality of the radio operator: other images, social stigmas, procedural problems, and embarrassments had to be overcome. This problem might be resolved if the radiotelephone were located in a central place and not in a mission or a school, perhaps somewhere that villagers helped to provide and that would be seen as "theirs."[24]

A related problem is the intervention of telephone operators. The involvement of an operator constrained telephone use in Papua New Guinea in two ways: users who only spoke a local tribal language had difficulty communicating with the operators establishing their call, when necessary, and, where this was not the case, the fact that the operator could listen in on the conversation inhibited use. This latter factor is a significant constraint throughout the developing world.

Finally, the practice of metering calls and billing the individual responsible for the telephone assumes that the person can control how others use the telephone. In societies in which considerable emphasis is placed on communal sharing, subscribers or vendors have difficulty restricting use by others or find doing so unacceptable. As a result, some choose not to obtain a telephone. Such problems have been observed in both Papua New Guinea and several countries in West Africa.

Conclusions about Telephone Access and Use

The information and analysis reviewed in this and the previous two chapters clearly show that telephones are an important means for facilitating economic activity. This was found to be true for business, residential, public coin box, and PCO telephone service. In developing countries, on a national basis, within urban areas, and even within small towns and villages, the business community and, to a lesser extent, government dominate access to telephone service.[25] Even when a large proportion of telephone lines are connected to residences, the evidence suggests that, overall, telecommunications services are mainly used to facilitate and coordinate directly productive or distributive activities of the economy. Telephones have a critical role in carrying out tertiary (or information) sector activities—government administration, finance, trade, commerce, services, transport—and are intensively used by managers and white-collar organizers in other sectors. Such groups tend to gain the first access to telephone service and tend to use it most frequently. Agriculture, at the other extreme, although it is one of the least-intensive users of telephones, appears to make contact with a relatively wider variety of economic activities than do most other sectors.

Telephones tend to be used primarily by individuals who function above the subsistence level, and they may be relatively widely used in regions or countries in which a significant proportion of the population has at least some formal education and is engaged at least to a minimal extent in market-related activities above the subsistence level.

The limited empirical evidence on the use of business and residential subscriber telephones suggests that it would be an oversimplification in developing countries to associate investment in business telephones wholly with "productive" uses and investment in residential telephones solely with "unproductive" uses. Although in most instances business telephones are overwhelmingly used for work-related purposes, they also meet some of the personal and family needs of workers.

Conversely, although residential telephones in most countries examined are used more than 50 percent of the time to communicate with family members, kin, and friends, they are also used extensively to participate in the economy by performing work, gaining access to goods and services, and pursuing the organization and creation of economically relevant activities. They also allow the relatively higher-income and skilled people in rural areas to keep in touch with friends and family and to update information associated with creating and maintaining jobs. Finally, the use of both subscriber and PCO telephones for emergency purposes seems to be of universal importance in developing countries, perhaps even more so than in industrial countries.

Although population groups functioning at a subsistence level do not make significant direct use of telephone facilities, little evidence supports the view that telephones are in any sense superfluous consumer goods for the high-income sectors of the population. Hence, opposing investment in telecommunications for distributional reasons is at best shortsighted and could, in the long run, retard the extent to which productivity could be increased through division of labor and could affect the level, distribution and spatial organization of economic activity.

Assuring increased access to PCO and coin box telephones is, of course, an investment decision that can generally be financed and implemented rapidly by most telecommunications administrations in developing countries.[26] Assuring that potential high-benefit users receive first access to subscriber telephones and that the expanded PCO telephone program is adequately financed is at least partly a matter of deriving proper pricing policies. Hence, in the following three chapters alternatives for pricing policy are reviewed, and a framework for telecommunications tariff policies is suggested.

Notes

1. See, for example, Saunders and Warford (1979) and Saunders and Dickenson (1979).

2. Both the subscriber and the PCO surveys are presented in Economics Study Cell (1981).

3. Individuals still in school were not included in the population percentage categories. Also in the survey in Papua New Guinea, 75 percent of the PCO users were men, and users in the urban population over thirty-five years old were found to use public telephones relatively little.

4. A more complete description of the results of this study is presented in appendix C.

5. Mayo and others (1987).

6. Long trips, however, tended to have other objectives in addition to placing a call.

7. Telephone density in Papua New Guinea and Costa Rica was approximately 1.4 and 8.9, in 1976 and 1975, respectively.

8. National Council of Applied Economic Research, India (1978).

9. Nicolai and Wellenius (1979).

10. Nordlinger (1986).

11. One of the problems that arise in analyzing the various survey data on PCO use is the imprecise scope of categories used. However, the rough equivalencies used in table 12-4 are thought to represent reasonably well the situations as they were reported.

12. Answers relating to the purpose of the most recent call were considered to be more reliable than those requiring individuals to recollect general use over a longer period.

13. Nordlinger (1986).

14. Preferences for alternative means of making the calls varied considerably from one village or country to another, presumably reflecting, among other things, the relative costs of alternative means of communication and purposes of the calls.

15. Presumably, if another telephone had been available in the vicinity, it might have been a more common alternative mode of communication. In the United States when a switching center fire left 90,000 telephone customers without service for twenty-three days, relatively few residential subscribers increased their use of other modes of communication: 48 percent used emergency street phones set up by the telephone company, 33 percent (virtually everyone with a daily occupation) made calls from work, 10 percent wrote more letters, and less than 2 percent communicated by telegram. See Wurtzel and Turner (1977), pp. 246–61.

16. Mayo and others (1987).

17. In contrast, the Bangkok results indicated that only about 2 percent of families would not replace telephone calls made from the home by other means of communication should their residential telephone not be available. This suggests that the telephone may generate far less communication in large cities, which have fairly widespread alternative means of communication.

18. Nordlinger (1986).

19. These three categories are not necessarily mutually exclusive, and there are, of course, other relevant ways to categorize constraints on telephone usage. Bryan and Evans have, for example, suggested eleven constraints, some of them related to overcoming a "threshold" to use and some to overcoming "barriers to increasing frequency of use." See Bryan and Evans (1979), pp. 172–87.

20. In small towns and cities in many developing countries waiting several hours for a booked long-distance call to be connected is not uncommon.

21. Chan-Kil (1979).

22. Related to this, Bryan and Evans have noted that for relatively lower-income persons desiring to subscribe to telephone service, the lack of a suitable permanent structure for a home or shop can be a constraint to gaining a subscriber telephone. Bryan and Evans (1979) p. 174.

23. Littlechild (1979), p. 37; Meyer and others (1980), app. C; Alleman (1977), chap. 4; and Taylor (1980), chaps. 3 and 4.

24. Bryan and Evans (1979).

25. As shown in chapter 10, in small more-rural places with very limited telephone penetration, government or quasi-government telephone access sometimes exceeds that of the business community.

26. Some proponents of rural telephony argue that governments should subsidize rural telecommunications from general government revenues. Although in some instances rural facilities may need financial subsidies at least during their first few years of operation, it is not clear that these subsidies can or should come from general government tax revenues, where they must compete with sectors that do not earn revenue, such as education, health, nutrition, roads, and so forth.

Part V
Telecommunications Tariff Policy

Chapter 13

An Introduction
to Tariff Policy

GENERAL AGREEMENT EXISTS in the literature on public enterprise tariff policy that, in the absence of good reasons to the contrary, prices charged should reflect the costs of providing services. There is, however, some disagreement among accountants, economists, lawyers, politicians, and so forth as to how those costs should be defined. Also, given the level and structure of telecommunications tariffs in both industrial and developing countries, in many instances there are apparently "good reasons to the contrary"; prices charged for individual services generally do not reflect historical average, current average, or incremental costs.

Objectives of Public Enterprise Tariff Policy

This situation exists partly because telecommunications tariffs in practice attempt to satisfy different simultaneous objectives, which are often inconsistent with one another. These objectives can be grouped into three broad categories: those that relate to attaining financial goals, primarily aimed at assuring the financial viability of the telecommunications operating entity but also possibly at contributing to general government revenues; those that promote an efficient allocation of a country's resources; and those that promote an equitable allocation of resources.[1] These objectives must be analyzed in light of a complex set of constraints facing policymakers; political and social acceptability and the administrative feasibility of establishing any particular tariff policy framework are among the primary considerations.[2]

255

Financial Viability

Subject to foreign exchange scarcity and, on occasion, unrealistically low tariffs, most telecommunications operating entities in developing countries can easily generate sufficient resources from users of the services to cover operation and maintenance costs as well as debt and interest payments. Likewise, they can usually generate a sufficient return on assets to attract needed capital and to cover a reasonable proportion of the future costs of expanding the system from internal cash generation. Subsidizing the sector from general government tax revenues is thus recognized as being inappropriate for several reasons.

First, the sector as a whole does not require government financial subsidies.[3] As outlined in chapter 1, in developing countries a large gap typically separates the supply and demand for telephone services: the number of potential subscribers on waiting lists sometimes exceeds the number of telephones in service, and the average waiting time for a new connection can be as long as ten years. Likewise, business-hour traffic generated by telephone subscribers usually exceeds the network's capacity. Given this excess demand, the market power (degree of monopoly) of the operating entity, and the fact that the short-run demand for telecommunications services is relatively price inelastic, particularly in urban areas, most telecommunications entities find it relatively easy to set their overall average charges at levels that cover their current costs and a reasonable portion (typically from 40 to 80 percent) of their expansion costs. In many instances they also generate substantial tax, interest, and dividend revenue for government.

A second reason for not subsidizing telecommunications services relates to the inefficiencies involved in interfering with consumer choice by transferring, through general taxation, resources from the private sector to the public sector.[4] It can be argued that the costs of the distortions in resource allocation resulting from such a subsidy are likely to exceed the benefits.

A third argument against government financial subsidies relates to management. A case can be made that, if the management of telecommunications entities perceive that government subsidies will be forthcoming whenever costs exceed revenues, much of the incentive for cost-consciousness and efficient management is lost. Throughout the developing world, efficient management of public utilities is being undermined in this way (problems relating to management objectives and efficiency were outlined in chapter 3).

Finally, from an administrative point of view, a stable and reliable source of revenue facilitates least-cost investment planning by telecommunications entities and supports their prudent financial management. Telephone tariffs can provide such a reliable source of revenue; in many instances, government tax revenues prove to be less reliable. Numerous needs compete for tax revenues in developing countries, and financial subsidies for telephone service tend to be one of the easiest expenditures to cut during a national fiscal squeeze.

Efficient Allocation of Resources

Price can be used to influence consumers so that the resources used to provide telecommunications services are not wasted. The price that consumers willingly pay for a good or service can signal at least the minimum value of that good or service to the consumer. If the price paid for telecommunications services exceeds the cost of production, this is a signal to expand output. If new output can only be sold at a price below the cost of producing it, this is a signal to reduce production.

To promote an efficient allocation of resources in operating a public utility sector dominated by what are essentially monopoly organizations, as is the case of telecommunications in most developing countries, traditional welfare economics suggests that price should be set equal to the incremental or marginal cost of expanding output, or, if capacity is fully used, at a level sufficient to "clear the market."[5] If, on average, the price of a telephone call is set equal to the incremental cost of providing an additional telephone call, then consumers will indicate whether or not the value to them of an additional call is worth the cost. In effect, the responsibility for determining whether or not the telephone system (or some portion of it) should be expanded is therefore shifted from central planners or company officials to the ultimate consumers of the service. Hence, if prices charged for different telecommunications services reflect the incremental costs of providing those services, then signals are given about which services should be developed most quickly.

In the short run, if supply cannot be expanded rapidly enough to meet demand, price can be increased to ration the supply of services (telephone lines or calls) to persons who value them the highest and, therefore, presumably derive the largest benefits from their use. Several problems are associated, however, with relying entirely on price to

promote efficient resource allocation. In particular, for monopoly telecommunications organizations, the tariff analyst must deal with conceptual problems of estimation, problems of demand forecasting and
measurement, market distortions, the presence of externalities, and
declining unit costs.[6] These issues are discussed in chapter 15, which
reviews some of the specific problems of estimating the marginal costs
of providing telecommunications services.

Equitable Allocation of Resources

Most developing countries exhibit highly skewed income distribu-
tions and large differences in regional income.[7] Hence, some potential
telephone users might not be able to pay telephone tariffs that cover
the full cost of telephone service. This is particularly true in smaller
towns or more remote low-traffic areas where the cost of providing
service is generally higher than in larger urban areas, where incomes
are low, and where the initial volume of traffic may not be large.[8]

As a result, providing selected telecommunications services at
prices below cost may sometimes be desirable in order to spread the
benefits of rapid communication widely and reduce the number of individuals excluded from at least minimum access because of low incomes. Even in higher-income areas such as North America, universal
telephone service has been a primary goal of pricing policy for many
years, and cross-subsidies between local and long-distance service and
some urban and rural areas have been common.[9] A certain degree of
cross-subsidies are also the norm in most other countries in both the
industrial and developing world.[10]

Traditional Approaches to Telecommunications Pricing

Telecommunications services include several "outputs" that are
usually charged for separately. Therefore, telecommunications tariff
structures have typically included all or most of the following
elements:

a. A nonreimbursable fee for initial connection to the network
b. A monthly, bimonthly, or quarterly rental, access, or subscription fee for terminal equipment and continued connection to
   the network, sometimes including a limited or unlimited number
   of local calls[11]

c. Charges for local calls not included in the rental (increasingly, the length of local calls and time of day, as well as the number of calls, are being introduced into the tariffs)[12]
d. Charges for long-distance and international calls, varying with duration, distance, time of day, and other factors, such as the use of operator assistance rather than subscriber dialing, person-to-person, and so forth
e. Miscellaneous charges for business telephones, PCOs, private branch exchanges, data communication, telex, and so forth.

In several developing countries, subscribers must also contribute capital that gives them an equity share in the telecommunications entity. Sometimes this is treated as debt, and subscribers are reimbursed when service is terminated, with or without adjustments for inflation or interest. Such arrangements are discussed further in chapter 14.

Where a regulatory body exists, the traditional approach to pricing is to establish an overall rate of return that yields an adequate flow of funds to cover the entity's operation and maintenance and a fair return on capital. If, based on an examination of historical costs, the total revenue earned by an entity generates a reasonable rate of return, when compared with the opportunity cost of capital, the regulatory authorities can conclude that the overall level of charges is reasonable. A second concern of regulators is to ensure that a monopoly telephone entity does not abuse its market power. Abuse in pricing practices can occur when an entity has excessively high prices or gives preference to some users either by charging other customers excessively or by cross-subsidizing its competitive services. A third concern of regulators is to provide incentives for the telecommunications entity to extend access to telephone service. Within the framework of achieving an overall revenue requirement, both regulators and telephone operating entities have traditionally favored holding down monthly subscription charges (network access charges) in order to make basic service affordable and thus achieve the maximum penetration feasible among users with lower incomes.[13]

Four key principles are traditionally taken into account in setting rates: (a) rate averaging across a company or a system, (b) value of service, (c) costs, and (d) usage. Rate averaging across a company means that all customers pay the same price for the same class of service throughout the company's operating territory. An example is when, within a particular exchange, all subscribers pay the same local exchange access rate (monthly rental charge), regardless of their use or distance from the switching center. The somewhat ambiguous value-of-service principle (defined as such by common usage) reflects the idea that a prospective buyer will pay a price that is related to the value derived from the service and that telephone services are more valuable to some classes of customers than to others. Thus, basic local flat-rate charges are higher at locations with more subscribers in the local calling area than in those with fewer, and rates for business subscribers are often higher than those for residential subscribers. Costs are taken into account by charging more for higher-cost services. Thus, operator-assisted long-distance telephone calls are priced higher than direct-dialed calls. Usage is a rating factor that can be associated with both value and cost. For example, subscribers attach value to use and are willing to pay more for it, and increased use may, in turn, increase costs. Charges may be reduced for Sunday, weekend, or night service because such discounts contribute to improved network use at a low incremental cost to the system and thereby help reduce peak-traffic demand.

Type of User

Telephone subscribers are sometimes categorized on the basis of the purpose for which their telephones are most likely to be used. Hence business and government users may be placed in one tariff category, whereas residential users are put into another.[14] Sometimes separate categories are defined for more specific groups of nonresidential subscribers, such as businesses, industries, professionals, nonprofit organizations, and government.

There are two reasons for the basic business-residential dichotomy, one based on demand factors, the other on cost of supply. First, telephones are generally assumed to be more important to business firms than to private residences because they are, for competitive reasons, concerned with finding the least-cost means of communicating rapidly. Business firms also tend to produce a much greater volume of communications. Therefore, the value of telephone service is assumed to be greater for business firms than for private residences, that is, the demand for business telephones is relatively more inelastic to price changes, and in any case such charges are tax deductible in many countries.[15] If this is so, businesses can be charged somewhat higher fixed monthly fees for their subscription without greatly distorting the business demand for telephones, and the number of residential connections can be increased through lower monthly fees.[16] Hence, it is argued that new residential subscribers benefit because they have a telephone, and all existing subscribers, including business firms, benefit because more subscribers are connected to the network and can be called.

The second reason for the business-residential dichotomy is that, although business and government calls are made primarily during peak business hours, many residential calls occur during nonbusiness hours when they do not burden the overall capacity of the system to handle traffic, if residential lines are intermixed with business lines in the same exchange area. Thus, since the marginal cost of an off-peak residential telephone call is in some instances relatively low, it is argued that residential telephones should be encouraged through lower monthly fixed charges.[17]

In industrial countries, a third argument in favor of lower residential telephone rentals or connection charges is sometimes put forth. This argument suggests that for reasons of equity residential users (families) should be subsidized by business firms, which can better afford to pay more for telephones. This argument is essentially incorrect, since business firms will pass on at least part of the increase in telephone charges to the final consumers of their goods and services, and it should not be given much weight without a detailed analysis of who ultimately bears the burden of higher charges to business firms. To the extent that upper- and middle-income groups have a greater proportion of residential telephones than do lower-income groups, and to the extent that business firms sell goods and services to lower-income groups who do not have residential telephones, it could be argued that when business service subsidizes residential service, lower-income groups subsidize the residential telephones of groups with higher incomes. This may be a more powerful argument in favor of relatively higher charges for residential telephones in poorer countries, where residential services tend to be in the hands of higher-income groups. An important alternative to subsidizing the access of residential users to telephone service across the board is to target subsidies toward either individuals or regions.[18]

Finally, special interests—doctors, fire fighters, politicians, and so forth—frequently claim, usually without much supporting analysis, that social concerns qualify them for favorable tariff treatment. International organizations have even suggested charging different tariffs for different categories of users. For example, in the 1970s UNESCO petitioned telecommunications operating entities to set low concessional tariffs for journalists transmitting news. One argument against this kind of subsidy is that such costs should be borne by governments, not telecommunications subscribers and users.[19]

Although subsidies for special-interest groups may be neither a desirable nor a practical option, the rapid spread of new communications technologies and services complicates the issue further. Very high capital investments are required to introduce new communications technologies, and, where tariffs are related to costs, small and noncommercial users may not be able to use the new services, particularly in a developing country.[20] Concerns have been raised that these investments may only benefit the relatively few individuals and large businesses that have sufficient means to use the new technologies. Some authors argue that, over the long run, potential noncommercial users of new services (including such services as data banks and satellite distribution networks) may risk being marginalized through their inability to participate in increasingly important developments in communications technology.[21]

Size of the Local Service Area

In many countries the size of the fixed monthly access fee that must be paid to retain telephone service varies according to the size of the local telephone service area, with higher fees being charged in larger service areas. In fact, distinguishing telephone tariffs by size of service area is recommended by CCITT.[22] The rationale for such a pattern is once again based partly on the demand-oriented concept of value of service and partly on the cost of supplying service.[23]

From the value-of-service perspective a telephone may be more valuable to any one subscriber if more people can be called without incurring extra distance-related charges. Hence, subscribers and potential subscribers in larger local service areas are assumed to find telephone service to be relatively more valuable than those in smaller areas. Also, it is argued that large settlements have more types of communication events for which local telephone service is an effective medium, and telephones tend to be a lower-cost substitute for alternative means of communication (travel to order supplies, travel to arrange meetings, and use of messengers).

From the point of view of telecommunications service entities, which have traditionally generated their highest profits and contributions to fixed costs from long-distance service, any enlargement in the inexpensive local calling area would reduce long-distance revenues. These would normally be replaced by compensatory increases in monthly charges for local network access.

A case for increasing monthly access charges as the local service area increases in size is sometimes made from the point of view of the cost of providing service, although empirical evidence supporting such a contention is limited. As the size of the service area increases, so do the requirements for switching capacity and connections between exchanges. Also, expanding service in heavily built-up urban areas can be costly in terms of land, buildings, duct construction, and work on outside plant. As outlined in chapter 2, however, growth also reduces the average cost per subscriber added, because of economies of scale and improved network use as well as the opportunity for technological innovation. Also, small service areas are generally located in remote rural areas, where low population density increases fixed investment costs for each customer served. The net result of these opposing trends varies from one situation to another.[24]

Distance Called

Traditionally it has been accepted that charges for calls should increase with distance.[25] Figure 13-1 outlines the typical relation between distance and call charges and shows that although charges do increase with distance, long-distance charges tend to increase at a decreasing rate.[26] The principle that telephone charges should increase as the distance called increases is once again based on both cost of service and value of service.

With regard to value of service, the number of alternatives to telecommunications for speedy and convenient communication tends to decline as distance increases, and the cost of such alternatives increases. Nevertheless, there is evidence that demand for telephone calls is more elastic as distance increases.[27]

On the cost side, a long-distance telephone call is generally more costly than one within a local exchange area simply because it includes local service area costs (on both the sending and receiving ends) plus the costs of long-distance transmission and switching. In the early days of telecommunications, when long-distance transmission was mainly through open wire lines, the associated costs were more nearly proportional to distance. With more modern technology, however, incremental (and average) long-distance costs tend to flatten out as distance increases. This is particularly so in international services using communication satellites, where cost is essentially independent of distance. Costs are also a function of traffic volume. Shorter-distance calls over low-traffic routes can be much more costly on a per-call basis than longer-distance calls over high-traffic routes,

Figure 13-1. *General Relation between Price and Cost for Local and Long-Distance Traffic*

Average cost and price per unit call

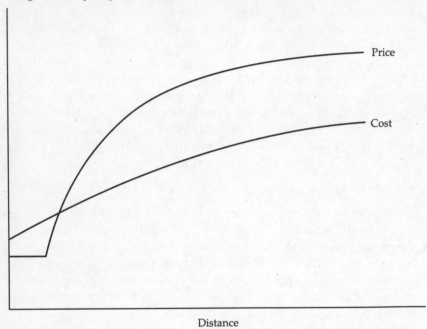

Distance

particularly as microwave and fiber-optic transmission is implemented in an increasing number of situations.

Even though long-distance tariffs usually increase at a decreasing rate with distance, tariffs typically overshoot substantially the cost of supplying high-volume long-distance service.[28] As illustrated in figure 13-1, the tariff/distance curve traditionally levels out at a price that is well above cost, both for domestic long-distance and international service.[29] This situation began to change during the 1980s in the United States, as competition in the provision of long-distance services increased. Significant reductions in both domestic and international long-distance charges have occurred in several other countries as well, including Canada, Japan, Mexico, and the United Kingdom.

Time at Which the Call Is Made

Since the use of telephones is not constant throughout a twenty-four-hour period, different charges for identical calls are frequently levied based solely on when the call is made.[30] In some countries this applies both to local and long-distance calls, although more often only long-distance calls are metered. Although not universal among developing countries, such peak pricing of long-distance calls has become widespread in Western Europe, North America, Japan, and Australia, and peak pricing of local calls is becoming increasingly accepted in many countries.[31] The adoption of time-of-day call metering can have a moderate effect on the distribution of traffic at different times of day.[32]

Once again, arguments based both on cost and on demand factors support having higher call charges during daytime business and early evening hours than during nighttime and weekend periods. From the demand perspective, it is generally argued that business and government demand for telephone service is relatively inelastic during daytime business hours, since the ability to complete a call is more valuable during these peak hours than during evening and nighttime hours when most offices are closed.

From a cost-of-service point of view, it is argued that consumers who make increased demands on the telephone system when it is operating at or near capacity should pay the incremental costs that must be incurred when the system needs to be expanded. Also, with peak pricing, there is some possibility of reducing the peak, shifting or smoothing the traffic load, and hence saving resources by reducing or postponing future expansion. Given this, calls that are made at off-peak times and that do not strain the system's capacity to handle traffic should then be charged only the incremental operation and maintenance costs they impose on the system and should not be required to bear the costs of system expansion.

The problems usually encountered with time-of-day pricing relate to technical metering difficulties and with improper choice of peak periods. For example, if a short peak period is selected, the call charge would have to be relatively high to recover capacity costs from the smaller number of calls. And, having a high call charge and a short peak period increases the likelihood that the peak will shift to adjacent lower call-charge periods. If the peak period is lengthy, however, subperiods exist when demand is low but charges are high, and there is thus no incentive to shift from high- to low-demand subperiods.[33]

Finally, to avoid confusing the subscriber too much, it has been suggested that no more than three price levels should be in effect during a twenty-four-hour period.[34] Having three or four call-charge periods creates the problem of educating and then periodically reminding telephone users which prices are in effect during which periods. Such an argument is, however, not totally convincing. If the size of the price differences among the periods were at all substantive, consumers would find it in their best interests to remember and use them to their best advantage.

The costs of introducing time-of-day metering into different types of existing local exchanges vary considerably. The costs tend to be relatively high when metering is being added to existing step-by-step and rotary exchanges (where they still exist) but is considerably less costly when being added to crossbar facilities. With modern digital electronic exchanges, the cost is very low. For example, the incremental capital cost required to time local calling on older nondigital exchanges can range from as little as $2 to more than $50 per line, depending on the size of the local exchange, the type of switching equipment, and the extent of record-keeping required.[35] During the mid-1970s the New York Telephone Company installed metering equipment in its large crossbar exchanges for $15 per line.[36] Record-keeping and billing costs for time-of-day metering for local calls are estimated to range between $0.001 and $0.002 per call. In general, the larger the exchange and the more modern the equipment, the lower the metering cost.

Notes

1. The objectives of policies governing public utility pricing may be categorized in many ways. Meyer and others list eight categories for telecommunications: universal service, static efficiency in resource allocation, equity for different kinds of users and services, financial self-sufficiency, preemption of uneconomic entry, consistency with expected technological change, administrative simplicity, and historical continuity. Meyer and others (1980) pp. 75–81. Baumol (1968), pp. 108–23, lists seven.

2. For a thorough examination of innovation and the relation between the theory and actual implementation of tariffs, see Mitchell and Vogelsang (1991), chap. 12.

3. This does not necessarily imply that there should never be any financial cross-subsidies within the sector. Also, different forms of subsidies, for example, per-unit subsidies, general deficit financing, and so forth, have different effects.

4. The same argument would apply under either private or public ownership of the telecommunications entity.

5. For a brief discussion of welfare economics and of marginal cost pricing in Britain and North America, respectively, see Littlechild (1979), chaps. 9 and 14; Mitchell and Vogelsang (1991) chaps. 3 and 4.

6. Saunders and Warford (1977).

7. World Bank (1990)

8. In rural areas, fewer people can generally be reached without incurring toll charges than in urban areas; hence the value of telephone service in those areas might be less, thus justifying lower charges. The value of a telephone depends on many things, however, such as remoteness, emergency needs, cost of alternative means of communication, loneliness, and level of economic activity.

9. U.S. Congress. *The Communications Act of 1934*, as amended, 50 Stat. 189, sec. 1.

10. For examples from Western Europe, see Mitchell (1979). Additional information on tariff structure issues can be found in OECD (1990) and Mansell (1990).

11. Since the regulatory authorities in many countries have decided that competition in the supply of equipment on the customer's premises better serves the public interest, many subscribers can purchase telephone instruments from sources other than telephone companies. Hence, the monthly charge can be separated into at least two parts: one part for a subscription to the network, which everyone pays, and a second part for rental of telephones, which subscribers who purchase their own equipment do not pay.

12. Historically, local calls have not been metered in North America. The current trend, however, is toward charging for individual local calls, resulting in the progressive introduction of local measured service. In general, the metering of, and charging for, calls by number and duration is referred to as usage-sensitive pricing in North America. See Schultz and Barnes (1984); Alleman and Emmerson (1989). Park and Mitchell (1987) argued that changing the technology for providing local service could reduce the cost of local usage sufficiently to make local measured service inefficient.

13. Kahn (1970), pp. 63–65.

14. This is generally the practice in North America and England, although less so in the remainder of Western Europe. Of the eighteen countries whose tariffs were surveyed in chapter 2, eight differentiated their connection fee and monthly rentals by category of subscriber.

15. The few existing empirical estimates of price elasticity of demand for telephones tend to show that business demand for local and long-distance calls may be more price inelastic than residential demand. However, the evidence on connection charges and monthly rental or subscription fees is somewhat mixed. See Taylor (1980), pp. 168–74; Meyer and others (1980) armex C; Alleman (1977) chap. 4; Littlechild (1979) pp. 34–38; and Wilkinson (1983).

16. When the monthly rental or subscription charge includes an unlimited number of local calls, the rental for business telephones may be higher, although the average price paid per call is lower, since businesses tend to use their telephones more intensely.

17. It may, however, be more adequate to reflect such time-of-day, traffic-related effects in the price for traffic (call charges) than to manipulate rentals or subscription charges on this account. Differentiating rental charges among classes of subscribers does not create incentives to shift calls at the margin from peak to off-peak hours.

18. The argument offered was that telecommunications should not charge the mass media on the basis of a so-called return on investment, since the exchange of media information is not commercially motivated but is directly related to the social and economic development of a country. Presumably, this argument was not meant to apply to the commercially oriented press of North America, Western Europe, Japan, and Australia. It also may not reflect the fact that in some developing countries, newspapers, television, and radio broadcasting are private business enterprises.

19. Helling (1981) pp. 27–29.

20. For instance, in 1986, one application planned for Brazil's domestic satellite was the national transmission of educational television programs for the Ministry of Education's FUNTEVE network via a dedicated transponder. However, switching FUNTEVE program distribu-

tion from existing means (terrestrial microwave, at a late-night discount) to the satellite would have increased the network's transmission costs by a factor of ten. Under these circumstances, use of the satellite for educational broadcasting was unrealistic without substantial tariff concessions. See Nettleton and McAnany (1989). Even in industrial countries, small users may find new technology services too expensive to use. For example, a basic on-line data base in the United States, Mead Data Central's Lexis Financial Information Service, costs its subscribers $39 an hour for access, plus $6 to $35 per individual search, plus a monthly service charge of $50. Only a minority of very well-informed small businesses can be efficient enough in their searches to avoid incurring expenses higher than the profitability derived from the information they retrieve. See Race (1990).

21. CCITT (1965).

22. In the United States, as in several other countries, pricing according to the size of service area has not been uniform. Although service areas are fairly consistent in size within states, local tariffs seem to be almost random among states. Meyer found that "in general nonuniformities in local telephone rates can be discovered almost at will, and it is difficult to explain many of the disparities by cost differences or density differences or other characteristics." See Meyer and others (1980) p. 86.

23. In the mid-1960s, a study in Texas showed that the average cost per main telephone decreased as the number of main telephones in the exchange area increased. At some size between 50,000 and 100,000 main telephones, however, unit costs leveled off and started to rise because of increasing loop lengths and more complex switching problems, which resulted in the need for more exchange and trunking equipment. See Bowers and Lovejoy (1965). See also the discussion of economies of scale in chapter 2 of this volume.

24. CCITT (1965). Mitchell and Vogelsang found that this is still generally true in the United States for interstate calls, even though competition and sharply reduced costs for transmission technology have made prices less sensitive to distance and have shrunk differences between peak and off-peak rates. However, the introduction of optional tariffs means that in some cases, simplified, uniform per-minute rates that do not vary with distance are being offered during night and weekend hours, particularly for residential and small business users. For an example of this kind of tariff, see the discussion of AT&T's experimental 1983 Optional Calling Plan in chapter 8 of Mitchell and Vogelsang (1991).

25. In the United States, which has a diverse and decentralized regulatory situation, large anomalies exist in the relation between distance called and call charge. Generally, a 100-kilometer call among states is priced the same irrespective of the location of the calling and called points. However, a 100-kilometer call within one state may cost a subscriber more or less than a 100-kilometer call in a neighboring state and more or less than a 100-kilometer call between states. Another factor is the purely geographic basis used to determine long-distance zones; customers who live near the boundaries of a particular area code often have to pay long-distance charges to call nearby locations in the neighboring area code.

26. No doubt other reasons contribute to this increasing inelasticity of demand at longer distances; for example, a relatively high proportion of long-distance calls are made by business and government users. See also Ministry of Supply and Services, Canada (1988).

27. Telex charges are often uniform throughout the country, reflecting what is sometimes a star-shaped network in which distance is a less meaningful factor of cost.

28. For a comparison of trunk call revenues and costs in Sweden, see Mitchell (1979) pp. 4–7.

29. Many countries in which the supply of telephone service is roughly equal to demand tend to have three distinct periods of peak use: between 10 a.m. and 11 a.m., around 3 p.m., and in the early evening. In many developing countries in which demand exceeds supply, the daytime peak begins much earlier, ends later, and does not dip as much around mid-day.

30. Mitchell (1979) and Mitchell and Vogelsang (1991).

31. In-depth analyses of the application of peak pricing in telecommunications have only become available in the past ten years, even for industrial countries; see Park and Mitchell (1987) and Sherman (1989). The importance of peak pricing has been long recognized in other sectors, however, notably electricity. See, for example, Boiteux (1949) and as translated (1960). The classic article on peak-load pricing is Steiner (1957).

32. In some instances large multiexchange areas could also experience the complication of having different exchange areas experience call traffic peaks at different times. For a brief discussion of this point, see Selwyn and Borton (1980).

33. See Mitchell (1979).

34. Alleman (1977) pp. 118–27; Mitchell (1976), pp. 26–27; and Mitchell (1978).

35. Mitchell (1976 and 1978).

36. Microprocessor-based electronic devices, which collect and send telephone billing information from different kinds of switching facilities to central offices, have lowered the cost of installing metering equipment, record-keeping, and billing for older switching equipment. See, for example, Rettig (1981). However, adding this kind of equipment to very old facilities may be shortsighted; doing so should be compared to the option of using the capital expenditure to obtain all the advantages of newer switching equipment, which already includes this feature.

Chapter 14

Tariff Policy for Economic Efficiency

PRICES THAT ENHANCE ECONOMIC EFFICIENCY should be a major consideration in establishing a tariff policy for any country. This is particularly
true in the developing world, where improving the standard of living
is an important concern and resources often are very scarce. An economically efficient allocation of resources maximizes the value of output in an economy for any given income distribution. Since the
consumption of resources is influenced by prices, economists have devoted much attention over the years to specifying the set of prices
that will be the most economically efficient for a given situation.[1]
This chapter explores issues relating to economically efficient pricing
of monopoly telecommunications services.

Pricing

As outlined in chapter 13, if an efficient allocation of resources is
of primary concern, prices for monopoly telecommunications services
should not be based on historical accounting costs nor on the somewhat arbitrary value of service designations that may not explicitly reflect significant differences in demand elasticities. The traditional
analysis of historical accounting costs does not reflect current resource choices, and, since the approach is backward looking, the level
and mix of costs assigned to any particular service depend partly on
when that service was instituted or expanded and on how large the
changes in prices and interest rates were during the intervening
period.

A more appropriate approach (which applies when output can be expanded to meet demand) is to set prices on the basis of the forward-looking marginal or incremental costs of expanding output in order to maximize economic efficiency.[2] These costs are used because they are tied to decisions to change the level of output. Thus, they can be avoided and are the only economic costs relevant to future consumption and investment decisions that affect the allocation of resources in an economy.[3]

Applying the concepts of marginal cost to the pricing of telecommunications services is complex and encompasses several aspects of telecommunications tariffs. For example, the rationale behind establishing fixed monthly fees for gaining access to the network is that costs associated with local loop infrastructure do not generally vary with the number of calls made or with the duration of a call.[4] On the other hand, switching costs and long-distance transmission costs vary with the volume of traffic and are, therefore, reflected in usage charges—that is, charges per call and charges that vary according to duration of the call. A further example of the application of marginal cost pricing occurs when demand for use of the telecommunications network varies during the day. In this case, network use can be improved by providing discounted prices during off-peak hours, as discussed in chapter 13.[5] These discounted prices should be no less than the short-run marginal costs of providing off-peak service.[6] Where networks will be expanded to meet demand, prices should be set no less than long-run marginal cost.[7]

Externalities Arising from Connections to the Network

Marginal costs should, in principle, also include costs (or benefits) that are "external" to the market.[8] In the field of telecommunications services, one of the most important externalities is that the value of telecommunications networks increases as the number of subscribers grows. A positive external effect is created each time a new subscriber is connected to a network, because the telephones of existing subscribers increase in value as the potential number of contacts increases.[9] Theoretical problems arise when telephone installation immediately follows payment of the financial costs involved. Where no queues for service exist, the connection charge to an individual subscriber should, ideally, equal the cost of connection *less* the benefits (or disbenefits) accrued to persons or firms wanting (or not wanting) to be in contact with that subscriber. This suggests that the optimum connection price might in some instances be less than the connection cost and implies that a financial subsidy, which presumably should vary by class of subscriber (theoretically, by individual subscriber), may be necessary to allocate resources efficiently in these circumstances.

However, the implementation of any subsidy scheme is never without cost. The particular external benefits discussed here accrue to other telephone users, and therefore, it may be argued, for reasons of equity, other telephone users should provide the subsidy. But there seems to be no completely satisfactory way of arranging this. A fixed monthly rental or subscription charge to finance the subsidy would not be appropriate, as it would produce exactly the same problems, and, for reasons discussed in the next section, a surcharge on call charges might also be inefficient. One option would be a subsidy from general tax revenue. However, such a subsidy scheme would be particularly undesirable in most developing countries, because of a shortage of public funds, and because a government subsidy might have unfortunate consequences for the autonomy, management, and long-term planning of the telecommunications entity.

Externalities Arising from Calls

Similar theoretical problems are encountered in applying the rules of marginal cost pricing to calls, if it is accepted that a telephone call benefits not only the party who pays for the call but also the party called. In such a case, the optimal price of a call could be something less than marginal cost, and the justification for expanding the system should be determined not by just the revenues generated by callers but also by the benefits accruing jointly to both callers and callees. For example, if the value of a telephone call were the same for both parties, the price would be set equal to half the marginal cost charged to each party. Some judgment as to the size of the external effects is therefore necessary if marginal cost pricing of this joint product is to help determine the size of the investment.[10]

The empirical problems of attempting to sort out the separate benefits of a joint product are immense; certainly the financial costs people are willing to pay to receive calls or the extent to which externalities are internalized by such things as mutual arrangements to return calls or reverse charges cannot be accurately identified. A further complication is that some calls made are not beneficial—they create external disbenefits rather than benefits.

If, overall, external benefits outweigh external disbenefits, the case

for including externalities in telecommunications tariffs is strengthened in the frequent instances in which competing services, such as
public radio, transport, and the postal service, are subsidized at least
partly because they generate external benefits. Other forms of communication have similar characteristics: the recipient of a letter does
not pay; face-to-face communication is usually at the expense of the
traveler, and so on.[11]

On balance, given the current state of knowledge about external
benefits, the most practical solution is probably to ignore the
externality problems discussed here; they are probably even less of an
issue in developing countries, where the external disadvantages associated with unsatisfied demand dominate the situation. There is, however, a clear need to research the issue of monopoly communications
pricing so that judgments about second-best policies can be made
more satisfactorily.

Ramsey Pricing

One of the most problematic results of marginal cost pricing is that
where an industry is operating under conditions of declining average
costs, as many analysts believe is the case in the telecommunications
services sector, marginal cost pricing will not generate enough total
revenues to cover costs (see chapter 2). As a result, either the operating entity will go out of business, or a subsidy will be required. Subsidies, either from taxes on income or from sales, also have important
disadvantages as a response to the problem. They distort economically efficient pricing in other sectors of the economy and do not provide an operating entity with suitable incentives to run its operations
efficiently.

Instead, many analysts have proposed marking prices above marginal costs according to "ability to pay" or "what the traffic will bear"
or "value of service." These concepts are comparable to marking
prices above marginal costs inversely with the price elasticity of
demand—an approach known as Ramsey pricing. The intuitive rationale for this approach is that since prices above marginal costs produce demand (and therefore output) below the optimum, output
reductions are minimized if prices are marked up most where demand
is most price inelastic, thereby minimizing the loss of economic
efficiency.[12]

Ramsey pricing is, however, problematic. First, price elasticity of demand for telecommunications services is not entirely determined
exogenously by the market. It is, in fact, very significantly affected by

government policy on sector competition and on authorizations for users to meet some of their own telecommunications requirements by means of radio communications. Therefore, price elasticity is not an independent factor guiding government tariff policy but is, in part, a result of government policy.[13]

Second, the categories of customers with very inelastic demand for telecommunications services are likely to include persons with no alternative, or only very expensive or less desirable alternatives, to the services offered by the operating entity. Charging much higher tariffs to the most vulnerable customers often cannot be supported when public policy fairness is considered.[14] In practice, the concept of charging more where demand is more price inelastic is usually applied to broad groups of customers, such as business subscribers.

Price Rationing

One of the most important elements of the economic inefficiency that exists in the telecommunications sectors of many developing countries is the inability to expand networks sufficiently to meet the demand for service (at prices that cover costs). Thus, in many countries, a shortage of telephone lines is indicated by waiting times of months or years to obtain service; this is often accompanied by heavily congested network traffic during daytime business hours (see chapter 1).

In this situation, administrative forms of rationing, such as giving priority to select groups, tend to be arbitrary and cumbersome and to invite management or administrative irregularities. Even an apparently nondiscriminatory form of rationing such as queueing (meeting demands in order of application) is almost always too rigid to allow an economy to function efficiently.[15] Further, in a situation of unserved demand, setting the price of telecommunications services on the basis of calculated long-run marginal costs is not necessary. Instead, price can serve as a rationing mechanism. Scarce telephone lines can thus be allocated to customers who will derive the highest value from their use by charging prices above cost to the point where the level of unmet demand is almost eliminated and additional funds are generated for the enterprise.

To the extent that telephone lines are fungible (can be transferred from one user to another, usually within one exchange), the optimum way to set prices so as to ration the service is to increase the monthly line rental. This approach serves to ration the total supply of available lines, including both new and existing lines. An extension of this approach is to permit trading in "telephone rights" (that is, local network access) either officially, through a market exchange, or unofficially, in order to ration the scarce supply of local access lines in an economically efficient way. However, where monthly rental charges cannot be increased beyond a certain level, the scarce supply of new telephone lines is commonly rationed by charging high installation prices to new subscribers.

Similarly, where network traffic is congested, increased peak-hour call charges are appropriate. A salient advantage of this approach, as noted in chapter 13, is that decisions about the importance and use of scarce telecommunications services relative to other goods and services are left in the hands of customers.[16] One important variant of call charging in response to network congestion is to charge for unsuccessful call attempts. This would tend to reduce the common practice of dialing repeatedly, which contributes to network congestion. Although industrial countries have generally chosen not to charge for unsuccessful call attempts, developing countries should perhaps consider this approach. From the viewpoint of economically efficient rationing of scarce network capacity, such a charge should apply both if the call is blocked by the network and if the called line is busy. However, charging for calls blocked by the network does not give the telephone company the best incentive to improve its call completion rate and may well be seen by customers as an unfair practice. Nevertheless, even charging only for unsuccessful call attempts when the called line is busy would reduce network congestion.

Call Metering

The practice of charging for individual calls is a prerequisite for any form of efficient call traffic pricing. To keep tariff schedules administratively simple and easily comprehensible, however, call charges can never completely reflect the exact cost of making each call. Nevertheless, groupings into general categories or along specific routes can roughly reflect costs and have been found to be administratively workable and easily understood.

Following from this, particularly in developing countries, special attention needs to be paid to the disproportionately large burden placed upon a network during peak hours by heavily used subscriber lines. Because they are busy a relatively high proportion of the time, these heavily used lines engender an unproductive circle of unsuccessful attempts to establish connections and worsening traffic congestion.[17] In such instances, call metering and pricing for both local and long-distance calls made during peak periods should be seriously considered.[18]

In the short run, however, before metering equipment can be installed, a temporary approach to this problem might be to use a multiple tariff, in which the charge per time unit or per call for each telephone line increases during the billing period after a certain amount of use has occurred. The objective of such a policy would be to provide incentive to restrict the amount of traffic per subscriber line so that the line would not be continuously tied up, thus restricting incoming calls.[19]

Automatic toll ticketing may also be desirable in some instances in developing countries. The cost of introducing it is low—about $10 to $20 per direct exchange line for larger crossbar switches that are already installed and virtually nothing for new electronic digital switches—and the benefits can be very large. With automatic toll ticketing, the telecommunications operating entity can collect better traffic data to use in network planning and can expect to reduce subscriber complaints and related administrative problems. Given these benefits and the low equipment costs, one South Asian country in 1980 estimated that the automatic toll ticketing equipment that was projected to be installed in its (crossbar switched) network would have a payback period of much less than two years. Automatic toll ticketing can also have broader national benefits. It has the potential to make subscribers more conscious of their calling rates and habits and to help them pinpoint opportunities, as well as waste and inefficiency, in their telephone use.

Differentiating between Business and Residential Subscribers

The practice in some countries of charging business subscribers a higher monthly rental fee than residential subscribers is also subject to special question in a developing country. As noted in the previous chapter, in situations in which telephone service is generally available on demand, the rationale for a residential-business dichotomy is based partly on the concept of value of service and its underlying premise that business demand for telephone service is more price inelastic than residential demand. Hence, it is argued that increasing the price of business telephones may not significantly retard their use, whereas reducing the price of residential telephones can increase the penetration of telephones throughout the country and thus benefit subscribers and others.[20]

In the typical developing country, however, in which demand greatly exceeds supply, any attempt to stimulate residential consumption through lower prices may not be justified when business demand is not being met.[21] In particular, the highly skewed income distribution in most developing countries suggests that the final burden of telephone charges should be analyzed before accepting proposals to impose tariffs that are different for businesses than for residences.[22]

Long-Distance Telephone Tariffs

The practice still common in developing countries of pricing high-traffic density, long-distance calls above their cost cannot, other things being equal, be defended on the grounds of economic efficiency when supply can be readily increased to keep up with demand. Although no definitive analysis similar to that in figure 13-1 has been undertaken for developing countries, the partial data that are available closely match the patterns found in many industrial countries. Short-distance and local calls tend to be subsidized by high-density long-distance and international traffic, which, on average, generates revenues well above incremental and average costs. In fact, in many developing countries investment in the provision of international services typically yields incremental financial returns on the order of 100 percent annually. As outlined in chapter 1, although the sector as a whole typically obtains financial returns in real terms of between 15 and 30 percent on new investment, incremental financial returns on domestic long-distance services, although lower than for international services, are also usually much higher than the sector average.[23] As an example of this, an analysis of the expected incremental financial returns on the 1981–85 telephone investment program in one large South Asian country suggested that investment in the larger urban areas was expected to yield an annual return of about 37 percent, and in smaller towns and rural areas it was forecast to yield 16 percent, whereas new investment in higher-traffic long-distance facilities was expected to yield almost 50 percent.

The exact implications for economic efficiency of such tariff and investment policies are unknown for several reasons. First, reliable data are lacking on the elasticity of call traffic demand in various situations in developing countries. Second, the elasticity of demand is not known for the many goods and services whose prices partly reflect the high charges that producers pay for long-distance telephone calls. Third, little is known about the calls that were not made, nor the inefficiencies incurred, because of inadequate service or high call prices. In general, however, a convincing argument can be made for the existence of a global underconsumption of long-distance telephone and telex communication in developing countries, brought about partly by the practice of charging prices that are significantly higher than costs for long-distance calls and partly because the resulting revenues are not sufficiently devoted to investment in additional long-distance capacity.

Arguments against overcharging for traffic for long-distance telephone, telex, data, and so forth include the overwhelming business orientation of their use and the likely distortions in resource use associated with using second-best alternatives for communication—often travel (rail, road, air) or postal services—that themselves may be financially subsidized by government. Furthermore, the implications of such a policy on regional development, national trade, and transport and energy consumption are unknown. Finally, any equity implications of long-distance price-cost differences will vary greatly among countries and among regions within countries, depending on which groups have access to telecommunications services. In general, however, residents of smaller towns and villages outside the main urban areas tend to make a disproportionate number of long-distance calls at the higher tariff rates.

Given the above, telephone administrations in developing countries should closely examine the extent to which they overcharge on or underprovision long-distance networks and should think through some of the implications of so doing. In many instances, they might consider reducing long-distance tariffs over heavily used routes and perhaps making long-distance tariff structures less sensitive to distance (see chapter 13 for a discussion of the sensitivity of costs to distance).

Subscriber Financing

In several developing countries, subscribers are required to contribute capital to the telecommunications operating entity. This method, called subscriber financing, compels new subscribers to purchase bonds or shares of stock in the telecommunications enterprise in order to provide funds for network improvement or construction (it also serves to ration scarce supply). It works best where there is a large unsatisfied demand for service and no other acceptable source of sufficient capital is available. These schemes have been implemented in different forms in several countries. In Taiwan, it was an irrecoverable charge; in Japan and Thailand, a bond; in Argentina, Bolivia, and Brazil, a contribution to equity capital; and in Mexico, a contribution to either the equity or the debt capital of the telecommunications authority.

Subscriber financing was used with spectacular results in Japan, where it was a key ingredient in the growth of the Japanese telephone system from 550,000 lines in 1946 to 42 million lines in 1981.[24] For periods of several years, such capital subscription or connection charges assisted the telephone systems in Brazil and Taiwan, among others, to expand at a rate of approximately 20 percent or more a year. Furthermore, during those years the rapid expansion of the system was attained without drawing on government capital.[25] Indeed, such expansion was achieved in Brazil and Taiwan with substantial net financial contributions from the telecommunications authorities to government.

An Investment Strategy to Complement Efficiency Pricing

An analysis of tariff policy in developing countries in which telecommunications will remain as a government-regulated monopoly, and in which waiting lists and system traffic congestion are significant during daytime business hours, could in many countries yield a three-pronged strategy: set prices in the short run near a market-clearing level, speed up the overall rate of investment in the sector, and put an adequate amount of investment in the near term in public telephone facilities (PCOs and coin box telephones) both in the larger cities and in smaller cities, towns, and rural villages.[26]

When deciding to set prices in the short run at a level that comes close to clearing the market, a relevant question is, what does "close to clearing the market" mean? Roughly, one answer might be prices that bring about perhaps no more than a six- to eight-month waiting period for connections in most areas and a completion rate of approximately 40 to 60 percent for the first attempted call during busy hours. For the United States the normal call completion rate during busy hours is about 75 percent, and for Australia it is about 70 percent. In a large city in South Asia the figure is currently estimated to be 5 percent. The suggested six- to eight-month waiting period and the successful call rate of 40 to 60 percent must be considered only starting points for discussion.[27] In some instances attaining an acceptable successful calling rate during busy hours may be difficult because telephones are used almost continuously, thereby generating busy signals when a high proportion of the available telephones are simultaneously trying to call one another.

As outlined in chapter 13, a large unsatisfied demand for calls and connections at prices in excess of costs indicates that a rapid expansion of the sector is justified; the incremental benefits perceived by telephone subscribers and users exceed the incremental resource costs. As expansion takes place, the price may be gradually lowered so that demand and supply remain roughly in balance. Traditional economic welfare theory argues that the expansion should not be halted until the price has fallen to the level of marginal production costs, and supply equals demand, or, if as is more likely in this declining-cost sector, to a level at which a lower price would result in financial losses for the telecommunications entity.

Not increasing the rate at which telecommunications services are expanded can cause another problem, in addition to the risk that communication-related inefficiencies would multiply throughout the national economy. In the absence of competition, the high profits, which keep rolling in because market-clearing prices are relatively high, have in several countries contributed to significant inadequacies and inefficiencies in the management of telecommunications entities. The financial incentive for efficient management is lost. Costs within the sector rise, and resources allocated to the sector may be increasingly misallocated—all the result of management inefficiencies. Limiting the natural growth of the telecommunications system, typically by government regulation of entry, prices, profits, investment, and management control, would in these circumstances typically require further government action. Such action could include levying a special unit tax to avoid the excessive profit that would otherwise accrue to the telephone authority and to ration the decreed level of service so that higher-value uses are maintained.

Given a short-run market-clearing price strategy for subscriber telephones, increasing investment in PCO or coin box facilities throughout a country may be desirable for several reasons.[28] First, PCO and coin box facilities are generally the least expensive way to provide wide telephone access to the most people. Second, the high market-clearing prices charged in the short run for subscribers' telephones may for a time prohibit their acquisition by smaller business and agricultural establishments that occasionally have an urgent need for rapid two-way communication. As shown in chapter 12 public telephones provide a means through which these smaller commercial entities and government administrators in towns and rural areas can satisfy their highest priority communication needs. Third, as reflected in the evidence reviewed in chapters 1 and 7, public telephones in market centers and villages near highly traveled routes can increase the efficiency of a country's transport system by providing points at which drivers can receive new instructions from headquarters or report problems or breakdowns. Finally, evidence reviewed in chapter 12 and appendix C also suggests that such facilities provide a way for the general population to report or discuss emergency or health needs or simply to communicate with family or kin who have left the area.

X-Efficiency

Another aspect of economic efficiency, separate from the allocative efficiency discussed earlier, is known as X-efficiency, a term coined by Harvey Leibenstein.[29] The concept incorporates the ideas of motivation, incentive, and skill or, on the other hand, inertia, apathy, and ineptitude into the economic analysis of efficiency. At a practical level, the concept looks at how efficiency can be increased by working smarter or harder, by reducing costs, or by increasing quantity and quality of output.

The basic X-efficiency hypothesis is that neither firms nor individuals are always as productive as they could be, and, as a result, costs are not always kept to their minimum. Leibenstein linked the concept to competition, observing that competitive pressures engender efforts to reduce costs. All other things being equal, X-efficiency will be higher in a competitive than in a monopoly environment because each environment has different incentive characteristics.[30] Many authors have found evidence to support the concept of X-efficiency.[31] In some cases, the losses from X-efficiency can be larger than those from allocative inefficiency.[32]

Although the terminology of X-efficiency rarely appears in discussions of policy initiatives dealing with privatization and competition in the telecommunications sector (see chapters 3 and 16), the concept has clearly been a major motive in the process. For example, according to Sir Bryan Carsberg (director of the United Kingdom's telecommunications regulatory body, OFTEL), the main objective behind the privatization of British Telecom and other public enterprises in the United Kingdom during the 1980s was to increase efficiency; to promote competition in the interests of creating incentives to become more efficient; to allow freer participation in various markets; . . . and to foster efficiency by an appropriate form of regulation in areas where competition cannot be effective.[33]

As an increasing number of countries around the world move to increase competition and privatization in the telecommunications sector, X-efficiency can be seen as a significant source of increased potential productivity and improved performance. X-efficiency considerations have clearly become important in influencing decisions about the sector's organization and policy (as discussed in chapter 16), as well as in designing and selecting tariff regulation modalities for the sector (as discussed in chapter 15).

Notes

1. Traditional normative welfare economics suggests that welfare will be maximized for natural monopolies if the price of each service is set equal to its marginal cost, if output is expanded to meet demand at those prices, and if several related assumptions are made. For a brief discussion of welfare economics and marginal cost pricing in telecommunications, see Littlechild (1979), chap. 9.

2. Kahn (1970), pp. 65–66. Kahn defines marginal cost as "the cost of producing one more unit; it can equally be envisaged as the cost that would be saved by producing one less unit." Although marginal cost and incremental cost are often used synonymously, marginal cost, strictly speaking, refers to the additional cost of supplying an infinitesimally small additional unit of output, while incremental cost sometimes refers to the average additional cost of a finite and possibly large change in output.

3. Lewis (1949 and 1968) For more on marginal costs and economic efficiency, see chapter 13.

4. The costs of subscriber network access consist of the subscribers' terminal equipment (if rented from the telephone operating entity), jack, drop wire, cable-pair to the local exchange (switch), and line termination equipment at the switch. See Littlechild (1989), app. 1.

5. A discussion of varying charges according to when a call is made is found in chapter 13. Current peak-load pricing theory focuses on uncertainty and nonprice rationing; it is also concerned with devising better ways to handle the pricing of services so as to accommodate the lumpy investments in capacity that occur as the sector expands. See Mitchell and Vogelsang (1991), p. 23.

6. Short-run marginal cost is simply the change in total variable costs caused by producing an additional unit; Kahn (1970) pp. 71–72. To the extent that depreciation does not vary with use (that is, for a fiber-optic transmission system), depreciation costs are not included in short-run marginal cost.

7. Long-run marginal costs include capital costs. Setting prices equal to long-run marginal costs is equivalent to setting prices so that the resulting net present value of an investment program is zero. This assumes that the discount rate used to calculate net present

value is appropriate in that it reflects the opportunity cost of capital for projects of comparable risk.

8. Air pollution is an example outside telecommunications of a negative externality resulting from the production of some goods and services.

9. Of course, sometimes private contractual arrangements take this into account. A business firm may pay for the home telephones of key employees or children may pay for their mother's telephone so that they can keep in contact. For a more detailed discussion of externalities as they relate to telecommunications, see Squire (1973) and Littlechild (1979) chap. 12.

10. It might be argued that everything comes out in the wash, because of the identity of the number of calls received and made, and that price should equal marginal cost after all. This would allow the financial equivalent of marginal cost to be recovered from telephone users. Nevertheless, if price is to be used to allocate resources efficiently, every decision to make a call should result from a conscious comparison of benefits and costs. A price based on the marginal (capacity plus operating) cost of the telephone enterprise alone may in some instances not allow this to be achieved.

11. It has also been argued that there can be no connection-related externalities unless there are call-related externalities, and, if the information content of a call is separated from the call-delivery of that information, then at least for business firms, the cost of the call will be reflected in the fees charged by the caller, which will be paid by the person receiving the call; hence, there may be no market failure. See Jonscher (undated).

12. Hazelwood (1950–51 and 1968); Baumol and Bradford (1970); Littlechild (1979) pp. 128–89; and Culham (1987).

13. For example, if government policy favored introducing as much competition as possible in the provision of telecommunications services, and if that policy had not been achieved in certain areas, it would seem perverse for government policy to support charging the remaining monopoly customers a markup higher than cost.

14. Writing about the suitability of Ramsey pricing concepts to pricing of telecommunications services in the United Kingdom, Bryan Carsberg (director general of telecommunications) wrote, "I am much more doubtful about the usefulness of the economic approach for determining the proper balance between local call charges and long-distance charges . . . I am keen to ensure that the economic approach should not be used as an excuse by BT [British Telecom] to load the burden of any internal inefficiency onto those callers who have most need of telephone service." Culham (1987).

15. Especially since in developing countries waiting times to get new connections typically are several years. If for political or other reasons the official telephone tariffs cannot be used to ration demand effectively, a secondary or black market might be encouraged to do so. In several developing countries telephone brokers use newspapers to advertise the availability of, or demand for, telephone lines.

16. Thus, high-value users who might otherwise be unknown to telecommunications or national planning or development authorities can increase their likelihood of gaining access to a telephone, telex, and so forth.

17. As noted previously, peak-period pricing in Europe has shifted traffic patterns. In several instances, in fact, sharply reduced off-peak rates stimulated calling at off-peak hours to such an extent that capacity was exhausted and overload conditions occurred. See Mitchell (1979).

18. One objection to call metering is that some telephone users pay more and use the telephone less. Nevertheless, it can be argued that pricing by number, time, and duration of the call tests the value of telephone use. Parties who use the service place a value on it at least equal to the price they are paying (calls or duration are no longer free), and to the extent that busy signals are reduced by eliminating overly long or low-value calls, total satis-

faction from the system increases. Even if the average length of calls is relatively short before metering, metering could significantly compress the distribution of calls by length.

19. A variant approach has been tried in India. There, rather than immediately installing peak-period metering equipment for local calls, P&T instituted a two-part call charge. The first 1,750 calls made (or pulses generated) by a subscriber during a quarter (billing period) were billed at the standard call charge of 3 paisa. All calls in excess of 1,750 were charged a 33 percent surcharge and billed at 4 paisa for each call. The objective was partly financial and partly to encourage heavy telephone users (who are generally business firms calling mostly during peak periods) to apply for additional telephones and to reduce the amount of time spent on each subscriber trunk dialing call; the network is highly congested, and all calls are bulk metered, with each pulse counting as one call. This solution is imaginative and could be effective if waiting lists were shorter, additional telephones were more readily available, or heavy users received priority telephone allocations.

20. In economic terms, such a tariff policy tends to maximize the economic measure of welfare—the sum of consumer and producer surplus. See Mitchell (1978) and Baumol and Bradford (1970).

21. The current practice in the United States and several other countries of charging more for the rental of touch-tone telephones, partly on the assumption that the demand for them is more price inelastic, may also be undesirable in developing countries, since touch-tone dialing permits more efficient use of the network.

22. There have been attempts to redistribute income directly through the monthly charge to rent a telephone. In Colombia, for example, both the monthly rental and the initial capital contribution required of residential subscribers are related to the tax rate value of the dwelling in which the telephone is to be installed.

23. Two alternative explanations justify such a pricing and investment policy: (a) government has set revenue targets relatively high, and this has generated too few calls and too little investment, or (b) short-run financial constraints or other investment priorities have limited investment in long-haul facilities, and hence the high charges for long-distance calls are appropriate in the short run to ration the fixed capacity. Both explanations probably have some validity.

24. The Japanese financing program reached its peak between 1963 and 1972, when it generated 32 percent of the investment funds. The public accepted the program, which remained in effect until 1982. See Gellerman (1986).

25. From the point of view of the telecommunications operating entity, capital subscription charges only generate local currency; in most countries, considerable foreign exchange is also needed for the system expansion.

26. It is sometimes argued that there are a few cases in which exceptions to the high connection fee should be granted to meet distributional or equity objectives of the country's development policy. For example, education, health, and emergency services might be favored with lower connection fees and rentals, and, if it was thought desirable, differences in regional incomes or urban-rural incomes might be partially compensated by having higher connection fees for larger cities and lower fees for smaller cities and towns. This is, in fact, done in several countries that have instituted significant capital subscription requirements in addition to connection fees. It can be argued, however, that governments should subsidize these services directly and should not force other telephone subscribers to bear a disproportionate amount of the subsidy burden, particularly by charging lower rates for telephones used for public service activities.

27. For an interesting discussion of some of the economic issues involved in determining an appropriate grade of service, see Littlechild (1979), chap. 11.

28. If direct subsidies from government are not feasible, which is the situation in most developing countries, such facilities can usually be financed out of the financial surpluses generated from urban subscribers and the long-distance network, even if in the short run

some of them lose money. In an interesting simulation example, data for a Latin American country were used to simulate the possibility of generating extra revenue from urban services through price increases and using it to subsidize rural services. The results of the simulation, which were partly predetermined by the assumptions made, showed that during 1978–87, the extra revenue from urban services more than canceled out the reduction in revenue from rural services and predicted that by 1987 rural traffic would be 30 percent higher than it would have been without the increased cross-subsidy. See Walsham (1979).

29. Although the terminology of X-efficiency was new at the time, the concept was not. See Leibenstein (1966 and 1973).

30. For example, if a monopoly firm fails to produce a required level of service in any given time period, the consequences to the firm will be minimal since its customers cannot easily go elsewhere. In a competitive environment, such mistakes or behavior send customers elsewhere and lose business.

31. Scherer (1970) writes that "X-efficiency exists, and it is more apt to be reduced when competitive pressures are strong than when firms enjoy insulated market positions." Also see Frantz (1988).

32. Hollas and Hereen (1984).

33. Carsberg (1986).

Chapter 15

Tariff Policy in Practice

ALTHOUGH PRICING BASED ON MARGINAL COSTS is theoretically the ideal approach from the viewpoint of allocative efficiency, several difficulties are associated with its practical implementation. Several areas in which such difficulties arise in the telecommunications sectors of developing countries were discussed in chapters 13 and 14. These include demand forecasting, income distribution, the need for telecommunications operating entities to achieve adequate financial returns, and the presence of externalities.[1] This chapter examines additional problems involving the estimation of marginal costs and reviews several approaches to tariff analysis and regulation that are followed in practice.[2]

Estimating Marginal Cost

Traditional economic welfare theory implies that marginal or incremental costs of expanding the system should be used at least as a benchmark to develop a tariff structure for telecommunications monopoly organizations when demand can be met. Therefore, it is important to have a forward-looking cost analysis for each telecommunications product or service. Unfortunately, many conceptual as well as practical problems are involved in defining the incremental costs to the system of even just expanded telephone service. Problems are also associated with measuring those costs.

For an incremental cost analysis, the output of the telephone authority (more telephones, private branch exchanges, local calls in big cities, local calls in small towns, long-distance calls over low-traffic routes, long-distance calls over low-traffic routes, long-distance calls over high-traffic routes, data transmission, television transmission, telex, and so forth) is difficult to measure. Even when output can be adequately defined, difficulties of measurement arise because investments might not be made in frequent, small increments. Investment streams for individual services or products can be "lumpy." This results in conceptual and practical problems in determining the proper time horizon and the accompanying incremental quantity of output for the analysis.

These issues are discussed briefly in following sections, but first the real costs, which are incurred by the economy as a whole, must be distinguished from the financial costs, which are paid by the telecommunications entity itself. This involves the concept of shadow pricing.

Economic Input Costs

Market distortions frequently exist in developing countries. Hence useful estimates of the real economic costs of producing various goods and services are obtained by adjusting the financial costs and prices according to economic analysis. Competitive market prices generally indicate the opportunity cost of inputs for the production of telecommunications services when those inputs are produced and supplied under near-competitive conditions or are freely imported. In some circumstances, however, it is possible to identify inputs whose prices do not reflect their real value to the economy; therefore they should be shadow priced.[3]

In particular, overvalued local currencies (resulting from artificial support such as foreign exchange controls and protective measures) and wages for unskilled workers determined by minimum wage laws or labor agreements (when unemployment and underemployment are widespread) tend to require the use of shadow values so that the resource cost to the economy of using foreign exchange, unskilled local labor, or local production in the telecommunications sector can be properly estimated. As in the calculation of internal rates of return, which was discussed in chapter 8, transfer items such as import duties should be excluded from cost estimates. Such adjustments are especially important for the telecommunications sector in developing countries, which is typically capital intensive and relies heavily on imported equipment and materials.

Interest rates should also reflect opportunity costs of borrowed funds in the economy as a whole. This is not always reflected in the rates that enterprises actually pay, particularly when they are publicly owned, as is often the case with telecommunications entities in developing countries. Finally, income generated by the telecommunications operating entity for the public sector may have a premium placed on it, relative to consumption, when public savings are inadequate.[4]

The foregoing adjustments allow a set of shadow prices, called efficiency prices, to be derived. A more detailed discussion of shadow pricing in the telecommunications sector is presented in appendix D.

Capital Indivisibility and Incremental System Costs

Problems associated with setting price equal to marginal cost are particularly apparent in the presence of capital indivisibility, a condition that is often encountered in telecommunications expansion programs, particularly in smaller developing countries.[5] For example, if the objective is to meet demand fully, switching and transmission capacity might be installed to make up for local deficits in supply and to meet some portion of future demand. The initial costs of subscriber, exchange, and transmission facilities are relatively high in relation to operating and maintenance costs, and scale economies can usually be realized with some initial overprovisioning. Even where extending capacity in fairly small increments is technologically possible, fluctuations in the availability of finance can mean that capacity must be extended in somewhat larger lumps. Strict marginal cost pricing in these circumstances may require periodic, large changes in price.

If in such circumstances price were set equal to marginal system cost, price would equal short-run marginal cost when capacity is less than fully used, and, when demand increases so that existing capacity is fully used, price would be raised to ration the fixed capacity. This procedure would continue until subscribers, system users, or parties demanding new access or service reveal their willingness to pay a price for additional service equal to short-run marginal cost plus the annual equivalent of marginal capacity cost.[6] At this point—where existing capacity is fully used and consumers are paying a price that equals long-run marginal cost—investment in additional capacity can be justified. Once the investment has been carried out, however, price would need to be lowered again to the point where it reflects only short-run marginal costs; then the only real costs (or opportunity costs in terms of alternative benefits forgone) are for operation and maintenance. Price therefore plays two distinct roles: obtaining efficient use of resources invested in existing capacity and providing a signal to invest in additional system capacity.

Alternative Approaches to Estimating Marginal Cost

In view of these difficulties, several approaches have been developed to estimate or approximate marginal costs in the telecommunications services sector. In the simplest case—a telecommunications system in which investment roughly keeps up with demand and occurs simultaneously in several parts of the network—the investment streams associated with one or more telecommunications service outputs may be relatively smooth. In such an instance a measure of marginal cost that has been used widely in electric power pricing—total long-run incremental cost (TLRIC)—seems to be an appropriate method.[7] TLRIC is defined as

$$\text{TLRIC}_t = \frac{C_{t+1} - C_t}{Q_{t+1} - Q_t} + \frac{rI_k}{Q_{k+1} - Q_k}$$

where t = year or period for which the estimate is being calculated, C_{t+1} = operation and maintenance costs in year or period $t + 1$, Q_{t+1} = quantity of telecommunications service output in years or period $t + 1$, I = the next major capital expenditure, r = the capital recovery factor, or the annual payment that will repay a \$1 loan over the useful life of the investment with compound interest (equal to the opportunity cost of capital) on the unpaid balance,[8] and k = the year or period in which the very next major investment expenditure is completed.

During the period t through k, the term $rI_k / (Q_{k+1} - Q_k)$ remains constant, reflecting the annual equivalent of marginal capacity cost for the next increment of investment. After the investment has occurred in year or period k, k is redesignated to be the next year or period in which a large investment will be completed, so that the price reflecting TLRIC will be adjusted in a stepwise fashion over time.

Other analytic methods have been developed to estimate incremental costs in the production of local exchange services. Four methods can be employed: econometric methods, engineering planning models, optimization models, and engineering-economics process models.[9] Each method has strengths and weaknesses. For example, the econometric method is based on historical data describing expenses and output but cannot easily account for technological change nor for new methods of production. Engineering planning models, in contrast, handle the costs of new technology well but require extensive data that may not be easily available to planners in developing countries. Optimization models can yield accurate information on specific design problems, but the relatively narrow focus of this approach makes it difficult to use as a general model to obtain marginal costs of access and local use. The engineering-economics model avoids some of these problems and works with the major factors that determine incremental costs without the need for highly detailed data. A particular advantage of this kind of model is that it can be used to estimate the average incremental costs of supplying an increased level of service for a specified set of demand and other conditions.[10]

Tariff Analysis in Practice

Tariff analysis is an extremely complex undertaking, due to the existence of multiple objectives for tariff policy as well as to the problems that affect the allocation and estimation of costs in concept and in practice. As a result, many approaches to analyzing tariffs for the telecommunications sector have been developed. Each approach is tailored to the specific needs of a particular country and its telecommunications system, and none is clearly superior. Two approaches are described in this section.

In the first example, TLRIC methodology was applied to the tariff structure of a telecommunications operating entity in a large Asian country, as shown in table 15-1.[11] The TLRIC-derived marginal costs were compared with actual tariffs charged by the telecommunications administration in order to determine the extent to which tariffs actually reflect costs of operation.

Connection costs were estimated based on the cost of connecting a subscriber to a distribution point plus an allowance for inside wiring. Subscription charges were estimated by adding together the annualized investment cost, the operating and maintenance costs for an exchange line, plus an allowance for local exchange equipment related to line termination. Local call charges were estimated from the annualized cost of traffic-sensitive local exchange investments plus operating and maintenance costs. These costs were spread across expected peak local traffic.

The comparison in table 15-1 shows that tariffs were significantly out of line with costs. Connection charges were generally too high, and subscription charges were far too low. Although charges for local calls on automatic lines appeared to be in line, no charge was made for local calls on manual lines. Trunk call charges were much too high.

Table 15-1. TLRIC *Analysis of Existing Tariffs in an Asian Country*
(U.S. dollars)

Tariff component	Estimated LRMC[a]	Existing tariff
Connection charge		
Automatic	65.04	66.37–442.48
Manual	152.21	Mostly over 176.99
Monthly subscription		
Automatic	14.51	1.77–3.10
Manual	11.68	0.88–1.77
Local call charge *(per pulse)*		
Automatic	0.05	0.07
Manual	0.12	No charge
Long distance charges *(per three-minute call)*		
Automatic (distance in kilometers)		
25–100	0.07	1.99
100–300	0.10	2.99–2.39[b]
300–600	0.19	3.98–2.99[b]
600–1,000	0.24	3.98
Over 1,000	0.32	5.97
Manual (distance in kilometers)		
25–100	0.17	0.99
100–300	0.22	1.19–1.49[b]
300–600	0.28	1.49–1.99[b]
600–1,000	0.32	1.99
Over 1,000	0.35	2.99

a. Estimated LRMC (long-run marginal cost) was used to illustrate an economic (as opposed to a financial) incremental cost estimate.

b. Exact comparisons with estimated LRMC prices were not possible since the existing tariff used different categories for distance.

Table 15-2 shows the extent to which these existing tariffs satisfied five typical objectives of tariff policy. In particular, the economic efficiency of the tariff structure, as indicated by the relation of the tariffs to LRMC was contrasted with other objectives such as efficient network use and generation of tax revenue.

The economic efficiency of this particular tariff structure was found to be bad to very bad, yet many other policy objectives were well met. The most extreme case appears in trunk, or long-distance, calls, where economic efficiency and assistance in rural development were very bad, due to high tariffs, even though the ability of this tariff to pro-

Table 15-2. Existing Tariffs Evaluated in Terms of Policy Objectives

Charge	Economic efficiency[a]	Efficient network use[b]	Service rationing[c]	Assistance to rural development[d]	Generation of tax revenue[e]
Connection charge					
Automatic	x (+)	n.a.	vv	x (+)	v
Manual	x (+)	n.a.	vv	x (+)	v
Monthly subscription					
Automatic	x (−)	n.a.	x (−)	v	xx (−)
Manual	x (−)	n.a.	x (−)	v	xx (−)
Local call charge					
Automatic	x	v	n.a.	v	x (−)
Manual	x	x (−)	n.a.	v	x (−)
Trunk call charge					
Automatic	xx (+)	v	n.a.	xx (+)	vv
Manual	xx (+)	v	n.a.	xx (+)	vv

n.a. Not applicable.

Note: Performance ratings for objectives are v, good; x, bad; (−), tariff too low; vv, very good; xx, very bad; (+), tariff too high.

a. Economic efficiency is defined here as the convergence between long-run marginal cost and tariff revenues.

b. Efficient network use refers to the degree to which peak-load pricing encourages off-peak use, suppresses excess peak demand, or both.

c. Service rationing refers to the ability to ration scarce access of new subscribers to network connection.

d. Assistance to rural development refers to the ability to keep prices low to increase access in poor and rural areas.

e. Generation of tax revenue refers to the likelihood that sufficient revenues will be generated to enable transfers from the telecommunications authority to other government departments.

mote efficient network use was good, and its ability to generate tax revenues was very good. Even so, network efficiency might be improved further by introducing lower, off-peak charges to take advantage of spare capacity, and additional revenue could be raised by increasing access charges, with little detrimental effect on economic efficiency and network use. The local pulse rate for automatic service was found to encourage efficient use to a certain extent but was too low, since the network continued to suffer congestion during busy hours. The manual local service, available at no charge, was clearly extremely inefficient, and these circuits were badly congested as a result. As regards the goal of rationing access to service, the performance of the tariff was contradictory: high connection charges discouraged new applications, whereas low subscription charges encouraged subscribers to hold onto existing lines. Both charges would have to be high in order to ration effectively. Assistance with rural development was facilitated through cross-subsidy of local charges but probably was counteracted completely by very high trunk charges, which are incurred more frequently by rural subscribers than by urban users. These tradeoffs highlight the complexity of determining the appropriate tariff structure in practice, where conflicting policy objectives must be balanced.

A quite different approach is used by the Canadian telecommunications regulatory authority (CRTC) in its cost inquiry phase III methodology.[12] This approach was developed specifically to identify the sources and recipients of cross-subsidies among broad categories of existing services and the existence, if any, of a cross-subsidy of competitive categories by monopoly categories. The approach uses historical accounting data instead of prospective long-run marginal costs. For a regulatory agency, such data have the particular advantage of being auditable; another advantage is that the phase III methodology is much less complex and less charged with conceptual problems than theoretically more appropriate forward-looking marginal costs.

The phase III methodology establishes eight broad categories for the allocation of costs and revenues (see appendix F for the definitions). In 1989, these categories were access (A); monopoly local (ML); monopoly toll (MT); competitive network (CN); competitive terminal—multiline and data (CT-MD); competitive terminal—other (CT-O); other (O); and common (C).[13]

Of particular importance is the creation of "access" (that is, subscriber network access) as a separate category.[14] As a result of an immense amount of work over several years, standardized procedures were developed to allocate almost all revenues earned and costs incurred in the provision of the carrier's services in a given year on the basis of causality. "Common" is a separate category used for costs that cannot be causally related to a particular service category.

The resulting comparisons of revenues and costs or, alternatively, of net revenues per category permit the direction and magnitude of broad cross-subsidies to be assessed quickly. In this way, the results provide important information relevant to assessing possible changes in the balance or structure of tariffs for telecommunications services. For example, the analysis shows whether competitive carrier services are cross-subsidized by monopoly services, an important concern of regulators. Furthermore, the analysis indicates the magnitude of the cross-subsidy from monopoly toll to subscriber access categories. To meet efficiency goals, the appropriate prices can be adjusted to reduce or eliminate cross-subsidies. Phase III methodology results for Bell Canada in 1987 are shown in figures 15-1 and 15-2.

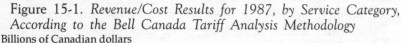

Figure 15-1. *Revenue/Cost Results for 1987, by Service Category, According to the Bell Canada Tariff Analysis Methodology*
Billions of Canadian dollars

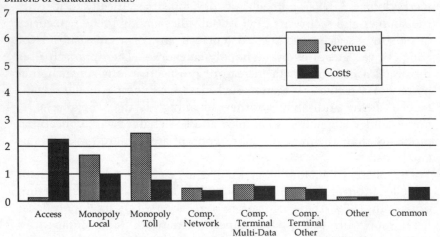

Service categories
Source: Canadian Radio-Television and Telecommunications Commission.

Figure 15-2. *Revenue Surplus/Shortfall Results for 1987, by Service Category, According to the Bell Canada Tariff Analysis Methodology*

Billions of Canadian dollars

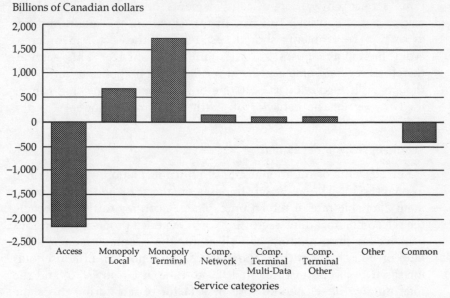

Service categories

Source; Canadian Radio-Television and Telecommunications Commission.

Tariff Regulation

Until the early 1980s, telecommunications operating entities in most countries were either government departments or public enterprises. Privately owned telecommunications entities mainly existed in North America and the Philippines. Under these circumstances, regulation used to play a less prominent role in policy implementation in most countries than it does today, because governments exercised direct control over the telecommunications operating entities. Starting with the privatization of British Telecom in 1984, a global trend toward increased privatization has occurred in many countries. In some, the telecommunications operating entity was reestablished as a commercially oriented government-owned corporation (for example, in Fiji, France, the Netherlands, and Sri Lanka); in others, the operating entity was privatized (for example, in Argentina, Chile, Japan, Malaysia, Mexico, and New Zealand).

Despite attempts to foster competition in the telecommunications sectors of many countries (often occurring as a complement to policies of commercializing or privatizing the dominant telecommunications service entity), significant elements of natural monopoly or market power continue to exist. In this situation, the question arises as to how governments should best regulate the sector in order to avoid monopoly abuses and optimize performance. This section, therefore, reviews some aspects of regulation as it applies to tariffs, although tariff regulation is only one option among many that can be used to control the behavior of a firm with market power.[15]

Rate-of-Return Regulation

Key elements of traditional rate-of-return regulation were described in chapter 13. The approach imposes on the regulated firm a maximum allowable rate of return on capital. According to this approach, the telecommunications sector in North America produced enormous benefits for the economy, as well as for the telephone system. The ratio of households with telephones to the number of all households increased from one in three in 1920 to nine in ten in 1970, and the total number of telephones rose by a factor of ten during this same period.[16] Other evidence indicates that the productivity gains in the U.S. telecommunications industry between the 1950s and the 1980s were consistently higher than those in the U.S. economy as a whole.[17]

Despite these benefits, rate-of-return regulation has well-known drawbacks as well. These include weakened incentives for the regulated firm to maximize technical efficiency because of the cost-plus character of rate-of-return regulation, incentives (under some circumstances) to use excessive amounts of capital relative to other inputs, incentives to engage in predatory, below-cost pricing in competitive markets, and the costly and time-consuming administrative procedures connected with the regulatory process.[18]

Price-Cap Regulation

Because of the drawbacks of rate-of-return regulation, there has been widespread discussion of whether alternative regulations might improve economic performance. Most of this discussion has focused on price caps.[19]

Under a price-cap approach, the regulated firm is required to ensure that the aggregate price level of a basket of services will rise by no more than a certain amount. Typically, this amount is calculated using the formula $CPI - X$, that is, the general rate of inflation, as defined by the consumer price index, minus the estimate X, which is the gain in productivity in the telecommunications sector that is greater than the gain in the economy as a whole.[20] The formula is then fixed for a specific period, for example, five years. Thus, price-cap regulation differs most significantly from traditional rate-of-return regulation in that it sets limits on the rates that carriers may charge for their services, while eliminating almost all limits on the profitability of such services during the period in which the fixed price-cap formula is in effect. However, rate-of-return regulation and price-cap regulation are not totally different because both depend on forecasts of rates of return. Thus, even under the price-cap approach, the all-important estimate of the productivity gain X is partly determined by future rates of return calculated under various scenarios.

The main advantage of the price-cap approach is that it forces a telecommunications enterprise to pass on a minimum assumed productivity gain of X to its customers in the form of reduced real prices and creates an incentive to increase productivity, which is reflected in profits that can be retained. By the same token, the enterprise is compelled to absorb any losses resulting from inefficiency. Additionally, the price-cap approach eliminates the inappropriate incentive that may exist under rate-of-return regulation to use excessive amounts of capital inputs, or to "gold plate." Finally, the approach is believed to be somewhat simpler to administer than rate-of-return regulation.

Another advantage of price caps applies in some developing countries, where telephone rates have the potential to be politically controversial. Once a price-cap tariff adjustment formula has been fixed for a predetermined period, tariff adjustments can be taken out of the political arena to some extent and implemented automatically by the regulated firm. This gives the telecommunications entity greater security in its future revenues and allows it to make investments that will meet demand and generate profits.

However, price-cap regulation does not solve all tariff regulation problems. The productivity gain factor can be difficult to determine, and carriers' profits are greatly affected by small variations in this figure. Exogenous cost changes are also problematic, since the generalized price indexes such as the current price index may not accurately measure all the cost changes occurring outside the firm's control.[21] Finally, price-cap mechanisms can create an incentive for reducing the quality of service and result in a cumulative deterioration in network infrastructure.[22] In a sense, price-cap regulation shifts some of the focus of regulation from profits to quality of service.

Price-cap regulation is currently the most prominent type of incentive regulation. It is designed to provide telecommunications enterprises with an incentive to achieve efficiency gains and in this sense is linked to the discussion of X-efficiency in chapter 14. The method was first applied to British Telecom when it was privatized in 1984 and has subsequently been applied in the United States to AT&T, to major local exchange carriers, and in many state jurisdictions, as well as in Mexico and other countries. Developing countries have not had widespread experience with the method. Because the track record of the price-cap approach is still incomplete, there is some concern that the incentives for improving efficiency will also provide incentives for reducing the quality of service. Further, if the price-cap method is not administered well, it can lead to profit levels that some would consider excessive.

Notes

1. For a more general discussion of demand forecasting in telecommunications, see Littlechild (1979), chap. 3. For a discussion and review of the factors underlying telephone demand, see Taylor (1980) and Mitchell and Vogelsang (1991), chap. 4.

2. The objective of this book is not to discuss the premises underlying the portion of normative economic theory known as welfare economics. A good introduction to this subject in the area of telecommunications can be found in Littlechild (1979), chap. 9.

3. Unless otherwise noted, all costs are assumed to be in constant prices with respect to a given base year, that is, net of inflation.

4. For a general discussion of shadow pricing, see Little and Mirrlees (1976) and Squire and van der Tak (1975). For further elaboration of the use of shadow pricing in public enterprise pricing policy, see Munasinghe and Warford (1978).

5. This section relies heavily on Saunders, Warford, and Mann (1977).

6. In theory, the total cost of an additional lump of investment should be attributed to the very last unit of output. In practice, however, the output considered as incremental has to be averaged over a particular period of time. In fact, the investment can be justified when consumers show their willingness, over a given period, to assume the financial (and presumably the economic) burden of additional capacity investment measured over the same period. See the definition of the capital recovery factor (r) in the TLRIC estimation method.

7. Turvey and Anderson (1977). Where capital indivisibility is a major problem, as when inadequate data or price inflexibility do not permit the TLRIC approach to be used satisfactorily, a simple shorthand approach that has been widely used in World Bank public utility projects is the "average incremental cost" method, which is defined as the discounted present worth of system costs divided by the similar discounted value of incremental output. For a description, see Saunders, Warford, and Mann (1977).

8. $r = I(1 + i)^n / (1 + i)^n - 1$, where I is the investment cost, i is the appropriate interest rate, and n is the useful length of life of the investment.

9. The material in this paragraph is drawn largely from parts of a paper by Mitchell (1989).

10. For a more detailed discussion of these methods, see H. Okazaki (1984); Roojsma (1985); Yoshita and Okazaki (1985); and Skoog (1980).

11. Tomlinson and Wellenius (1987).

12. The Canadian federal telecommunications regulatory agency is the CRTC (Canadian Radio-Television and Telecommunications Commission). For an explanation of the CRTC cost inquiry phase III methodology, see CRTC (1984); Bigham and Wall (1989); and Bigham (1990).

13. Terminology often differs between British and North American descriptions of identical items in telecommunications systems; for example, subscriber apparatus (United Kingdom) = terminal equipment or customer premises equipment (North America); subscriber local line (United Kingdom) = local loop (North America); exchange (United Kingdom) = switch or central office (North America); long-distance (both) = trunk (United Kingdom) = toll or interexchange (North America).

14. Kahn and Shew (1987).

15. These include approval of interconnection arrangements, authorization to provide competitive services, authorizations for customers to provide their own telecommunications links, and nonsectoral antitrust legislation.

16. Selwyn and Lundquist (1989).

17. Federal-Provincial-Territorial Task Force on Telecommunications, Canada, (1988).

18. Johnson (1989).

19. See Littlechild (1983); Rohlfs and Shooshan (1988); Selwyn (1988); Director General of Telecommunications, OFTEL (1988); Hartley and Culham (1988); and Beesley and Littlechild (1989).

20. In the United Kingdom, the equivalent of the consumer price index is known as the retail price index. In the United States, the Federal Communications Commission has proposed a price-cap formula based on the price index of the gross national product. See U.S. Federal Communications Commission (1989).

21. See Johnson (1989) and Bethesda Research Institute (1988).

22. Johnson (1989).

Part VI

Mobilizing Resources
and Promoting Efficiency:
Alternative Approaches

Chapter 16

Restructuring the Telecommunications Sector

FOR TELECOMMUNICATIONS, THE 1980s was a decade of revolution. The traditional modalities of the sector's organization were successfully challenged in industrial countries and gave way to structures that were vastly more complex and dynamic. A rapidly growing number of developing countries undertook or began preparing to undertake a major overhaul of their telecommunications policies. A rapidly expanding body of literature addresses the issues and options for changing the structure of telecommunications and draws lessons from the practical implementation of sectoral reforms.[1]

This closing chapter provides a link between the approach taken in this book, which examines telecommunications from the standpoint of costs and benefits, and alternative approaches that emphasize policy, structure, and regulation of the sector. To this end, we shall briefly examine several questions: How has the perception of the role of telecommunications in development been evolving in recent years? Why are structural changes taking place in the telecommunications sector, and what directions are they taking? What are the main issues that need to be addressed for these changes effectively to help overcome past constraints?

Evolving Views on the Role of Telecommunications in Development

Telecommunications is a huge business that moves at a breathtaking pace, commanding the attention of politicians, government au-

303

thorities, transnational corporations, and the world's capital markets. *The Economist, The New York Times,* and other major periodicals regularly carry editorial material and news on telecommunications. It is easy to forget that twenty years ago, only a few analysts perceived the importance of telecommunications and began to make the case that it should be a basic input in economic development. How did this turn-around occur?

Until the early 1960s, telecommunications were often considered to be a service mainly consumed by the well-to-do for mostly trivial purposes. During the 1960s and 1970s, however, policymakers gradually began to recognize telecommunications systems as essential infrastructure for economic development. It was shown that telecommunications services were used mainly in connection with a wide range of economic production and distribution activities, delivery of social services, and government administration. They also contributed to improving the quality of life and to achieving social, political, and security objectives. When they were available, telecommunications benefited a broad cross-section of the urban and rural population from many different income, educational, and occupational strata. These features generated high social and private returns (with appropriate tariffs) from telecommunications investment and considerable capacity to mobilize resources.

In the 1980s, information came to be regarded as a fundamental factor of production, along with capital and labor. The information sector accounted for one-third to half of GDP and of employment in OECD countries in the 1980s, and this figure is expected to reach 60 percent for the European Communities by the year 2000. Information also accounts for a substantial proportion of GDP in the newly industrialized economies and in the modern sectors of less-developed countries.[2] In the 1980s, economic activity became increasingly intensive in information, and the globalization of capital flows, trade, manufacturing, and other activities advanced. This produced a strong demand for better, more varied, and less costly communications and information services. Growth in demand became intertwined with rapid changes in telecommunications technology fueled by advances in microelectronics, software, and optics. These changes have greatly reduced the cost of transmitting and processing information, altered the cost structures of telecommunications and many other industries, created new ways of meeting a wider range of communication needs at lower cost, reduced the dependence of users on established telecommunications operating companies, and increasingly integrated information and telecommunications technologies and services.[3] These interrelated processes, illustrated in figure 16-1, are still well under way and show no signs of abating.[4]

In this context, telecommunications are now widely considered to be a strategic investment that is key to maintaining and developing competitive advantage at the level of the nation, the region, and the firm. Telecommunications constitute the core of, and provide the infrastructure for, the information economy as a whole. From the standpoint of the user, telecommunications facilitate entry to the market, improve customer service, reduce costs, and increase productivity. They are an integral part of financial services, commodities markets, media, transportation, and tourism, and they provide vital links among manufacturers, wholesalers, and retailers. Moreover, industrial and commercial competitive advantage is now influenced not only by the availability of telecommunications facilities but also by the choice of network alternatives and the ability to reconfigure and manage networks as corporate objectives change. Countries and firms that lack access to modern systems of telecommunications cannot effectively participate in the global economy. This applies to the least-developed countries of Africa and Asia as much as to the middle-income countries in Latin America, East Asia, and Central and Eastern Europe, which aspire to join the ranks of industrial countries in the next decade or so.

Changing Sector Structures

As the views on how telecommunications relate to the economy have evolved, major changes have occurred in the organization of the sector. Telecommunications used to be regarded as a natural monopoly and a relatively straightforward public utility. Economies of scale, political and military sensitivities, and large externalities made telecommunications a typical public service.

Industrial Countries

In most industrial countries, telecommunications services were provided on a monopoly basis by government departments or state enterprises (the main exceptions were Canada, Finland, and the United

Figure 16-1. *Driving Forces of Reform in the Telecommunications Sector*

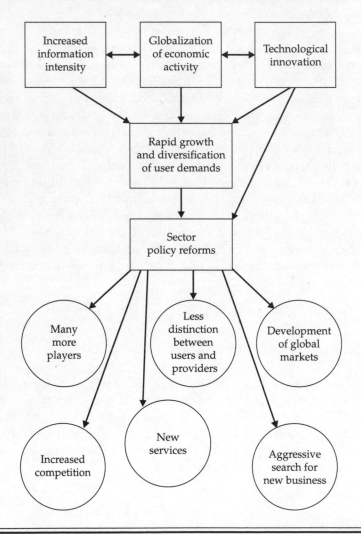

Box 16-1. *The Precursors: Liberalization and Privatization*
in the United Kingdom

The liberalization of the telecommunications sector and the eventual privatization of British Telecom was a flagship project for the Thatcher government's program of free market economic reform and one of the world's pioneer efforts to privatize a public telecommunications authority. In order to introduce competition, the markets for consumer equipment and cellular services were opened to new providers in 1981. In 1984, Telecom was privatized through the sale of 51 percent of shares in the domestic and foreign markets. A duopoly was set up in the area of basic fixed-link telecommunications between Telecom and a new operator, Mercury Communications, Ltd., initially a joint venture between Cable and Wireless, British Rail, and Barclays Bank.

Also in 1984, the Office of Telecommunications (OFTEL) was created as an independent, nonpolitical, regulatory body, outside the Department of Trade and Industry (the ministry responsible for telecommunications policy and licensing). OFTEL was headed by a powerful director general for telecommunications, with authority to enforce and amend rules, recommend licensing, and develop strategies. The regulator was intended to be strong enough to control the dominant operator (Telecom) and help Mercury develop as a competitor, but without intervening excessively in the market and thus undermining the benefits of competition.

By 1992, the British telecommunications market had been thoroughly transformed into one of the most dynamic in the world. In particular, Telecom now has new drive and vigor and is very profitable. It has implemented a massive domestic investment program and successfully entered into new venues such as foreign investments and information systems and services. More than 2 million telephone lines have been added, and over 90 percent of households now have telephone service. With some 2 million shareholders, Telecom ownership is widely distributed, including 96 percent of its employees. On the other hand, Telecom has experienced problems with quality of service, and its high profitability has resulted in public clamor for lower tariffs.

OFTEL, while remaining a relatively small organization with about 200 staff, has succeeded in playing a strong, proactive role in shaping British telecommunications policy. Price-cap regulation, pioneered by OFTEL and increasingly adopted in other sectors and countries, has reduced tariffs in real terms by about one-third. Policy has evolved flexibly in consonance with actual development of the competitive environment. Effective competition with Telecom and Mercury has developed in important markets, especially from cellular operators and a large number of cable television operators. Competition in basic network services has, however, been growing slowly under the duopoly system, and Telecom still controls 95 percent of voice telephony in Britain. Mercury remains quite small in size and scope and is competing mainly in digital local and long-distance services for businesses in large cities, still aided by significant pricing protection from OFTEL. A review of the duopoly policy in 1990 resulted in a regulatory decision to open the core telecommunications business to more competition. So far, this opening has not attracted new investors in new basic networks and services.

Source: Drawn largely from King (1989); Laidlaw (1991); Carsberg (1991); and Gillick (1991).

States). These monopolies generally succeeded in building and profitably operating national infrastructures, meeting the demand for basic telephone service, and introducing more advanced services. In the 1980s, however, major structural changes took place, driven by the twin forces of technological innovation and qualitative changes in demand. Deregulation and divestiture of the Bell system in the United States were followed by privatization and the introduction of competition in the United Kingdom, Japan, and, more recently, Australia and New Zealand (see box 16-1). By the early 1990s, virtually all OECD countries were in the midst of, or had completed, restructuring their telecommunications sectors. Overall, these reforms increased the system's ability to satisfy user needs, greatly broadened user choices, increased productivity, and reduced prices.

Developing Countries

In developing countries, telecommunications services were initially run by foreign private companies and colonial government agencies. During the 1960s, most telecommunications operations were nationalized and taken over by the public sector. Unlike the PTTs in the industrial countries, however, these state monopolies generally fell short of meeting needs. There continued to be large unmet demand for telephone connections, congested call traffic, poor quality and reliability of service, limited territorial coverage, and a virtual absence of modern business services. In the 1970s, attempts to overcome these shortfalls focused on obtaining a larger share of the limited public funds and external development credit and aid for telecommunications. Moreover, enterprises in creditworthy countries borrowed funds and, in some cases, imposed mandatory subscriber financing. These efforts resulted in sustained, rapid growth and modernization of telecommunications services. In Brazil and Costa Rica, for example, annual growth rates reached or exceeded 15 percent. Overall, however, the sector did not have access to enough capital, nor did it use its scarce resources efficiently. Deteriorating national economies and tight international credit in the early 1980s further constrained investment and led to escalating shortages of supply and lower quality of service, even in countries that earlier had done fairly well.

Starting in the mid-1980s, a growing number of governments in developing countries realized that sectoral arrangements had to be overhauled. This movement toward change was driven by the same factors underlying reforms in the industrial world, namely technological innovations and increased demand, amplified by five additional factors:

• State monopolies had reached the limit of their ability to accelerate the supply of telecommunications services. In particular, governments realized that they could not provide the huge amounts of capital required to catch up with demand. For developing countries in Asia and the Pacific, Latin America, and Africa to catch up, as a group, with existing unmet demand for basic telephone service by the year 2000, they would have to invest about $25 billion to $30 billion a year, throughout the 1990s, or in real terms, about five times the average level of investment achieved in the 1980s. To this figure at least $10 billion a year should be added for upgrading facilities in Central and Eastern Europe and in the former U.S.S.R., and this does not include the unknown, but rapidly growing, amount needed to develop more advanced business services and facilities. Although reorganizing the enterprise and improving management allow telecommunications entities in some developing countries to use their scarce resources more efficiently, in the context of inadequate government policies, these internal changes alone have limited potential.

• In recent years, many developing countries have begun to adopt market-oriented economic strategies, including measures to liberalize trade, promote competition, deregulate financial and capital markets, reduce restrictions on foreign investment, and restructure public enterprises. In order for these broad economic reforms to be effectively implemented, adequate telecommunications infrastructures urgently needed to be developed. The constituency for telecommunications development quickly grew to embrace a crowd of powerful and vocal users, including multinational companies, domestic and foreign investors, industrialists, traders, and bankers, all demanding the services they needed to succeed under the new economic strategies. Telecommunications became a central theme in multilateral and bilateral negotiations on trade in services. Suddenly, governments felt the need to become involved in telecommunications.[5] The new economic strategies also provided a context in which the new models of sectoral organization became politically acceptable. In several countries (such as Argentina), the state-owned telecommunications operator was the first target of efforts to restructure state enterprises.

• Popularly elected governments found that public dissatisfaction with service and, in many countries, extensive corruption of telephone company personnel resulted in widespread public support for major reform initiatives.

• Parallel changes in the telecommunications sectors of industrial countries raised international awareness of a wide range of sectoral policy issues and options and demonstrated the viability and increasing political desirability of alternatives to state monopoly.

• Telecommunications operating companies in industrial countries, repositioning themselves in their own changing domestic and regional markets, aggressively pursued new business opportunities in developing countries. Also, foreign banks sought to shift their exposure in highly indebted developing countries from nonperforming loans to new investment opportunities. Telecommunications investments in developing countries appeared to be particularly promising.

Main Directions of Change

Although fairly universal policy issues and options face governments attempting to reform the telecommunications sector, their relative importance, the sectoral solutions adopted, and especially the strategies to implement them are highly country-specific.[6] Yet, all telecommunications reforms so far involve some degree of change along each of four directions: (1) commercializing and separating operations from government; (2) containing monopolies, diversifying supply of services, and developing competition; (3) increasing private sector participation; and (4) shifting government responsibility from ownership and management to policy and regulation. This is true for developing as well as for industrial countries.

Commercializing operations. Telecommunications operations are being reorganized along the lines of commercial companies. There is widespread agreement that regardless of who owns them, telecommunications operating entities perform best when they are run as profit-driven businesses. Achieving this goal with state-owned entities involves either transforming them into companies or otherwise creating for them conditions that approximate the freedoms, incentives, and discipline of a commercial enterprise. The continuum along which state operating entities are separated from governments begins with the reorganization of government departments into state enterprises; state enterprises are then transformed into state-owned companies, which are transferred, in turn, to mixed state/private ownership and then from mixed to totally private ownership. Whatever the starting point, moving farther along this continuum tends to free the entity from rules that may be appropriate for government administration but not for a dynamic, high-technology, capital-intensive business. All reforms are moving in this direction, although only some have completely privatized state operations. At the same time, improvements are also being undertaken in internal organization and management, such as reorganizing the enterprise into cost and profit centers, subcontracting functions that can be performed efficiently by other organizations,[7] establishing or improving commercial accounting and management information systems, and emphasizing customer service, cost awareness, financial discipline, and staff performance.

Diversifying supply and developing competition. Allowing new enterprises to supply services and networks can attract fresh sources of capital and management to the telecommunications sector, develop rivalry among service providers regarding performance and price, and generate cost benchmarks to guide pricing of monopoly suppliers. A single monopoly operating enterprise, whether state-owned or private, is increasingly unable to meet equally well the large, varied, and rapidly changing demands of all types of users. Diverse alternatives are available. They include (a) monopolies divided by regions; (b) joint ventures in which the established operator and other enterprises provide specific new services or facilities (for example, very small aperture terminal, packet-switched data, cellular); (c) build-operate-transfer and related arrangements with foreign operating companies, equipment manufacturers, and investors for creating parts of the public network; (d) selected specialized networks licensed to meet the needs of major communication-intensive sectors of the economy (such as banking, tourism, mining) and to attract resources from them for investment; (e) independent public telephone companies licensed in unattended areas (such as new industrial estates, residential developments); (f) licensed extensions of the public telephone network (for example, cellular, rural subscriber radio); and (g) articulated rules that govern voluntary commercial relations between dedicated and public telecommunications networks.

In the context of broad economic liberalization, an essential element of reforming the telecommunications sector is a policy on competition. Competition, or a credible threat of competition, is likely to spur established operating enterprises to focus attention on customers, improve service, accelerate network expansion, reduce costs, and lower prices. Competition also widens user choices and accelerates the introduction of new services and facilities. Elements of competition can be effectively introduced in the early stages of reform and then extended by stages to affect most or all segments of the market. Modalities include competitively awarded concessions to a monopoly or duopoly, regulated entry, and unrestricted competition.[8] Technological changes are making competition possible in a widening range of market segments. Competition in the provision and maintenance of terminal equipment belonging to the customer and value added services is beneficial in virtually all countries. The introduction of some competition in long-distance networks makes sense because the increased volume of traffic is reducing the importance of economies of scale. At present, large economies of scale still exist in local networks, which makes competition in the provision of wired local services viable only in exceptional situations, such as highly developed urban business districts. New radio technologies (for example, cellular, personal communication networks, mobile satellite), although still more costly than wired telephones, already offer competitive alternatives for business and high-income residences under certain circumstances, such as when conventional telephone lines are very scarce or perform poorly and when rapid installation is highly valued (for example, when existing networks must be upgraded quickly, as in Eastern Europe, or when disaster relief is crucial).

Increasing private sector participation. A larger role is being accorded the private sector in many countries. Given the right policy environment, private enterprise and investors have responded vigorously to business opportunities in telecommunications. As private sector participation increases, the telecommunications sector can be expected to attract new sources of capital, management, and technology and also to contribute to the development of the private sector overall. More and more developing countries are privatizing their state telecommunications enterprises. This has mainly involved selling the telecommunications provider to a consortium of domestic and foreign investors and foreign operating companies. Privatization by itself does not, however, guarantee improvement of the sector and is not always feasible.

Many other avenues are available for private participation in telecommunications. Existing enterprises may divest or outsource construction, maintenance, transportation, routine design, billing and collection, directory services, operator assistance, and other functions traditionally undertaken internally. State-owned enterprises can be reorganized under company law, and shares can be sold gradually to the public even if the state retains a controlling interest. Bonds and other debt instruments can be floated in domestic and, sometimes, foreign markets. Franchising, leasing, revenue-sharing, build-operate-transfer, and other arrangements offer options of growing interest. Private financial entities can be established to invest in profitable new telecommunications ventures; for example, facilities could be operated by other entities, such as the existing or a new telecommunications company, under lease, revenue-sharing, or some other arrangement. Subscriber financing schemes can raise a large proportion of the local funds needed for expansion when funds are in severely short supply, and they also can promote the participation of users in company ownership and debt. Most of the options to diversify supply and introduce competition are suited to private enterprise.

Separating and developing regulation. Policy and regulatory functions are being separated from operations. As long as telecommunications operations remain in the public sector, the political system provides, however imperfectly, for reconciling diverse objectives such as commercial efficiency and broader national and regional development. As operations move away from direct government control and the number of participants in the telecommunications business increases, however, this arrangement breaks down and the functions of policy, regulation, and operation must be developed separately. At this point, the nature and extent of additional sectoral reforms that can be undertaken are constrained by the development of institutions capable of effectively formulating and regulating the implementation of policy. In particular, regulation is essential when public sector monopolies are privatized, and an effective regulatory capacity must be in place at the time of sale. Regulatory frameworks and institutional arrangements must be developed early to monitor operator performance, to prevent abuses of market power, and to promote competition. Alternatively, temporary arrangements could be made to oversee the transition from public to private ownership and to deal with regulatory matters that require immediate attention while a more permanent capability is being built.

The Beginnings of Telecommunications Reform in Developing Countries

By 1992, major reforms of the telecommunications sector had been completed or were well under way in at least fifteen developing countries, and a comparable number were in preparation. The pace and scope of structural change have varied considerably among regions. The reform movement got to an early start in Latin America. Privatization of state telecommunications enterprises was completed in Chile (1987), Argentina (1990), Mexico (1990; see box 16-2), and Venezuela (1991), and in 1992 reforms were being prepared in Panama, Peru, and Uruguay and being considered in Bolivia, Nicaragua, Paraguay, and other countries.

Following the collapse of the communist regimes in Central and Eastern Europe and in the former U.S.S.R., rapid progress was made in outlining broad strategies for reforming the sector in this region as well (see box 16-3). By 1992 all of the new governments were developing

Box 16-2. *Privatization in a Large Developing Country:*
The Case of Mexico

At the time of the devastating 1985 earthquake in Mexico City, telecommunications in Mexico had already reached a state of near crisis. In 1989, the government responded by announcing a comprehensive plan to modernize the sector. This plan sought to establish a new regulatory framework that would promote efficiency, competition, and private investment; to privatize Teléfonos de México (TELMEX, the mainly state-owned monopoly telephone company) while regulating price and quality of service; and to introduce competition in local and long-distance telephone services as well as in cellular and other new services.

These far-reaching objectives were largely achieved by 1991. New telecommunications regulations enacted in 1990 established a comprehensive, modern sectoral framework in which policy formulation, licensing, and regulatory functions were exercised by the government and networks and services were provided largely by the private sector in an increasingly competitive marketplace. The telephone tax was abolished, and tariffs were increased on average and rebalanced to approximate international practice. The Secretaría de Comunicaciones y Transportes (SCT), the ministry responsible for telecommunications, was reorganized to become the telecommunications regulatory agency. Operating responsibilities were transferred from the SCT to Telecomunicaciones de México (TELECOM), a new autonomous state enterprise that became the domestic and international satellite carrier. TELMEX was reorganized along the lines of a modern commercial company, labor contracts were renegotiated to allow management greater flexibility and facilitate technological innovation, and capital structure was changed to prepare for privatization. A controlling interest of 20.4 percent was sold in 1990 to a consortium of private domestic and foreign investors in association with U.S. and European telecommunications operators. The remainder of state-owned TELMEX shares were sold to employees and the public at large, in domestic and foreign capital markets, in successive tranches in 1991 and 1992.

The new TELMEX has a monopoly over all basic services and networks for six years, after which competition will be allowed. During the reserved period, TELMEX's concession establishes stringent requirements for growth and quality of service, including quantified obligations to extend service to rural areas. Tariffs are subject to declining caps as well as limits to increases in residential charges. The market for customer premises equipment, private networks, and value added services has been liberalized, and a growing range of network choices is available to businesses. In each of the country's nine cellular regions, one TELMEX subsidiary and one independent operator compete on a duopoly basis. The foundation for competition in long-distance services after 1996 is being laid in several ways. The nine independent cellular operators have formed an association to coordinate their relations with TELMEX and the government, and three of the companies have linked themselves via a small digital microwave network and tandem exchange.

TELMEX has emerged from the reform as a very successful, viable private company. It has met or slightly surpassed its concession targets in nearly all areas. Growth in telephone lines exceeded 12 percent for 1991, quality and reliability of service have improved in most parts of the country (and a major network improvement program is under way in the metropolitan area of the capital), a new digital overlay network has been completed for large businesses using voice and advanced data and other services, substantial productivity gains have been realized, and prices have been further realigned toward costs while remaining within the mandated aggregate limits. Profits have risen dramatically, and over 60 percent of the 1991–96 investment program will be financed from internally generated funds. Share prices have increased more than four times in value.

Regulation has made slower progress. The SCT's response to several regulatory challenges has been good so far, but its new organizational structure was only approved in 1992, and there is an acute shortage of managers and senior professional staff able to deal with the rapidly increasing amount of regulatory issues. The agency may be overwhelmed by the technical demands of monitoring TELMEX's performance and dealing with regulatory issues of the North American Free Trade Agreement being negotiated.

Source: Based on Casasus (1991) and on internal World Bank sources.

Box 16-3. *Catching up with the Industrial World:*
Central and Eastern Europe

The telecommunications sector was much neglected by the former communist regimes in Central and Eastern Europe (CEE). The current technology lags fifteen to twenty years behind world practice, and there are on average only about 11 telephone lines per 100 inhabitants compared with 30 to 50 lines per 100 inhabitants in Western Europe. Furthermore, service is poor quality and unreliable. Today, CEE countries face three challenges as they develop their telecommunications networks: they must greatly accelerate the rate of growth and increase the scope of services; they must begin implementing the latest digital technologies; and they must adapt their institutions to a newly competitive environment.

In response to these challenges, most CEE countries aim to double or triple investment and to achieve an annual growth rate of 10 or 12 percent to reach an average density of about 30 telephone lines per 100 inhabitants by the year 2000. The offices administering posts, telegraphs, and telephones (PTTs) are being gradually commercialized and have been, or will soon be, separated from regulatory and policymaking functions in the ministries. Most countries plan to follow the framework for introducing competition and structural reform set out by the 1987 European Commission's Green Paper on Telecommunications.[1] Existing domestic manufacturers of equipment are being strengthened by joint ventures with foreign companies that try to give them a significant role in the growth efforts.

Implementation of these strategies has barely begun. A rapid explosion in demand, resulting from efforts to develop market economies, has put increasing pressure on countries to liberalize faster, but the legacy of socialist institutions and business processes has made this difficult to put into practice.[2] In particular, regulatory structures and policies are still very underdeveloped.[3] Financing is another stumbling block; estimates of what the region will need over the next decade range between $3.5 billion and $6 billion a year, amounting to some 2 percent of GNP, and foreign exchange, in particular, will be difficult to obtain in sufficient quantities.[4]

Overall, these goals are probably just barely feasible, if operators use every means available to them to reduce costs, increase revenues, attract outside debt and equity, and control financial planning. In addition, an integrated approach covering policy, regulation, investment financing, and enterprise organization and management is essential for each country, so that all the efforts to change occur in harmony. Finally, significant help from the West is required, both in financing the changes and in providing technical and management expertise, to bring reforms to fruition.

Notes

1. As Europe draws toward closer integration under the single European Community market of 1992, policy consistency in the area of telecommunications is being strongly influenced by the 1987 EC Green Paper on Telecommunications and by subsequent EC legislation. Common goals are (1) an open EC-wide market for terminal equipment, (2) an open market for value added services and liberalization of data services, while accepting national differences in basic network and voice telephony, (3) separation of regulation and operations, (4) development of consistent technical standards to promote a single terminal market, (5) mutual recognition of service licensing to promote a single services market, and (6) definition of Open Network Provision to give new service providers fair access to the facilities of the network infrastructure and basic services.

2. East Germany is, of course, an exception. Technically, however, it faced perhaps one of the worst situations; one-quarter of its switches were installed before 1934, none were installed after 1965, and many were fully depreciated two or three times over. Unification of the Federal Republic of Germany and the German Democratic Republic was completed in October 1990. Unification of the two PTTs occurred by the introduction of West German legislation (itself recently reformed in 1989–90 along the lines of the 1987 EC Green Paper) into the East German PTT. The government of Germany made the political decision to bring East German regions up to the general level of West German telecommunications by the year 2000, and a massive investment program was immediately begun to achieve this goal. The total cost is estimated to be around DM60 billion, which will undoubtedly exert a significant drag on other investment.

3. This is the case even in Hungary, where reforms of the sector are most advanced. Reforms have been under way since 1989, when the Hungarian Telephone Company (Matav) was separated from the former PTT (Magyar Posta). A ten-year investment program aims to give Hungary a density of 28 lines per 100 inhabitants by 2000, requiring investment of $4.5 billion; a separate three-year program aims to install business services and a digital overlay network at a cost of $1 billion. Less attention has been given to matters of regulation.

4. Even without foreign exchange, much can be done. Between 1973 and 1983, the Bulgarian PTT implemented a concerted effort to expand and upgrade its network, based largely (about 80 percent) on internally generated funds. Municipalities, cooperatives, and state-owned enterprises also helped construct local networks, which were then transferred to PTT ownership. In spite of institutional and structural limitations, Bulgaria achieved a density of 17 lines per 100 inhabitants, nearly twice the regional average. The program was derailed by problems with technology transfer and funding cutbacks after 1983. The total cost was estimated at Leva 1.6 billion.

Source: Based largely on four papers: Muller and Nyevrikel; Neumann and Schnoring; Nulty; and Ungerer, all in Wellenius and Stern (1991). See also Whitlock and Nyevrikel (1992).

the details of implementation, and Hungary, Poland, and Ukraine, among others, were building state-of-the-art business networks through joint ventures (with, for example, foreign operating companies and through other investment and management modalities).

Reform has been slower and more limited in Asia: there have been corporatization (1987) and subsequent partial privatization (1990) in Malaysia, corporatization and liberalization of nonbasic services in Indonesia (1990), decentralization of operations in India (1985, partial) and China (1988, extensive), and reorganization of telecommunications departments to become state enterprises in Sri Lanka and Fiji (1990; see box 16-4). The pace appeared to be picking up in 1992, however, with Pakistan preparing for privatization, Thailand embarking on a series of major build-operate-transfer ventures, and the Indian government exploring options to restructure the telecommunications operating entity and possibly open the market to new service providers.

In relative terms, the least-developed countries have the most to gain from sectoral reforms, yet efforts to overcome telecommunications shortages in Sub-Saharan Africa have so far been confined largely to improving the performance of state entities (see box 16-5). Although some countries have explored new modalities, including joint ventures with PTTs and the introduction of some competition, most governments still hesitate to consider broader reforms and privatization. This mainly reflects their concern with limited resources, especially skills to prepare and implement such programs. Slow progress is compounded by small markets, extreme paucity of existing facilities, disproportionate size of social needs, and sometimes unfriendly economic policies, which make investment in these countries less attractive to foreign operators and investors than opportunities in other regions. Nevertheless, telecommunications privatization is now under way in Côte d'Ivoire and Guinea, other governments have expressed interest in their experience, small private ventures (for example, cellular) led by local entrepreneurs have been successful in a few countries, and regional organizations are starting to examine broad aspects of telecommunications restructuring.

Issues and Lessons

The initial experiences that developing countries have had in reforming their telecommunications sectors are generally encouraging. Where such changes have occurred, telephone service has expanded and improved at a faster pace, productivity has increased, new services

Box 16-4. *Restructuring without Privatization: Sri Lanka*

In the mid-1980s, the telecommunications sector in Sri Lanka was so weak that it constituted a considerable drag on overall economic development. There were only 0.7 telephone lines per 100 inhabitants, and the quality of service was very poor. The monopoly operating entity, the Department of Telecommunications (SLDT), a government department within the Ministry of Posts and Telecommunications, was a civil service organization suffering from inefficiency, noncompetitive remunerations, and low staff morale. In response to general dissatisfaction with this state of affairs, in 1985 the government appointed a Presidential Commission of Inquiry to consider reorganizing the SLDT and the telecommunications sector overall.

The Commission recommended a radical restructuring that would transform the SLDT into an independent, autonomous, government-owned, but commercially oriented company and the creation of an independent regulatory agency, the National Telecommunications Commission. A foreign operating partner would contribute capital and expertise once the new company was established. In 1986, the Telecommunications Board of Sri Lanka was appointed to implement these goals. At the same time, an act of parliament was drafted to provide the legal framework for these changes, several licenses were issued to new operators in cellular and radio paging services, and the market for subscriber terminal equipment was opened to competition. However, opposition from organized labor and SLDT management, mounting political opposition to privatization of state enterprises generally, and growing civil unrest in parts of the country eventually brought this process to a halt. The Board was dissolved in March 1989.

In 1990, the government adopted a scaled-back version of the original plan. The operating functions of SLDT were transferred to a new, government-owned corporation, Sri Lanka Telecom (SLT), which is autonomous and commercially oriented, but under overall government control. A new telecommunications act was passed by parliament in 1991, which confirmed the establishment of SLT and created an independent regulator, the Telecommunications Authority, headed by a director general, answering to the minister of posts and telecommunications.

Addressing the concerns of the opposition was the key to implementing this limited restructuring. In response to SLDT management's concern for loss of seniority and imposition of outside control, new management was only hired in certain critical areas, such as finance and marketing. Organized labor feared layoffs, loss of pensions, and loss of the social status conferred by being in the civil service; therefore, the transition was made flexible, and employees were given the option of maintaining their civil service status in the new SLT (only 17 percent of personnel chose to keep their civil service status, and 80 percent chose to accept the new working conditions, which required greater effort and efficiency, but which rewarded these efforts with higher remuneration). Finally, a massive training effort was undertaken, with the assistance of the ITU and the United Nations Development Programme to introduce commercially oriented work practices into the new company.

Source: Based largely on Watson (1991) and Mendis (1989).

Box 16-5. *The Beginnings of Reform in Africa*

Telecommunications development faces its greatest challenge in Sub-Saharan Africa. Densities for the region are among the lowest in the world, and most rural areas have no access to services at all. Investment is sporadic, reflecting the overall acute shortage of investment capital for the region, and project implementation and subsequent maintenance are complicated by an equally acute shortage of skilled personnel. Another significant limitation is the prevalence of the traditional structure of posts, telegraphs, and telephones (PTT). In the past four or five years, however, some partial reforms have begun to show positive results.

Benin. Modest reforms were begun in Benin in the mid-1980s, in concert with a $70 million investment program funded by a group of five multilateral agencies. The investment plan required the Office of Posts and Telecommunications (OPT) of Benin to implement an internal institutional reform program, which included (a) reform of the financial functions, including bill collections, the postal savings bank, and separation of the OPT treasury from the state treasury, (b) modernization of administrative functions such as marketing, human resources, and procurement, and (c) internal preparation for the institutional separation of posts and telecommunications at the end of five years (1988 to 1993). At the same time, the government of Benin was encouraged to give greater responsibility and autonomy to the OPT by replacing its direct involvement in day-to-day management with a system of contractual agreements between the state and the enterprise based on multiyear development plans.

The results have been encouraging. As of early 1992, the internal reforms were working well, especially in the area of management, and the investment program was on schedule. Financial reforms have also been successful. The rate of uncollected bills fell from more than 50 percent to reach about 10 percent, and OPT's treasury was separated from that of the state. However, many significant problems remain: the government continues to interfere in daily management, and entrenched civil service structures have limited the changes in employment patterns.

Guinea. A more ambitious reform effort is under way in Guinea as part of a government effort to privatize state-owned companies, which began in 1985. In the case of the telecommunications sector, an interministerial committee was created in 1989 to supervise institutional reform, and extensive outside assistance in the form of technical and financial consultants is being used. It has been determined that the PTT should be separated into two parts: a mixed-capital, commercially oriented monopoly telecommunications provider, SOTELGUI, and a government-owned Office des Postes Guinéennes (OPG). This separation is to be finalized on March 1, 1993. In the meantime, three preparatory processes are under way: (a) a search for private sector partners for the new SOTELGUI enterprise, (b) operational and technical assistance aimed at developing structures for the new entities, and (c) engineering studies to support the investment program that will accompany the restructuring.

Madagascar. Reform of Madagascar's telecommunications sector is also progressing at a brisk pace with the assistance of outside consultants. A current plan calls for (a) separating operational functions from the ministry, which will retain only regulatory responsibilities, (b) unifying all telecommunications operators (domestic services are currently provided by a department within the ministry, and international services are provided via the Societé des Télécommunications Internationales de Madagascar, STIMAD) into a single, mixed-capital, commercially oriented monopoly, (c) drawing up a social plan to guide the new operator's personnel planning and employee relations, and (d) implementing an investment program to improve the network and introduce new services. Implementation is currently set for the end of 1993, but a great deal of preparatory work still lies ahead.

Niger. Reform plans in Niger are still being discussed, but the government is well aware that the current structure is not viable. Among the ideas being considered are (a) integrating the satellite services provider STIN into the Office of Posts and Telecommunications, (b) separating Posts and Telecommunications into two offices, giving each enterprise responsibility for achieving certain objectives, along with the autonomy needed to do so, (c) decentralizing managerial responsibility within each organization, (d) creating a Human Resources Department within the telecommunications operating entity together with a scheme for developing human resources, (e) improving internal management, such as procurement practices, maintenance, and financial management, (f) creating a separate management department for the capital city, Niamey, and its immediate surroundings, where the majority of telephones in the country are located.

Rwanda. A comprehensive restructuring program for the PTT in Rwanda was drawn up by outside consultants between 1989 and 1991. This plan called for separating posts and telecommunications into RWANDATEL, a mixed-capital company, and the Office National des Postes (ONP), a state-owned company. Both would be autonomous in management and administration, have a commercial orientation, and be granted a thirty-year monopoly. In addition, the policymaking and regulatory functions would be reserved for the ministry in charge, while all operational responsibilities would be invested in RWANDATEL. The law creating the ONP was passed in March 1991. The law establishing RWANDATEL is still under study but is due to be adopted in July 1992; it calls for a first stage, under which state control of capital would be 99 percent, progressing to a second stage, in which the participation of national and foreign capital would rise to 49 percent. The law creating a director general for communications, with regulatory authority over the two companies, will be promulgated after RWANDATEL is created. The restructuring effort is taking place in concert with a large-scale investment program funded by three multilateral lenders.

Source: Drawn from a series of papers presented at a seminar on telecommunications restructuring, sponsored by the World Bank, the ITU, and ACCT, held in Tunisia in May 1992. The papers include Jacques Hababib Sy, "African Nations and Access to Telecommunications Services: Political Economy and Legal Issues"; H. Vignon, "La restructuration des télécommunications au Benin"; Mamadou Pathe Barry and Mohamed Sylla, "Le cas de la Guinée"; Mamiharilala Kasolojaona, "Colloque sur la restructuration du secteur de télécommunications Malagasy"; Anonymous, "Restructuration du secteur des télécommunications dans Niger"; and Habyalimana Malien, "Colloque sur la restructuration du secteur des télécommunications de Rwanda."

have become available, and in some cases, international capital markets have been tapped effectively. Sectoral reform is not, however, without dangerous pitfalls. Although it is still too early to assess the extent to which these reforms have overcome past constraints, several key areas of concern have already emerged.

Privatization. Privatization of state telecommunications enterprises is not universally feasible, nor does it necessarily improve performance. In particular, the timing and method of privatization in a given country are determined largely by relatively narrow and somewhat unpredictable windows of political opportunity and by broader developments in national economic strategy. In each case, governments must clearly identify the conditions necessary for privatization to be successful and the extent to which these conditions are likely to be met. A growing body of global experience with privatization shows the risks involved (to governments as well as to investors and operators) and highlights ways in which these risks can be reduced.

International market considerations are also increasingly important. In the near future, several telecommunications enterprises are expected to be sold in both industrial and developing countries. Timing and preparation will increasingly influence the extent to which privatization succeeds.[9] Commercialization of operations, organizational and financial restructuring of enterprises, renegotiation of labor contracts, and improvement of available information on the enterprise are actions that can make a particular offering more attractive.

Privatization, moreover, is not a one-shot deal, but rather a complex process of introducing private capital and know-how into telecommunications operations. The process has various aspects: (a) separating operations from government and nontelecommunications activities (for example, posts, manufacturing); (b) restructuring the telecommunications operator as an independent state enterprise that is financially self-sufficient and financially autonomous from the government; (c) reorganizing the enterprise internally so that it runs as a business; (d) restructuring the telecommunications enterprise under private company law; (e) devising a privatization strategy (including decisions on controlling interest, employee stock ownership, tranching of stock sales, and residual state ownership) and changing the company's capital structure to enable implementation of this strategy; and (f) carrying out the sale. These facets may all be dealt with over a relatively short period of time (Argentina, Mexico, and Venezuela each did so in less than two years), or they may develop in stages over longer periods (as occurred in Chile and Malaysia). Some stages (for example, internal reorganization) may be left to the new owners.

Designing and implementing this complex process properly require strong and visible high-level political commitment, clear allocation of authority and resources to manage the process, and expert (including foreign) assistance on policy, regulatory, legal, and financial matters. Early in the reform process, the government must clarify its position on tradeoffs among the conflicting interests that inevitably arise from privatization, such as conflicts among existing operators, organized labor, prospective buyers, potential competitors, investment bankers, the treasury, equipment suppliers, large users, and the public at large.[10]

Regulation and competition. The single most troubling issue in recent reforms has been the slow progress made in developing regulatory capabilities. All major reforms have been predicated on the expectation that effective public regulation of the privatized monopolies—especially with respect to prices, service obligations, interconnection, competitive behavior, and access to the public domain (including use of the radio spectrum)—can be implemented fairly quickly. Yet building regulatory institutions in countries with little or no regulatory tradition in any sector is proving to be an arduous and slow task. Whereas some developing countries have carried out satisfactory privatization in a short period of time (Argentina, Mexico, and Venezuela did so in a little more than a year), and at least one country (Chile) completed privatization some five years ago, so far not one has a properly functioning telecommunications regulatory system (see box 16-6). The largest privatized companies are operating with little or no competition and in a regulatory vacuum in which critical regulatory responsibilities regarding licensing, pricing, technical and accounting standards, and performance monitoring, to name just a few, are not being properly discharged. In a market dominated by one operator and lacking effective and proactive regulation, competitors are unlikely to emerge and become firmly established, and numerous forms of anticompetitive behavior may become entrenched.

The search for better regulatory solutions merits high priority in the design of sectoral reforms. Like privatization, development of regulation is not a one-shot affair, and it is intertwined with the development of the telecommunications market. Lack of an initial constituency for regulation might be partly overcome by establishing a council or advisory board that represents users and other key interests and

Box 16-6. *Regulation and Competition: Chile Five Years After*

Chile began a slow but ultimately very successful process of reforming its telecommunications sector by adopting in 1975 a national development strategy oriented toward free-market economics. Two mainly state-owned companies provided most telecommunications services as monopolies: Compañía de Teléfonos de Chile (CTC) had over 90 percent of local telephone lines, and Empresa Nacional de Telecomunicaciones (ENTEL) was responsible for most long-distance and all international facilities. The Ministry of Transport and Communications (MTC) had overall responsibility for the sector, while the National Development Corporation exercised state ownership of CTC and ENTEL, and the Ministry of Economy approved tariffs. In 1977, the Sub Secretaría de Telecomunicaciones (SUBTEL) was created within MTC as the specialized regulatory agency for telecommunications, responsible mainly for technical and administrative matters. A national telecommunications policy was promulgated in 1978, which set the basic principles that guided subsequent liberalization and privatization of the sector. A new telecommunications law in 1982, and an additional law in 1987, established the legal framework for reforms.

In 1986, the government began to sell some of its shares in ENTEL to company employees and the public at large through the Santiago stock exchange. In 1987, the government sold 35 percent of CTC shares with a controlling interest to Australian investor Alan Bond. That same year, the remaining ENTEL shares were sold to various domestic and foreign investors, and 20 percent were sold to Spain's Telefónica. In 1990, Bond sold his share of CTC (now 50 percent, after additional investments in accordance with the terms of the initial sale) to Telefónica, which then became a major force in Chilean telecommunications. Concurrently with privatization, the market was made more competitive.

The reforms have greatly revitalized the telecommunications sector. Investment has been substantial. Telephone lines, which had been growing at a paltry 5 percent a year in the 1970s and 1980s, expanded more than 20 percent in the early 1990s; by the end of 1991, CTC had 0.9 million lines in service, 35 percent of them installed in the previous two years. Networks have been quickly modernized; over 70 percent of telephone lines are digital, and extensive optical fiber cable and satellite networks are being developed. Many new services have been introduced, and some market segments have become very competitive. The supply of customer premises equipment has been fully liberalized, cellular telephone service is offered by three companies in the main cities, and data and private networks are provided by several new carriers. CTC successfully floated $100 million in new shares in the U.S. market in 1990. Reflecting improvements and good financial performance, the value of ENTEL and CTC shares roughly doubled in one year.

Not all is well, however. CTC and ENTEL still have too much market power, and strong competition in the core telephone business is slow in getting started. SUBTEL is not strong enough nor sufficiently well staffed to serve as an institutional locus for the regulatory system. Excessive involvement in litigation has delayed solutions to regulatory problems. Two court battles have remained unresolved during the past three years. First, CTC's planned expansion of long-distance facilities is being resisted by ENTEL, and CTC, in turn, opposes ENTEL gaining increased direct access to end users; second, the government is seeking to force Telefónica to divest either its CTC or its ENTEL holdings in order to avoid possible monopoly consolidation. No practical means have been developed to safeguard social service obligations, especially in rural areas; cross-subsidies are largely ruled out by mandatory cost-based pricing rules, and attempts at direct government subsidy have had limited success.

Source: Drawn largely from Melo (1991) and Ramajo (1991).

sets an agenda for the regulator. Restructuring the sector so that strong competitors are present in key market segments from the outset would generate demand among dominant operators for regulatory action and shift the regulatory burden to the technical issues of interconnection and standards. However, since there are limits to how much competition a particular market can sustain, another possibility would be to divide the regulatory function into discrete tasks, some of which could be subcontracted.[11]

The quality and progress of the telecommunications sector will be tied to the level of development of the country as a whole. Arrangements that are possible and necessary in the more advanced developing economies, which participate vigorously in competitive global markets, may be neither affordable nor necessary in less competitive ones. In many cases, optimal regulatory arrangements may not be possible, and compromise solutions will be necessary.[12]

Continued technological change. Rapid technological changes are likely to bring greater opportunities for lower-cost and reliable expansion of networks to developing countries, giving them the chance to leapfrog intermediate stages of network development. However, these changes also make designing, procuring, and managing new networks much more complex. For example, wireless technology for personal communication has emerged as a strong challenger to the fixed network; the cost of optical fiber systems continues to fall even as capacity increases; the new synchronous format for transmission systems permits flexible and inexpensive access to data streams; and faster computer technology is increasing the call-processing capacity of exchanges significantly. These developments are redefining the optimal network structure and reducing costs.[13] Broadband ISDN (Integrated Services Digital Network) has also arrived, permitting integration of video, data, and voice, although standards have yet to be established. The eventual result of this proliferation of new systems may be either a multitude of competing technologies or a limited number of proven technologies that emerge through market dominance and de facto standardization. The growing number of network alternatives has increased the importance of strategic planning and the need for revised solutions to designing and planning networks. The challenge is to reduce overall cost while ensuring the development of an integrated network that is robust, manageable, and consistent with broader development goals. In this context, the prevailing approach to network planning, which is based on long-term master plans followed by piecemeal projects, is likely to produce a suboptimal outcome.

Technological innovation also raises the question of whether plans based largely on today's dominant technologies are capable of evolving and accommodating major systemic changes. Some innovations will lower and eventually remove entry barriers that currently block new competitors in certain areas. For example, radio technologies such as new analog and digital cellular telephony, personal communication systems, and satellite mobile services, which are all reaching the market, will accelerate the demise of the wired local telephone network as a natural monopoly. Other developments, however, will have precisely the opposite effect. For example, new generations of optical fiber cables are capable of providing ever-cheaper bandwidth at a negligible marginal cost of transmission, reviving arguments for a natural monopoly in parts of the system where they no longer apply.

The potential impact of technology on regulatory strategy is especially intriguing. Policymakers should use continuing technological changes to lessen dependence on regulation because these changes will enable a larger portion of telecommunications operations to be submitted to the disciplines of the market. This raises the possibility that good regulation, if and when it is finally in place, may no longer be important. If that were the case, setting up temporary, self-extinguishing arrangements to oversee the initial years following reform would be preferable to developing self-perpetuating regulatory bureaucracies. Despite further technological changes, however, established service providers will likely retain considerable market power, which will continue to need regulation. Furthermore, some analysts believe that despite growing competition, successive waves of liberalization will tend to increase the need for regulation in the future.

Convergence of telecommunications and informatics. Telecommunications are becoming an integral part of a much broader family of information technologies and applications. Distinct economic activities are increasingly being structured as different end applications of a common business, namely the organization and management of telecommunications and information resources. Examples include banking, stock trading, broadcasting, publishing, library and information services, and national statistical services, as well as traditional telecommunications and computing. As the supply of telecommunications facilities grows and diversifies, the structural focus shifts from the conventional dichotomy between networks and users to a new, less well-defined sector, which requires organizing and managing facilities that are available from many sources and suit particular business needs. In this context, the boundaries among businesses become blurred, and, in particular, the strict identity of a telecommunications company is no longer clear.[14]

Policymakers are just beginning to look into the implications of these trends for economic development. The effects that at least four areas have on telecommunications policy and regulation require attention. First, the structure of the developing telecommunications supply is likely to influence the pace and direction of the overall informatization of the economy. An initial step is to explore the possibility of organizing the telecommunications sector in a way that allows a wide range of suppliers (including users and companies that resell to others) to develop facilities and services. This approach enhances the potential for innovation but makes it more difficult to establish industry standards and thereby risks slowing the development of critical new services and limiting interconnectivity among user groups. Second, the choice of technologies to modernize and expand the main telecommunications network influences the development of the sector and the applications of information technology in the future. In particular, generalized networks such as ISDN may provide solutions that are cost-effective for certain categories of users but not for others, delay the development of facilities-based competition, and simply cost much more than other solutions. However, some observers believe that this kind of common framework, once established, would provide a more viable platform for multiple uses than diversified network structures, which could be especially valuable in developing countries, and could in the long run eventually reduce costs. Third, the question arises of the role that dominant telecommunications companies play in providing nontraditional services. In particular, some analysts emphasize the economies that can be achieved when these companies enter the market for value added services; for example, they can add intelligence and processing capabilities to their existing networks and make these facilities accessible to users. Others, however, feel that new services should be implemented by alternative providers, as part of an overall effort to build up an increasingly competitive environment. And fourth, the development of advanced telecommunications services is likely to have important externalities. The pace of informatization of an economy will be affected by policy decisions as to whether telecommunications companies should merely respond to demand or whether they should also play a proactive role in, for example, building up a critical mass of data terminals among small and medium businesses to facilitate the development of information services.

Closing Remarks: Looking Forward

About 80 percent of all developing countries have yet to address structural change in the telecommunications sector. Change is, however, inevitable. Not only has the traditional model based on state monopoly lost favor in an increasingly procompetitive, pro–private enterprise global environment, but its technological and economic underpinnings have also shifted substantially. The question is not whether to reform, but rather how and when. Governments that fail to guide change in an orderly manner risk having crucial policy decisions made for them by more assertive players, and such decisions may not be in the country's best long-term interest. These governments are also likely to find it increasingly difficult to sort out the new opportunities and pressures that both domestic and foreign sources are placing on their telecommunications sector. Holding the fort is, in this environment, a lost cause.

There is, however, no universal blueprint for reform. Although a wealth of relevant knowledge can be gained from the experience of other countries, particular models cannot be moved readily from one country to another. The timing and pace of reform, the solutions adopted, and the implementation strategies followed are all highly dependent on the conditions in a particular country, and they are intricately interwoven with each country's broad economic strategy and political processes.[15]

As more developing countries revise the structure of their sector, more varied solutions are likely to appear. So far, the most ambitious reforms have been driven by high-level government decisions to privatize state enterprises, and the rest—market structure, policy on competition, regulatory development—is left to emerge as best it can. This approach is probably not well suited to all developing countries. An important measure of success will be the ability of future reforms to attract investment capital creatively. To a large extent, the design of sectoral solutions may well revolve around this one issue. Future reforms will also be measured by the extent to which they secure a better balance among the three pillars of reform, namely private participation, competition, and regulation. Regulatory arrangements that limit reliance on new institutions will become increasingly desirable, and unorthodox solutions are likely to be tried. The interdependence between competition and regulation will be played up to increase reliance on markets whenever possible as well as to generate demand for effective regulation.

In closing, it is worth suggesting that pragmatism is a valuable ingredient in whatever approach is eventually taken. Choices among increasingly numerous solutions can and should be made only after assessing realistically their potential contribution to the sector's growth and performance, rather than by citing doctrine. In particular, the decision to allow private ownership of telecommunications operations can be decided for each country based on its access to capital, management, and technology, regardless of whether one believes that private is inherently better than public. Likewise, a policy governing competition should be a means to achieve specific objectives for service and the network, rather than a mere reflection of the assumption that market forces know best.

Telecommunications companies in many developing countries were, after all, initially private and subsequently nationalized to forward practical development goals, not just to satisfy nationalistic fervor. And markets are not infallible; they have also given us industrial pollution, crowded airports, and the massive collapse of financial institutions.

Notes

1. Two books based on World Bank seminars on telecommunications reform in developing countries (Kuala Lumpur, 1987, and Washington, D.C., 1991) present the main policy issues and options as well as country experiences in preparing and carrying out reforms. See Wellenius and others (1989) and Wellenius and Stern (1991). World Bank (1992) summarizes the main features and trends of telecommunications in the developing world as viewed by a number of Bank staff and managers through successive rounds of discussion. This chapter draws heavily on these publications, and some parts are taken verbatim from the last reference.

A short study commissioned by the International Finance Corporation gives examples of various modalities of growing private sector participation in telecommunications in developing countries. See Ambrose, Hennemeyer, and Chapon (1990).

An advisory group on telecommunications policy, established by the ITU secretary general in 1988 and chaired by the late Poul Hansen, produced an excellent report that was endorsed by the ITU's plenipotentiary conference in 1989. Chapter 5 of that report gives recommendations that apply to most developing countries and a checklist of policy, regulatory, and legal issues that any comprehensive reform process should address. See ITU (1989). A bibliography and the summary findings of two ITU surveys are presented in ITU (1991).

2. Loosely speaking, the information sector comprises all activities that involve the production, processing, and distribution of information and knowledge, as distinct from physical goods. It includes activities that primarily comprise the handling of information, such as banking and government, as well as the information components of other activities, such as accounting in a factory and management of a farm. The information sector thus includes activities traditionally counted under the primary, secondary, and tertiary sectors. The in-

formation sector has been quantified by a number of researchers in the United States, Europe, and Japan since the 1950s. Data for developing countries are more limited. However, several studies in the Asia and Pacific region in the early 1980s, using data from the late 1970s, give some indication of the information sector as a proportion of GDP: Singapore, 25 percent; Indonesia, 19 percent; Malaysia, 14 percent.

3. Nulty (1991).

4. This figure was originally published in Wellenius (1989). It is shown here as modified in ITU (1991).

5. Wellenius (1990).

6. A checklist of aspects that need to be addressed through sector policies, regulation, and legislation is given in ITU (1989), p. 37.

7. It has been proposed that subcontracting can also be an effective tool among different organizational units of the same operating company. See Bruce (1991).

8. The following are examples: international competitive bidding for a ten-year license to provide cellular services in a given region (more than one or two are seldom possible in terms of market size and radio spectrum capacity); competitive supply of subscriber terminal equipment (also called customer premises equipment; for example, telephone sets, private branch exchanges) subject to technical standards and type approval to ensure network compatibility; and unrestricted competition in the provision of services such as shared data processing, information, electronic mail, packet-switched data, and store-and-forward facsimile and telex.

9. A growing number of major telecommunications operating companies from industrial countries are entering ventures in developing countries as part of their strategies to globalize their business. Some are already involved in several developing countries, and the trend is likely to continue. However, there are limits to the pace at which these companies can divert internal human and financial resources to foreign ventures, and as the number of opportunities grows, firms can be increasingly selective.

10. For example, sale price can be enhanced by giving the new owners extended monopoly privileges; however, reducing service costs and promoting responsiveness and innovation require increased competition and regulation.

11. Novel options may include placing regulatory authority with an existing government department but contracting out critical functions. Some (untested) possibilities include retaining internationally reputable audit firms to monitor compliance with franchise and other obligations (or requiring the main operating companies to retain such firms to report periodically); contracting out recurrent procedures and conflict resolution to local management or legal consultants, probably with foreign associates for specialized assistance; adopting technical standards and type approvals from another country; and retaining the spectrum management agency of another country to set up a local branch backed by the agency's established norms, practices, processing hardware and software, and expertise.

12. As an extreme example, entrusting monopoly services to an experienced foreign operator or investor may quickly remove communication bottlenecks in critical productive sectors, even if the absence of competition and effective regulation keeps prices high and compromises the pace of future innovation. In some countries, having to pay high prices for communication, which accounts for only a small proportion of the total costs incurred by most businesses, may be less of a handicap to users than not having the services at all.

13. For example, large switching nodes fed by remote interface electronics close to the customer are becoming more cost-effective than current solutions based on smaller switches collocated with demand centers and connected to customers through cables. Such networks can have several forms, including fiber-to-the-curb and disaggregation of the switch itself.

14. To a large extent, this relates to the convergence of what used to be quite different technologies. For example, with digitalization of telecommunications systems, differences

among voice, data, text, graphics, and video can only be made at the user end, although they all are bit-streams for purposes of processing and transmission. Major telecommunications equipment now essentially consists of specialized computers and software, and its operation and maintenance are akin to those of a data processing center; the similarity among network management centers for telecommunications, power, and railway systems is striking.

15. Attempts to interpret events in the telecommunications sector in the context of the political process are starting to appear. One of the first published papers focusing on developing countries is Cowhey (1991). The practical importance of the subject can hardly be overstated, since it is essential to understand, for example, what and when reforms are possible in a particular country, how reforms are likely to mobilize support and opposition, and what welfare distributional effects they would have.

Appendixes

Appendix A

Applications of Telecommunications for the Delivery of Social Services
Heather E. Hudson

THIS BOOK HAS EXAMINED the role of telecommunications in economic development, with examples of applications for production, trade, and management. Telecommunications can also play an important role in facilitating the delivery of social services, particularly to rural and isolated populations. This appendix provides an overview of applications of telecommunications for health care and education, with examples drawn from numerous projects in developing countries.

The past quarter century has been marked by dramatic technological developments in computers and telecommunications and the growing importance of information in all aspects of human life. Access to information, and to the facilities that produce, store, and transmit information, is now considered vital to development, so that the classifications "information rich" and "information poor" may mean more than distinctions based on GNP or other traditional indicators of development.[1]

Reductions in infant mortality as well as in contagious diseases and other preventable health problems are priorities for health care throughout the developing world. Development planners also realize that their populations must become better educated as well as healthier to participate in an increasingly globalized economy. Yet developing countries have a chronic shortage of teachers and health care workers. Telecommunications and information technologies can help overcome these shortages by providing training and consultation to practitioners as well as instruction to learners, both children in school and adults in the community or on the job.

Changing Technologies

New communications technologies including video recorders, satellite receivers, microcomputers, and facsimile machines are often considered the tools and toys of the industrial world. Nevertheless, advances in communications technology now make it possible to extend reliable communications to any village or encampment, no matter how isolated. Satellite and radio technology can be combined to reach virtually any location, whether in the desert or the jungle or on a remote island. These telecommunications links can transmit voice and data, and often video as well, so that the power of new technologies can be harnessed for development. Examples of these technologies and services are discussed below.

Communication satellites. Communication satellites now make it possible to extend basic communication and broadcasting services throughout a country or region, including the most remote islands and villages. Satellites can be used to provide basic telephony as well as radio and television networks with multiple sound tracks for local languages, where appropriate. The advantage of a satellite system is that the earth stations can be installed wherever they are needed, no matter how remote the location. In contrast, terrestrial technologies typically extend out from urban areas as funds are found to build more repeaters or string more wire and cable. Satellite services are considered to be highly robust, since an outage at an earth station affects only that location, whereas a cut cable or damaged repeater affects the entire network past that point. In addition, satellite earth stations can be located in or near the communities they serve, whereas microwave repeaters are often located in inaccessible locations such as mountain tops, and cable may be buried in the ground or under water.

Small low-cost earth stations such as those used for rural telephony with domestic satellites and the VISTA terminals used with INTELSAT satellites can bring voice and data communications to isolated regions. These earth stations may be installed in any community or project site without being connected to the national network by expensive terrestrial links. They may serve the surrounding territory through line-of-site radio links.[2]

Satellites can also be used to transmit educational programming such as rural development information to villages, instruction to students from primary to university level, and professional training to teachers, nurses, and other development personnel. Such services need not be one way: teleconferencing via satellite includes simple audio networks that serve as an improved version of the multiparty high-frequency radio services available in many developing regions and enhanced systems that offer computer conferencing, audiographics (graphic communication over narrowband channels), freeze-frame video, or motion video using new bandwidth compression techniques.

New applications of radio technology. Advances in radio technology such as cellular radio and rural radio subscriber systems offer affordable means of reaching less-isolated rural customers. These technologies make it possible to serve rural communities without laying cable or stringing copper wire. For example, digital microwave radio links provide a local loop via radio instead of traditional copper wire.[3] Cellular radio systems may also be used to reach isolated communities without installing wires. In some cases, these technologies may be combined with satellite earth stations, so that the earth station becomes a hub with radio links to surrounding communities. The application of solar power to eliminate reliance on diesel generators and community power supplies also makes these technologies increasingly attractive for developing countries.

Electronic mail. Communication via computer is a means of exchanging information immediately. Microcomputer users throughout the world may now interact using various electronic mail networks. Messages may be sent from one computer to another through a host computer equipped with communications and message-processing software including so-called mail boxes for subscribers. These services are cheaper than voice communications and overcome the differences in time zone that hinder real-time communications. Users may dial into local nodes of packet-switched networks to reduce transmission costs. Specialized electronic mail networks have been established for users in developing countries.[4]

Computer conferencing. Another application of computer communications is computer conferencing, that is, the interaction of many users through a central host computer. Each conference member may share ideas with the others and respond to their comments. Participants may log on at their convenience, thus avoiding the need to accommodate different schedules and time zones.

Facsimile. A technology with widespread development applications is the facsimile machine, which enables any type of hard copy, including print, graphics, and handwritten messages, to be transmitted over a telephone line. Facsimile boards may now be installed in personal computers to allow a message created on the computer to be sent directly to a remote facsimile machine.

Audio conferencing. A thin route service with considerable promise for development applications is audio teleconferencing. Several sites can be linked together through a bridge at a switching point or through a common frequency assigned on a satellite audio channel. This service can be used to hold meetings of, for example, project personnel, to tutor students in rural areas, or to train staff in the field.

CD-ROM *(Compact Disk, Read-Only Memory).* Information in the form of data bases, full texts of journals, video images, and other graphics may now be stored on compact discs and retrieved with a relatively inexpensive reader attached to a microcomputer. The advantages of CD-ROM include vast storage potential, low cost, durability, and ease of use. In addition, CD-ROM can be used on a stand-alone basis, without the need for on-line access to data bases. Of course, the discs must be frequently updated to keep information current.[5]

Access to on-line data bases. Although data bases in CD-ROM format are proliferating rapidly, much specialized information may be more readily accessible through on-line data bases. Users with a computer terminal and modem can dial into a data base and search for information using key words or phrases. Relevant information is then displayed as a citation, abstract, or sometimes a complete text of documents. Users can then select the relevant information and download it into their own computer or print it out.

Data broadcasting. The flow of information within the developing world has been hampered by the cost of distribution and by the lack of access to telecommunications facilities in rural areas. VSATs (very small aperture terminals) now allow information from news services to be disseminated to virtually any location. News services transmit copy by satellite from a hub earth station that may be shared with other data, voice, and video customers. These micro earth stations may be powered using photovoltaics or portable generators.[6]

VSATs for interactive data communications. Microcomputers or terminals linked to mainframes via interactive VSAT technology can be used to collect and update information from the field. A VSAT network called NICNET operated by the Indian government's National Informatics Centre now links 160 locations and will be expanded in the next stage to more than 500.[7] Similar systems may be used for accessing centrally stored patient records from any clinic, collecting epidemiological data, and downloading information from centralized data bases.

Desktop publishing. Enhanced graphics capabilities of microcomputers now make it possible to produce newsletters and other printed material without typesetting. These features may be particularly valuable in countries where newspapers, texts, and development materials in local languages are scarce and costly to produce. Development agencies can now produce their own materials in-house. Storefront desktop publishers may also spring up, as they have in the United States, in association with copy shops. This approach enables many small users to share desktop publishing equipment and software.

These technologies and services may be combined to enable development workers to access and share information electronically in the following ways:

Electronic messaging. Facsimile transmission and electronic mail may be particularly viable alternatives to sending hard copies of correspondence and documents through the mail in areas where service is often slow or unreliable. These technologies can also be used to link staff in the field with one another and with headquarters.

Electronic meetings. Managers, development experts, and project staff may now stay in touch electronically rather than through face-to-face meetings. Audio conferencing allows individuals at several sites to participate in the same meeting, while computer conferencing allows group members to interact at their convenience by reading and contributing to discussions stored on a host computer. These electronic meetings do not offer the richness of face-to-face interaction, but they may be an important supplement to meetings where transportation costs severely strain limited travel budgets.

Access to data bases. Computer terminals or microcomputers with modems linked to the telecommunications network can provide access to data bases anywhere in the world. Agricultural researchers, for example, may consult the data bases of the Food and Agriculture Organization of the United Nations in Rome. Health researchers may search the data base of the National Library of Medicine in Bethesda, Maryland. Others may use specialized development data bases such as those for agriculture and energy in India and for development project management in Malaysia.[8]

Costs may be reduced even further if these searches can be localized. Data bases may be downloaded onto computers within the country on tapes, floppy disks, or laser disks, with updates transmitted at regular intervals using telecommunications. For example, the same VSAT technology used to transmit information from new services may also be used to download updates. The search then becomes local and saves the cost of connect time.

Dissemination of information. Information may be transmitted from the field and from regional centers to desktop publishing locations via telecommunications networks. For example, development workers and reporters in the field could send reports by facsimile, which would then be edited and published in urban newsletters. Posters and news sheets could be sent by facsimile to rural communities, and newsletters could be sent either directly to the communities or to regional centers, where they would be duplicated and dispatched to schools, clinics, and government offices in their territory. Information obtained from sources such as news services, data bases, and teleconferences could also be disseminated to development workers throughout the country or region via facsimile.

Applications for Health Care Delivery

Primary health care is being provided to an increasing degree in developing countries by paraprofessionals drawn from the local population. These paraprofessionals may be community health workers or health aides with a few weeks or months of training, medical extenders or assistants with about one year of training, or practical field nurses. The duties of primary health care workers vary widely by country but generally focus on preventive, promotional, and simple curative care. This includes early diagnosis and treatment of common illnesses, maternal and child health care, midwifery, family planning, treatment of injuries, and referral of patients to higher-level facilities, if available. Primary health care workers may also organize immunization and mass treatment programs; provide guidance and education on nutrition, family planning, and hygiene; monitor epidemics, water quality, and sanitation; and collect demographic and health information.[9]

The second level of health care varies according to the location and complexity of the health care system. It normally refers to services available at larger health centers and small district or regional hospitals. Such services may be dispensed by nurses, paramedics, or general practitioners. The third level of care is normally the most sophisticated technologically and has the greatest mix of health personnel and services. It usually includes medical specialists based in major hospitals in larger urban centers. In some countries, highly specialized hospitals treating specific health problems, such as leprosy or mental illness, constitute a fourth tier.[10]

The health sector in developing countries is thus characterized by both breadth and specialization. It is composed of many services that require practitioners with widely varying levels of expertise, such as physicians with specialist training, general practitioners, registered nurses, other nursing personnel, nurses' aides, medical technicians, paramedics, midwives, and various kinds of community health workers or health aides.

In most developing countries the shortage of trained medical workers is critical. Hence, to make the most efficient use of the limited number of trained personnel, individuals with the most training tend to be concentrated in a few locations. Because resources are generally inadequate, individuals living in towns, villages, and rural areas outside major urban centers have limited access to organized health care; the same is often true for lower-income inhabitants of urban fringe and squatter areas.[11] Health care in developing countries must be dispensed by individuals with less training and less backup than their counterparts in industrial countries. Partly as a result of this and partly as a result of inadequate infrastructure, the health care sector in developing countries encounters major administrative, quality control, and logistical problems.

Communications and the Health Care System

The benefits of a rapid and reliable system of communication to help with hierarchical, spatial, and logistical problems should be self-evident. The relatively large number of lesser-trained health workers has a critical need to communicate up the hierarchy of expertise; rapid and reliable communication is necessary for the system to function as a system. With rapid communication, the limited higher-level expertise can be focused more effectively on the highest-priority problems, thus increasing the scope and quality of health services available per trained doctor or nurse. Likewise, lower-level care can be adequate when backup and support are available, and the general population's perception of the quality of that care improves.

In Australia, Japan, and the industrial countries of Western Europe and North America, the value of telecommunications in health care has long been understood. Examples include telephone access to all medical personnel, computerized patient records that are accessible on-line, mobile telecommunications links with ambulances and other emergency vehicles, and applications for training and consultation. In developing countries, where resource constraints are more binding, telecommunications are also used to support health care, although at a less capital-intensive level. Nevertheless, the functions fulfilled are much the same. In particular, the application of telecommunications to primary health care services (known in some contexts as telemedicine) offers a key to maintaining the system: building morale and maintaining confidence, providing emergency assistance, allowing consultation, facilitating administration and logistics, maintaining supervision and quality assurance, and supplying education and training. The following are examples of how telecommunications can support these functions.

Emergency Assistance

Emergency communications offer the most dramatic evidence of why communications are a critical component of health care delivery because getting help quickly can save lives. Telecommunications play an important role in health services in many developing countries. Studies in Costa Rica, Egypt, India, and Papua New Guinea showed that about 5 percent of rural calls were made for emergencies and medical reasons.[12] These calls saved lives and reduced suffering, which are highly significant indirect benefits. During natural disasters, such as earthquakes and floods, and epidemics, communication systems are used to secure assistance and to coordinate the logistics of emergency relief.

Two-way radios are used in many developing countries to coordinate disaster relief activities. Portable satellite stations may be brought in on trucks or small planes to provide reliable emergency communications. Portable satellite terminals were used following the 1980 Mount St. Helens eruptions in the United States and the 1985 earthquake in Mexico City.[13] In the South Pacific, the World Health Organization has used the experimental PEACESAT satellite network to summon medical teams during outbreaks of cholera and dengue fever and to coordinate emergency assistance after typhoons and earthquakes.[14] Mobile services such as cellular telephones can replace regular telephone service when cables are damaged. Cellular phones were used to coordinate disaster relief activities following the 1989 San Francisco earthquake.

Emergency location beacons transmitting signals to a satellite enable rescuers to find ships lost at sea or downed aircraft. Telecommunications systems can also help prevent disasters. For example, following several severe cyclones that caused great loss of life in Bangladesh in the mid-1970s, the government implemented a cyclone early warning telephone system consisting of single telephone installations in several coastal areas previously without access to telecommunications.

Consultation

Many developing countries now use paraprofessionals to deliver basic health services, particularly in rural areas. These workers receive basic training in the treatment and prevention of common health problems but need supervision and assistance in diagnosing and treating uncommon diseases and coping with serious health problems. Telecommunications links between village clinics and regional hospitals or health centers can be used to provide consultation and supervision.

Rapid communication allows paraprofessionals to maintain regular contact with a medical doctor or nurse. More than 100 Alaskan villages are equipped with earth stations that are used for the dedicated medical network, long-distance telephone service, and television reception. The medical network is a shared audio conferencing system that includes health aides (who typically have only six weeks of training) and sometimes provides in-service training. Each day during a scheduled "doctor call," a physician at the regional hospital contacts each health aide to discuss current cases, to provide advice on diagnosis and treatment of patients, and to authorize evacuations if necessary.[15] The Alaskan satellite network also provides twenty-four-hour emergency monitoring; by pressing a button on a telephone handset, a village health aide can activate an alarm at the nearest regional hospital.[16]

In Guyana, rural health workers called "medex" use a two-way radio network to communicate with headquarters in Georgetown to check on the delivery of drugs and supplies and to receive advice on major health problems. They may also request emergency evacuations and follow up on patients referred to the hospital. The medex, who have received about one year of specialized training and are therefore more self-sufficient than their Alaskan counterparts, only use their radios for advice on difficult or complicated cases.[17] The Georgetown training staff offer refresher sessions and "grand rounds" over the radio. At night, chatting over the radio helps medex reduce their sense of isolation and boosts morale.[18]

Administration and Logistics

Another important use of a two-way communication system is to administer rural health services, including ordering drugs, checking on delayed shipments of supplies, transferring patient records, coordinating staff travel, and evacuating patients. In Guyana, for example, the dedicated radiotelephone network was used most frequently for administration. Medex with two-way communication facilities received drugs more quickly and kept a more complete supply than medex without them.[19] Also, administrative problems that took weeks or months to resolve by mail or in person were resolved within hours by radiotelephone.

Medical services in some countries use small planes to take physicians to field sites rather than bring patients to central hospitals. This approach, known as the "Flying Doctor," began in Australia in 1928. People living on remote ranches or stations used two-way, high-frequency radios to contact regional flying doctor bases. Doctors flew in to provide assistance, often piloting the small bush planes themselves. Fifty years later the "mantle of safety," the area served by the flying doctors, covered more than 80 percent of Australia.[20] Today, the flying doctor service is replacing the use of high-frequency radios with telephone service provided either terrestrially (typically via digital microwave links) or via satellite.

Similar health communication networks are also found in other parts of the developing world. Flying doctor services in several East African countries (including Kenya, Malawi, and Tanzania) use two-way radio networks to link nurses at rural clinics with headquarters and to coordinate the aircraft used to transport doctors to the clinics and to evacuate seriously ill patients. Two-way radio is a vital support for this service, with radiotelephone communications linking the headquarters with all field hospitals and clinics and with the airplanes themselves through a common frequency installed in the planes' radios.[21]

Supervision and Quality Assurance

The deprofessionalization of health care and the decentralization of services exacerbate the problems of providing a relatively consistent quality of health care service in all locations.[22] Hence, supervising medical performance and monitoring the use of services offered become critically important.[23]

Periodic visits by senior personnel are the most effective form of supervision, although in many countries difficult terrain or shortages of vehicles, fuel, and expert staff mean that such visits are infrequent. Where available, the telecommunications system has been used to supplement personal contact by scheduling regular periods for telephone contact with each health worker to monitor or check on procedures and, on occasion, to prompt the health worker about specific functions and duties. Such frequent and interactive contacts with health workers may help maintain the quality of service dispensed by individual workers and reduce regional disparities in quality of service.

Another facet of health care administration in developing countries is that the management expertise necessary to organize and operate a complicated system of health services for rural and semiurban populations is in very short supply.[24] Rapid means of communication allow the limited number of managers to work more effectively and over a wider area as well as to manage available resources rather than only react to problems as they emerge.

Morale and Confidence

An important factor in the acceptance of primary health care by rural populations is the knowledge that a doctor or nurse is on call and ready to back up the local health worker's advice to a patient. Rural health aides have reported that they feel more confident in their diagnoses and treatment of patients and that patients are, in turn, more willing to accept treatment that is backed up by a telephone consultation with a physician.[25]

A second important factor involves the morale of field workers dispensing primary health care. The ability to maintain daily contact with health workers in other villages and other sectors of the system reduces feelings of isolation and boosts the sense of security and the morale of health workers. In Guyana, for example, health workers stated that being able to contact one another by radiotelephone greatly improved their morale.[26] In Lesotho, rural field nurses were adamant that health staff who were assigned to more isolated regions should be provided a means of communicating with one another.[27]

Applications of New Technologies

In many industrial countries, several new technologies are being used to augment the functions of the basic telecommunications network for health service uses. A carryover from two-way radio technology is the conference circuit, which has been highly useful for in-service, continuing education. In Alaska, the public health service medical network consists of five dedicated audio channels. Four are assigned to the regional service units, so that two hospitals and their associated villages share a channel. The fifth is dedicated for communication between hospitals to ensure privacy at the secondary and tertiary levels. The system can be configured for special courses and seminars for all health aides, for all nurses involved in coronary care, or for other uses simply by having the target group select the assigned frequency at the scheduled time.[28]

A simple, low-cost add-on to individual telephones is the speaker phone; both the patient and the health worker, or groups such as expectant mothers, can listen in and participate in a dialogue. Hard copy can be transmitted (for example, between regional hospitals and administrative headquarters) by low-cost facsimile or by microcomputers equipped with modems.[29]

Computerized record-keeping systems are also used by many agencies in several countries, including the public health service in Alaska and the southwestern United States, to store and update patient data. These systems allow patient records to be accessed and updated at different locations, which is particularly advantageous both for migratory populations who receive treatment at various clinics and for patients transferred within the system. Computerized records can also be sorted and grouped according to any variable so that it is possible to monitor, for example, heart patients with a history of rheumatic fever and to provide up-to-date lists that itinerant health workers can use to treat children requiring vaccinations or women requiring periodic checkups.

Another narrowband technology, slowscan video, transmits still pictures over a telephone line. Slowscan has been used to transmit X rays and pictures of dermatological lesions as well as for teaching. Telephone channels can also be used to transmit electrocardiograms, heart sounds, and ultrasound. Such relatively low-cost add-ons to the voice network are generally most useful for communication between the secondary and tertiary levels of the health care system, for example, between the general practitioner or nurse at the regional hospital and the specialist at a major medical center. Some use has been made of full-motion video for consultation, but this has not been cost-effective, although lower-cost digital compressed video systems are now available that use a fraction of the bandwidth used by conventional analog video systems.

Recently, a nonprofit organization called SatelLife, based in Cambridge, Massachusetts, and the former Soviet Union, organized a telecommunications system called Healthnet. Designed to operate independently from existing telephony networks, its goal is to increase the flow of information about recent medical research from industrial countries to developing countries, as well as to provide an electronic mail network that serves subscribers in developing countries who could not communicate otherwise because telephone service is poor or nonexistent. By means of inexpensive radio equipment (based on ham radio technology) transmitting signals to a small, low-orbit, store-and-forward satellite, users can query colleagues, hold conferences via electronic mail, order and receive medical literature from sources such as the *New England Journal of Medicine* (available at no cost through a subsidization agreement), and request data base searches at low, subsidized rates. As of 1991, demonstration sites had been set up in Kenya, Tanzania, Uganda, Zambia, and Zimbabwe.

Applications for Education and Training

Telecommunications and information technologies can also be used to train health care workers, teachers, and other field staff, as well as to educate students in the classroom and adults at home or in the workplace. The vast majority of such applications are found in industrial countries, although some are beginning to appear in developing countries. In the area of vocational technical training, most telecommunications-based programs serve relatively high-level students, such as engineers interested in keeping abreast of changes in their field or businesspersons seeking additional training in subjects such as marketing or accounting that can be taught at a distance. In the area of distance education, as in vocational training, telecommunications media are nearly always used in combination with other, more traditional media, such as books and papers sent through the mail, audio and video cassettes, and occasional small meetings of students with a tutor or instructor.

Training and Continuing Education

Field staff usually have little opportunity to take refresher courses or discuss new techniques once they return to the field after training. Most health agencies in developing countries provide only limited continuing education through infrequent and costly (in time and travel) meetings or seminars. In a few countries, however, telecommunications systems are used to offer continuing education, either directly as a medium for interactive instruction or as support for health workers studying on their own. Audio conferencing systems have been especially useful for instruction. Two-way radiotelephone systems generally have a few channels that are shared by many locations, each using the channel in turn or at scheduled intervals. Although such systems lack privacy, they allow rural health workers at all stations to listen, participate, and learn. Satellite systems can also be used for conferencing by having one audio channel that is shared by all sites.

In Peru, the Rural Communication Services Project linked seven rural communities, three via satellite and four via very high-frequency radio and then via satellite, to the national network. More than 650 audio teleconferences concerning agriculture, education, and health were carried out during the project.[30] In Alaska, health aides listened to consultations between their colleagues and physicians using an experimental satellite network. This conferencing feature was later included in a much expanded network.[31] In Guyana, a physician using a two-way radio network presented a case in a grand-rounds format once a week and tested health workers on diagnosis and treatment.[32]

Television programs that offer continuing education to health staff and public health education both to patients in hospitals and to the general public are increasingly available in the United States. Channels offering educational programs and entertainment are received via satellite and distributed via internal cable systems to patients' rooms. Continuing education channels for health care professionals are also distributed nationally via satellite.[33]

Distance Education

Satellite technology can reach distant learners at their workplace, at school, or at home, no matter how isolated. In China, the Television University distributes programs via INTELSAT and terrestrial facilities to more than 1 million students at their work sites. Indian educators use INSAT to transmit adult education programs and supplemental materials for classroom use to villages equipped with low-cost antennas. In North America, the National Technological University transmits technical courses to participating companies so that employees can continue their education without leaving the work site. The Learn Alaska network distributes educational programs to village schools where teachers tape them on video cassette recorders for classroom use.[34]

Using telecommunications for distance learning may reduce the rate of student dropout and staff turnover at remote locations. The University of the South Pacific operates a satellite-based audio conferencing network linking its main campus in Suva, Fiji, with its agricultural college in Western Samoa and with extension centers in nine Pacific island nations. The system is used to administer extension service activities and courses, offer tutorials for students taking correspondence courses, and extend outreach services to the people of the region through, for example, consultation, in-service training, and seminars offered by the United Nations and other development agencies. The benefits of this experimental network have been significant. The savings in travel time and costs have been at least ten times the cost of using the network. Dropout rates of correspondence students in courses with effective satellite tutorials have also been reduced.[35]

In the Caribbean, the University of the West Indies has established a teleconferencing network called UWIDITE to link its campuses in Barbados, Jamaica, and Trinidad with extension centers throughout the region.[36] It uses a combination of satellite and terrestrial audio links.

Pilot Projects to Encourage Applications of Satellites

During the 1980s two ambitious initiatives were undertaken to encourage the use of satellite technology for development. The Rural Satellite Program, sponsored by the U.S. Agency for International

Development (AID), was designed to help countries use satellites to help solve development problems. Pilot projects were carried out in Indonesia, Peru, and the West Indies. INTELSAT's Project SHARE provided free access to INTELSAT satellites for health and education demonstrations and projects.

AID *Rural Satellite Program*

By the late 1970s, AID had identified major obstacles to using satellites for development. These included the small number of visible and ongoing projects that could serve as models, the lack of planning capacity and relevant experience in developing countries with access to satellites, and the lack of awareness among development agencies and telecommunications lenders of the role satellites could play in rural development.[37] In 1979, AID undertook an exploratory program to help the developing world test the use of satellite communications for development. The experimental aspect of the program was not in the technology, although certain hardware innovations were introduced; rather, it was in the applications of that technology for development.

The goal of the program was to help developing countries that already had access to a satellite to use it for development rather than plan new satellite capacity more appropriately designed for development use. The program concentrated on interactive narrowband telecommunications rather than on broadcast community television because of the experience with audio conferencing in Alaska and the South Pacific, the low cost of narrowband equipment, and the simple production techniques required.

In Indonesia, AID supported a project that used Indonesia's PALAPA satellite to enhance higher education by linking thirteen new universities in the Indonesian archipelago. These Eastern Islands Universities suffered from a severe shortage of specialized faculty, particularly in basic sciences and agriculture. The satellite audio conferencing system enabled a professor at one institution to teach students at several locations. The network was also used to train faculty, handle administration, and enable administrators to meet electronically between infrequent face-to-face meetings. Sites were also equipped with facsimile machines and electronic blackboards that could transmit material written or drawn on graphic tablets. At most locations, the sites accessed existing Perumtel earth stations through telephone lines.

The Peru Rural Communications Services Project was developed and administered by ENTEL Perú (Empresa Nacional de Telecomunicaciones del Perú) with support from AID. The goal of the project was to use satellite communications via INTELSAT to provide basic telephone service and teleconferencing to support development activities in an isolated region of Peru. The project provided public telephone service and audio conferencing facilities to seven communities in the Department of San Martín, a high-altitude jungle area east of the Andes. Satellite earth stations were installed in three communities, and four smaller towns were linked via very high-frequency radio to one of the earth stations.

The teleconferencing activities were developed in cooperation with Peruvian agriculture, health, and education ministries and incorporated a wide variety of administrative, training, diffusion, and promotional strategies. A total of 658 audio teleconferences were sponsored by the ministries and ENTEL during 1984 and 1985, involving almost 12,000 participant hours.[38] The health sector was the most active and successful user, perhaps because it had a stable organizational structure and a pressing need for communication to support its operations. ENTEL itself became a major user of teleconferencing for training purposes. At the end of the two-year period, ENTEL transferred responsibility for developing teleconferencing to its commercial sector, with plans to promote teleconferencing among government agencies and private business.

The third satellite project supported by the AID Rural Satellite Program was the University of the West Indies Distance Teaching Experiment (UWIDITE). This activity built on the experience gained in a previous experiment called Project Satellite, which used NASA's ATS-3 and ATS-6 satellites to link the Jamaica and Barbados campuses with the St. Lucia extension center in 1978. UWIDITE began with five dedicated teleconferencing rooms, one at each of the main campuses (Barbados, Jamaica, and Trinidad) and one at each of two extension centers (Dominica and St. Lucia). Each room was equipped with audio conferencing equipment and slowscan television; telewriters and microcomputers were added.[39]

The main applications of the network were for extension studies (courses for credit toward a university degree), extramural studies (special nondegree courses), and extension services to support agricultural development and information distribution. A typical weekly schedule included in-service classes for teachers, meetings of the UWIDITE coordinators, class sessions for Challenge Examinations (which allow students to take their first year of university in their home countries), continuing medical education classes, and medical consultations.[40]

INTELSAT announced Project SHARE (Satellites for Health and Rural Education) in August 1964 as part of its twentieth anniversary celebration. INTELSAT offered to provide free satellite time for health and education purposes; users had to provide their own earth station facilities and support for the projects. Project SHARE was initially limited to sixteen months and was twice extended through the end of 1987. Its activities involved sixty-five countries on five continents.

Project SHARE applications were divided almost equally between video events or video conferences that were one-time or limited events and those that were recurring and longer term. Of the twenty completed projects summarized in the final evaluation report, eleven involved video conferences and four used audio conferencing. Most of the video conferences were one-time events, whereas the audio conferences were generally ongoing. Three projects involved computer communications. One project, China's Television University, used television for distance education.

Several projects, including those by the University of the South Pacific, China's Television University, the World Health Organization, and the Pan African News Agency, found satellite to be a cost-effective means of meeting their goals. The University of the South Pacific, based in Suva, Fiji, took advantage of Project SHARE as a transitional step to achieving operational service using the INTELSAT Pacific Ocean satellite and existing INTELSAT earth stations in eight countries. The network is used for tutorials and administration and now provides facsimile, data, and slowscan video in addition to audio teleconferencing.[41]

China used Project SHARE as the first step in implementing a national Television University. The goal was to deliver university instruction to students at the workplace, because the shortage of places for qualified students was critical. Courses were initially transmitted over the terrestrial network used for broadcast television, but much of the country was not covered by this system. Using INTELSAT, educational programs were transmitted to fifty-three locations. Following the pilot project, China purchased two transponders and leased a third from INTELSAT. By mid-1988, some 5,000 TVRO terminals built in China had been installed, and the Television University had an estimated student body of more than 1 million.

INTELSAT's board of governors has authorized a follow-up to Project SHARE known as Project Access. More market driven, Project Access is designed to stimulate service to rural and remote areas while emphasizing the potential for commercial applications. It also provides free use of spare space segment capacity for educational, health, or closely related social services. Project Access is intended for areas where communications are limited or where access to new communications can be developed.[42]

Applications for Education in Industrial Countries

During the 1980s, industrial countries increasingly turned to tele-communications and information technologies for education and training. Their use of satellites for distance education provides valuable models for developing countries, as does their use of media for community education and development. The following examples are particularly applicable to developing countries that are attempting to increase access to education and practical training in order to enhance the economic opportunities and well-being of their populations.

Satellites for Distance Education

Canada, the United States, and Australia have pioneered the use of satellites for making education available, accessible, and affordable for populations living in remote or isolated communities.

Canada. Canada has used its own domestic satellite system since 1973 to deliver broadcasting and telephony services to its vast isolated northern areas. In the mid-1970s, it undertook a series of experiments on the Communications Technology Satellite or Hermes, a joint project of the Canadian Department of Communications and the United States' NASA. Several of these experiments were continued on Canada's own ANIK B satellite, and the most successful are now operational services. For example, TV Ontario uses an ANIK satellite to deliver educational programs previously available only in the most densely populated areas to remote communities throughout Ontario and directly to schools, libraries, cable television systems, master antenna systems, and low-power television transmitters.

In British Columbia, the Knowledge Network delivers educational programs for credit to isolated communities, where they are retransmitted over cable systems or viewed at community centers and colleges. An audio conferencing link enables students to interact with instructors located in the Vancouver/Victoria area.[43]

United States. In the United States, educational television programs are delivered to schools and workplaces via satellite. In rural areas, small high schools cannot offer a wide array of vocational and college preparatory classes. A satellite network called TI-IN, based in Texas, offers specialized courses in foreign languages, mathematics, and science, enrichment courses, and staff development programs. Participants interact with teachers using toll-free telephone lines. At the university level, the National Technological University offers graduate courses in engineering and computer science via satellite to employees at their workplaces throughout the country. Employers support the program because it enables their professional staff to improve their knowledge without having to attend graduate programs at distant universities.

Continuing education programs are also distributed to homes via satellite. One example is the Learning Channel, an outgrowth of the Appalachian Community Services Project begun more than a decade ago to provide educational opportunities to small communities in the region's isolated hill country. Now the Learning Channel disseminates adult education channels via satellite to local cable television systems throughout the United States.

Australia. The School of the Air originated in Australia to supplement the correspondence courses sent to children in the outback. Using two-way radios to contact students on remote farms or ranches, the School of the Air offered some instruction and enrichment but, most important, provided a link with other children and "real teachers." Distance education has expanded in Australia with the creation of AUSSAT, the domestic satellite system. A pilot project demonstrated that educational programs could be delivered to rural and remote areas via satellite and received using small satellite earth stations. Several states now operate distance education networks to reach both children and adults.

Applications for Education in the Community

Industrial countries have developed applications of telecommunications and information technology at the community level that may be adapted to developing countries. Two examples discussed here are telecottages in Scandinavia and community radio in Canada.

Telecottages. Telecottages are rural sites equipped with computer and telecommunications facilities that are used by community residents. The official name of telecottages is Community Teleservice Centres (CTSCs). Their aim is to provide small communities with access to telecommunications, data processing, and computer-assisted services. A telecottage may be defined as "a centre where IT [information technology] apparatus is placed at the disposal of the citizens of a specific local community within a marginal geographical location, so that communal use may be made of the facilities available. The purpose of the CTSC is to counteract some geographically determined disadvantages which affect the local community, whether they are of an economic, educational, or cultural nature, or whether those disadvantages concern employment, services, or other infrastructure facilities."[44]

Approximately twenty-five telecottages have been established in Denmark, Finland, Norway, and Sweden. Telecottages form part of a strategy of economic diversification that aims to reduce dependency on agriculture in some areas and on resource industries such as forestry and mining in others. Attraction of small and medium-size businesses may be hampered by geographical and sociocultural distance from markets and decision centers, lack of easy access to information, lack of appropriate services including training programs, and other factors. Locally available and affordable access to telecommunications and information technologies is seen as a means of overcoming these barriers.

Telecottages are located in schools, libraries, local government buildings, and other converted or shared buildings such as coffee shops. They may contain an office, classroom, meeting room, and work facilities as well as a small kitchen. Equipment in a telecottage may include personal computers, printers, telephones, a facsimile machine, and possibly video equipment. Telecottage services could include word processing and desktop publishing, bookkeeping for small businesses, and training and continuing education both in the use of the technologies and in other topics available on computer disks or videotapes or via teleconferencing.

An important feature of the telecottage is the facilitator, who takes care of the equipment but also serves as a computer consultant, trainer, and development worker. This role is similar to that of the animator in Canada.

Community radio in Canada. Canada's native peoples are scattered in hundreds of villages across the vast northern regions. Most villages are isolated from urban centers. In the far north, the only access is by boat or plane. In these villages, people generally speak their own languages, although children are now educated in English or French.

Community radio began in northern Canada in about 1970, as an experiment in giving native people the opportunity to communicate in their own languages. The Canadian Department of Communications funded a pilot project in the early 1970s using a community development approach. The Canadian Broadcasting Corporation supported community radio by facilitating access to its own radio transmitters. Pilot projects with federal and provincial support explored a range of community media models, including local access, television transmitters, portable video, and community radio. A strong influence was the National Film Board of Canada's Challenge for Change Program, which used film as a participatory community development tool. The French concept of animator was applied to community radio so that residents learned how to use the medium to inform the community but also to collect information and opinions and thus give the community a more powerful voice in its own development.

Community radio stations in Canada are nonprofit, controlled by the community, and responsible for programming and production. Some produce all their own programming, typically a few hours a day in the morning, at noon, and in the evening. Others also broadcast network programs from the Canadian Broadcasting Corporation and other networks. Community radio stations are found across the country, particularly in small towns and remote communities.

In 1983, the federal government recognized the importance of native broadcasting in its Northern Broadcasting Policy, which established the Northern Native Broadcast Access Programme. This program has provided more than Can$40 million to enable thirteen communications societies to establish production facilities, train broadcasters, and produce and distribute programming in native languages and oriented toward native themes and concerns. Many of the organizations distribute their programs via satellite.

Native broadcasters have combined the local and development-oriented approach of community radio with the distribution power of a satellite so that programs produced by native Canadians can be distributed via satellite to community radio stations that retransmit them to their own communities. In some regions, native organizations also produce native language television programs that are carried via satellite to supplement the national network programming distributed from southern cities. The Wawatay Television Network produces programs in Cree and Ojibway Indian languages that are transmitted to thirty remote communities in northern Ontario on the ANIK C satellite, using an audio subcarrier of TV Ontario, the provincial government's education channel. The Inuit Broadcasting Corporation now broadcasts five hours of Inukititut (Eskimo language) programming each week using time on the Northern Television Service channels of the Canadian Broadcasting Corporation. Programs are uplinked from satellite stations in the Northwest Territories and Arctic Quebec.

The Need for Integrated Planning

The examples in this appendix demonstrate that telecommunications technologies can be used to improve delivery of health care, education, and other social services. For these applications to occur, however, telecommunications planners must be educated about them and social service planners and practitioners must be made aware of the potential of the technology. In the glitter of all this new technology, we must not lose sight of the lessons learned from decades of work with instructional media. As Wilbur Schramm noted fifteen years ago, motivated students can learn from any media—whether large or small—including satellite-delivered television, audio cassettes, and workbooks.[45] Educators need to work with technologists to determine which mix of technologies is most appropriate for a specific application.

Demonstrations and pilot projects can be effective means of testing new technologies and extending awareness. Careful planning is required, however, to move beyond the experimental stage. Again, both telecommunications and social service professionals must be involved throughout the process. As the evaluators of the rural satellite project in Peru pointed out, "the transfer of sophisticated telecommunications technology is not, and probably will not be for the foreseeable future, a straightforward exercise. Even the most thorough plans require revision, endless patience, and dedication if they are to be implemented successfully. New and innovative communication strategies require extensive promotion as well as innovative management structures, especially when changes in standard operating procedures are involved."[46]

In many cases, the most difficult hurdle for telecommunications applications in the social sectors is their high cost. Particularly in the area of satellite applications, tariffs are commonly much higher than nonprofit organizations can afford to pay on an ongoing basis, even if the necessary equipment can be obtained from outside donors. For example, the Learn Alaska network, which was set up by the suddenly oil-rich state government of Alaska when the price of oil was high, suffered significant cutbacks as the price of oil fell. In fact, most socially oriented satellite projects depend on special tariff subsidies, and when subsidies are terminated, the projects often cannot continue without the full support of the government, as is the case with the Chinese Television University. Even where simple voice telephone applications are called for, social service entities in some developing countries find it difficult to pay related costs. For example, in Guatemala, the Ministry of Health does not have sufficient funds to pay the monthly telephone bill of its headquarters in Guatemala City.

Yet the rewards are well worth the effort. Telecommunications administrations and national planning agencies must remember that the telecommunications network is more than a financial asset and a source of revenue. It is a vital strategic resource for development. By meeting the challenges posed by new and converging technologies, developing countries can maximize the benefits of telecommunications as a strategic resource for national and regional development.

Notes

1. ITU (1984) and Hills (1990).
2. Hudson (1990).
3. Parker and others (1989).
4. International Development Research Centre (1989).
5. National Research Council (1990).
6. Parker (1987).
7. Blair (1988).
8. Stover (1984).
9. World Bank (1980), p. 57, and Evans, Hall, and Warford (1981).
10. In Latin America, instead of resembling a pyramid, the structure of health staffing "resembles a sand clock with a slip neck of technicians and two receptacles representing a cadre of auxiliaries and a cadre of professionals." See Golladay (1980), p. 10.
11. Dorozynski (1975) and Golladay (1980), pp. 5–11.
12. Hudson (1984).
13. Goldschmidt, Hudson, and Lynn (1980).
14. Hudson and others (1979).
15. Hudson and Parker (1973).
16. When feasible, arrangements can be made to evacuate critically ill patients. In such situations, it is not clear whether rapid two-way communication results in an overall increase or decrease in hospital admissions. Some difficult and expensive evacuations may not be necessary, whereas others, which might be marginal, may be easier to arrange. See Hudson and Parker, (1973), p. 7.
17. Goldschmidt, Forsythe, and Hudson (1980).
18. Hudson (1984).
19. Hudson (1984).

20. Page (1977).
21. African Medical and Research Foundation (1979).
22. Rockoff (1975).
23. Golladay (1980), p. 41.
24. Evans, Hall, and Warford (1981), p. 1126.
25. Hudson and Parker (1973). Information conveyed directly to the village concerning patients who are away for medical treatment helps allay the fears of family and friends.
26. Goldschmidt, Forsythe, and Hudson (1980).
27. African Medical and Research Foundation (1979).
28. Hudson (1990).
29. Hudson (1980).
30. Mayo and others (1987).
31. Hudson and Parker (1973).
32. Goldschmidt, Forsythe, and Hudson (1980).
33. Hudson (1990).
34. Hudson (1990). For more in-depth discussion of technology in distance education, see Tiene and Futagami (1987); Perraton (1986); Jenkins (1988); and Lockheed, Middleton, and Nettleton (1991).
35. Pierce and Jequier (1983).
36. Block (1985).
37. Block and others (1984).
38. Mayo and others (1987), p. iii.
39. Block and others (1984), p. 29.
40. Block and others (1984), p. 29.
41. INTELSAT (1988b).
42. INTELSAT (1988a), p. 29.
43. Hudson (1990).
44. Qvortup (1989).
45. Wilbur Schramm (1976).
46. Mayo and others (1987), p. iv.

Appendix B

Posts and Telecommunications as an Integrated Communications Sector

Douglas Goldschmidt

IN THE MAIN TEXT of this book the postal system was mentioned several times: chapter 3 discussed some of the advantages of organizational and management autonomy for postal and telecommunications administrations, chapter 5 reviewed joint postal and telecommunications consumption patterns by sector for several countries using input-output analysis, and chapters 11 and 12 outlined survey results on the substitutability of postal services for telecommunications services.

Traditionally, the postal and telecommunications sectors have for many purposes been viewed as one, both organizationally and analytically. The primary reason for this, in spite of their vastly different requirements of capital and labor inputs, is that both services facilitate the same general function: two-way communication on a spatial basis. Hence, in most countries telecommunications historically have been treated as an extension of the publicly owned postal monopoly.[1]

Both services have also traditionally been tied closely to transport infrastructure and facilities, and, as discussed in chapter 7, transport acts as both a complement to and a substitute for other communications services. In spite of this, the tendency has been to treat the transport sector as separate from the communication sector, particularly as modern communication has expanded beyond the physical movement of objects. Nevertheless, the functions performed by the transport, postal, and telecommunications sectors are increasingly merging and overlapping, and the interaction of these sectors may not be efficiently facilitated in the future within the traditional institutional and regulatory framework that exists in many countries.

Substitutability of Telecommunications and Postal Communication

The telephone user surveys presented in chapters 11 and 12 gave ample evidence of the substitutability of telephones and posts, and also travel. In Thailand a sample of residential telephone subscribers ranked writing a letter as the third most preferred means by which to set up appointments and communicate with kin and friends. The telephone ranked first, and a personal meeting or visit ranked second. Likewise, in both Costa Rica and Papua New Guinea, when rural public call office (PCO) users were asked how they would communicate if the PCO were not available, writing a letter was the first alternative in Costa Rica and the second (behind travel) in Papua New Guinea.[2] For a sample of telephone subscribers in two rural regions of Kenya, the telephone and postal service together accounted for almost 80 percent of all economic contacts that subscribers made; mainly telephone or only telephone accounted for about 51 percent of the contacts, mainly letter or only letter accounted for about 11 percent, and an equal combination of telephone and letter accounted for 18 percent.

The respondents in these surveys might not be typical of the general population, however, since they were by definition telephone subscribers or PCO users. The only information reviewed on postal use from a more general sample population in a developing country came from a survey of nine rural villages in Egypt. There, of almost 500 male inhabitants polled, about 20 percent had sent or received a letter during the preceding year compared with about 10 percent who had talked on a telephone.[3]

In the past two decades, in most developing and industrial countries, telephone use grew at a faster rate than postal communication. For example, figure B-1 presents the situation in the United Kingdom, where the number of letters sent each year has declined since 1970, while the volume of telephone calls has increased dramatically. Such changes do not result entirely from the substitution of telephone calls for letter writing. Although part of the change reflects a substitution of telephone calls for letters, the rapid increase in calls is probably more closely linked to the major expansion of telephone access in Great Britain.[4] The overall demand for telecommunications

Figure B-1. *Inland Letter and Telephone Traffic in the United Kingdom, 1950–76*

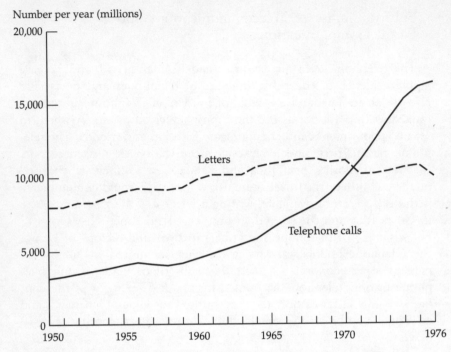

Number per year (millions)

Source: Smith (1978). Reproduced with permission of the Controller of Her Britannic Majesty's Stationery Office.

also increased as quality of service improved and as new services, particularly relating to data and message communications, were introduced.

Postal and telecommunications services are not perfect substitutes even for the transfer of information. Although information is moved spatially by both means, the postal service is not as efficient in facilitating two-way interaction.[5] For certain types of messages, speed is highly valued. For example, in the Papua New Guinea and both of the Indian studies reviewed in chapter 12, more than 75 percent of all calls made by the PCO users sampled were perceived as being necessary immediately or as having to take place during the day on which the telephone calls were made.

Electronic Message Transfer

Several chapters of this book cited instances in which postal and telecommunications functions merged. In the study of villages in rural Egypt, village administrators used telephones primarily to dictate messages, which were then hand carried to the intended recipient. Also, the general public sometimes used telephones to transmit telegraph messages, which were then written down and delivered; in Egypt such messages are called phonograms. Likewise, in Myanmar several provincial public telephone offices were closed each afternoon while (telegram) messages were dictated over the telephone lines for transcription and hand delivery at the other end; similar telegraphy systems have been observed in parts of the Philippines and in other developing countries. These, of course, are rudimentary examples of the ultimate convergence of the information transfer function of posts and telecommunications: electronic mail.

Electronic mail systems are being introduced throughout the industrial and developing world. These systems are similar to a traditional postal system in that documents are sorted and transmitted to specific addresses. The difference is that the documents (messages) are transmitted electronically over telecommunications networks rather than through transport networks, and they need not be printed on paper.

One form of electronic mail is the transmission of messages using telecommunication/computer interfaces. These systems use a variety of computers ranging from inexpensive home or small business microcomputers to the largest mainframe systems. Essentially, the electronic mail system is a software package that provides computer sorting, mailbox (storage), and transmission functions and an interconnect, which is accessible from as many other computers or terminals as desired.[6] Such small (or sometimes extensive), mostly privately owned and operated computer-telecommunications systems are becoming common in North America and parts of Western Europe within commercial enterprises of all types and sizes, as well as among universities and research institutions.[7]

In the public sector, the United States Postal Service, among other postal carriers in industrial countries, has begun to implement a national electronic message transfer system that integrates conventional systems of postal collection and distribution with computer, facsimile, and telecommunications facilities.[8] Increasingly, large-volume users will be able to send tapes or magnetic cards to local postal centers where the messages will be read and routed electronically; others will send bundles of nonenveloped message pages, which will then be facsimile encoded and transmitted. Ultimately, individuals with hard copy messages may be able to access the system through coin-operated facsimile machines (electronic mailboxes), which will be located in post offices or other public places (large stores, shopping centers, bus and railway terminals, and so forth). Individuals with terminals in their home or small business may be able to submit messages directly over local telecommunications links.[9] As a first step, the U.S. Postal Service inaugurated in January 1981 regular international electronic postal service (Intelpost) between the United States and London, and other Western European countries will be added to the system. In January 1982 an Electronic Computer Oriented Mail service (E-COM) was inaugurated that allows large commercial mailers to send computer-generated messages to twenty-five specially equipped postal facilities, where the messages are printed, stuffed into envelopes, and delivered as regular first-class mail.[10]

Given the rapid advancement of technology, in the foreseeable future, a major function of the current postal system in all countries—the transmission of bills, invoices, and funds—may be largely replaced by electronic banking services using telecommunication/computer interfaces. Not only will funds be automatically debited from accounts and detailed accounts maintained for many transactions, but customers will be able to use simple data entry devices, such as push-button telephone sets, for banking functions.

In fact, if falling data processing and transmission costs bring the cost of an electronic message below the cost of a first-class letter, electronic mail could substantially displace the manually carried first-class letter.[11] In particular, electronic message transfer could replace the first-class mail generated by business and industry. The system could transmit market information, product specifications, technical details, purchase orders, contract copies, and so forth. This could directly reduce market response and purchase lead times, which would permit companies to reduce inventory levels, while providing an equal or improved level of customer service.[12] If electronic mail systems are widely adopted, major savings in labor, paper, vehicles, and energy could also be realized.[13]

The Postal System in Developing Countries

It is likely that both public and privately owned electronic mail systems of varying degrees of complexity will soon be relatively widespread in many developing countries. Already telecopiers, data terminals, teletypewriters, facsimile machines, and so forth are found in the larger cities of many developing nations. The widespread development of domestic computer/communications networks is now under way; in fact, packet-switching systems being planned or implemented by many telecommunications authorities in developing countries will directly facilitate the development of electronic message systems. However, such systems are mostly for communications within and among the larger urban and commercial areas.[14] In the absence of reliable telephone systems in many smaller towns and rural places, it is unlikely that electronic mail will be a significant factor in less urbanized places during the next decade.

The absence of modern electronic mail services in rural areas will not, however, be a roadblock to development. In rural areas, where transport and other communication services are often highly inadequate, the traditional post office can continue to perform several functions, including the transport and movement of bulky items.

In fact, some of the characteristics that make the postal service difficult to manage—the need for an elaborate transport system and the large volume of labor required—offer a special opportunity. If properly managed, the postal systems in developing countries can provide a valuable scheduled movement of correspondence that does not require immediate response as well as of newspapers, supplies, freight, and even people. The postal system can also be an attractive vehicle for implementing and supporting a variety of rural development programs. Personnel operating a postal system have in some countries provided banking services in areas without banks, disseminated information on behalf of government and other public and private enterprises, maintained PCOs, and provided various social services by virtue of their physical placement in small towns and villages. No other communications media or civil service maintains such an extensive pool of labor in remote locations. The following sections cite brief examples of the postal system's rural potential to support agriculture, health, and nutrition extension activities and business; education and literacy; transport systems; rural banking; and the generation of employment.

Rural Agriculture, Extension Activities, and Commerce

The need to disseminate information remains crucial to rural development; without adequate systems to facilitate the movement of information, maintaining many of the traditional development activities is difficult and costly.[15] In the absence of a reliable postal service and at least minimal telephone access, rural development or extension agencies and private commercial enterprises tend either to use courier services or to operate their own relatively high-cost transport and communication systems.

Agriculture

The growth in recent years of agricultural extension programs in developing countries has increased the need to disseminate various types of printed information and administrative correspondence regularly to small towns, villages, and rural areas. Historically, this was also the case in industrial countries. For example, in the United States the postal system had a significant effect on the agricultural sector in the early 1900s. One poll taken in 1912 indicated that farmers were very responsive to published agricultural information. Of over 3,500 farmers polled, 40 percent stated that agricultural periodicals delivered through the rural free delivery postal service were most helpful. An extension agent with the U.S. Department of Agriculture noted at the time that, "So far as this survey is an indication, the agricultural press would seem to be the most efficient of our agricultural extension agencies in reaching the farmer."[16]

Health and Nutrition

The existence of postal delivery to rural towns can also help disseminate information on health, family planning, and nutrition programs. A letter carrier who often handles few letters and parcels, but who travels long distances, could easily distribute posters, pamphlets, drugs, and medical supplies to a local health worker as well as other materials to government agencies at very low marginal cost. In effect, the size of a government's rural health extension program could in some instances be increased by making the postal service an integrated part of the effort.[17] The fact that postal carriers are generally not used in such a way and are in many instances underemployed probably reflects bureaucratic structures that resist interagency cooperation, a general disinterest in or ignorance of the possibilities offered by the post, or a distrust of the postal service resulting from experiences with traditionally poor or nonexistent delivery services in rural areas and small towns.

Commerce and Trade

As with the telephone, many commercial or business-related benefits of the post are so obvious as to be transparent. Presumably, the ordering of supplies, remittance of bills and payments, and shipment of parcels and produce can all be facilitated by a reliable postal service. The post also provides a conduit of information for rural dwellers, which can be essential for the conduct of trade and business. In 1900 the postmaster general of the United States noted, with regard to the then experimental rural free delivery service, that

> Rural delivery has now been sufficiently tried to measure its effects. The immediate results are clearly apparent. It stimulates social and business correspondence, and so swells the post receipts. Its introduction is invariably followed by a large increase in the circulation of the press and of periodical literature. The farm is thus brought into direct daily contact with the currents and movements of the business world. A more accurate knowledge of ruling markets and varying prices is diffused, and the producer, with his quicker communication and larger information, is placed on a surer footing. The value of farms, as has been shown in many cases, is enhanced. Good roads become indispensable, and their improvement is the essential condition of the service. The material and measurable benefits are signal and unmistakable.[18]

More recently, a retrospective study of rural America documented that the introduction of rural free delivery service affected the way in which farmers conducted business and ultimately led them to employ more precise and careful business practices as the level of business and financial transactions they conducted through the post increased.[19]

A Caveat

The postal system, nevertheless, should not be given too much credit; the problems of disentangling the joint contributions of diverse inputs are immense. For example, the value of farms at the turn of the century in the United States likely increased at least as much from the improvement in the road system as from the introduction of the rural postal delivery system. Also, the introduction of the parcel post in the United States appears to have had a somewhat negative effect on stores in small towns. The introduction of the parcel post allowed large mail-order companies to market a diverse selection of merchandise at prices sometimes lower than those charged by small local stores. Although many local stores attempted to compete by diversifying their merchandise, the overall effect of rural parcel post seems to have been to injure local retailing.[20] Such competition was probably beneficial for the country as a whole, but it would be difficult to determine the net effects on a particular village or small town of losing its store or shop to such competition.

In any case, comparing turn-of-the-century North America with the current situation in many developing countries is partly invalid because so many changes have taken place in communication technologies. Today there is a much wider variety of communications media for conveying market and other information to rural areas. Hence, attempts to facilitate the flow of information among rural areas, markets, and urban centers in developing countries now require both users and suppliers to make many more decisions about the cost and effectiveness of posts, telecommunications, transport, broadcasting, and various other communication media, as well as about the value of time.

Education and Literacy

In the United States in the early 1900s, as in many other countries, postal deliveries were viewed as one way to promote an informed and literate electorate.[21] Early accounts of the rural free delivery system emphasized that the post improved the level of education of both adults and children.[22] This was partly reflected in the large increase in the number of newspapers delivered to rural areas after the introduction of the system. By 1911 more than 1 billion newspapers and magazines a year were delivered on rural routes—a volume greater than that of all other rural mail combined. By 1919 this number had doubled.[23]

In developing countries, the postal system has also helped promote literacy through the transmission of information among correspondents and, more important, through newspapers, periodicals, and some books. Literacy campaigns that do not include a variety of printed literature have been difficult to sustain. It has been argued that the audiovisual media sometimes seen in rural development projects, although useful, cannot provide the diversity or depth of information provided by the print media, and, even more important, printed material, through its convenience, gives learners more control over their learning; they are able to work at their own pace and to choose when and where to inform themselves.[24]

An efficiently functioning postal system can also facilitate the implementation of correspondence-type education programs. In developing countries, in particular, the demand for school places generally exceeds the capacity of national economies to supply them, and this has led to a search for alternative education methods that reach larger numbers of people at low cost. One alternative becoming increasingly accepted is distance teaching; usually a form of correspondence school relying on radio broadcasts, printed literature, and in selected instances periodic group discussions and telephone conversations.[25] Administrators of distance teaching programs in both industrial and developing countries generally prefer to rely on the postal system to deliver instructional material, to return exams, to deliver assignments, and even on occasion to answer individual student questions.[26]

When a reliable postal system is not in place, however, such educational programs can be impeded, which incurs additional costs; staff of the International Extension College, in London, an organization that promotes university and other forms of adult education in developing countries, have encountered many such cases. In Nigeria, for example, the postal service is considered unreliable enough that the local extension university must maintain its own private courier system.[27] In Mauritius, all course materials for the Mauritius College of the Air are delivered by courier service.[28] Similarly, private courier systems are needed to supplement the post both in Indonesia, to assure the reliable and timely conveyance of audiovisual and printed teacher training materials to rural areas, and within the University of the South Pacific, to disseminate material among campuses located on different islands.[29]

Postal Services and Transport

Postal systems by definition depend on transport facilities. In particular, mail delivery to rural areas and to towns within those areas in many countries is unreliable in the absence of adequate roads, navigable waterways, or small airstrips. Although the mail can be delivered by messengers traveling by bicycle, on foot, or by animal, such delivery inevitably is slow and haphazard.

In many countries the development of the postal system has been closely correlated with an improvement of travel conditions and the upgrading of roads.[30] One of the conditions attached to the establishment of the rural free delivery service in the United States was that local residents had to maintain their roads; the post office was deliberately used to promote the construction and maintenance of roads. An assistant postmaster general noted in 1912 that,

> The department [U.S. Post Office] very much desires that postmasters, rural carriers, and substitute carriers shall not only constitute themselves apostles of good roads and spread the propaganda, but that they shall, by their works, arouse interest and emulation in others. Many postmasters and rural carriers have been instrumental in forming good road clubs and associations, the results of which have been vast improvement in the condition of highways, and, in several notable incidents, the appropriation of enormous sums of money for the rebuilding and improvement of entire county highway systems.[31]

This emphasis on upgrading postal roads is apparent through the years in various reports of U.S. postmaster generals.[32]

Following a general improvement in roads throughout the United States, the practice of maintaining postal roads was abandoned. However, the case is suggestive.[33]

Given improved road conditions, the postal service can also act as a form of local freight and general delivery service as it does, for example, in rural Sweden. There, a service introduced in the early 1970s by the Swedish post office allows rural farming communities to contract with the post office for the house-to-house delivery of basic necessities. The Swedish system allows the customer to order goods by telephone or letter from any shop in either the locality where the rural postman begins his route or on the route he follows while making his rounds. The shop packages the ordered goods and either drops them at the post office or hands them to the postman while he is on his rounds. The parcels are delivered as mail, and the postman may also collect payment for the goods. The Swedish government considers that this service makes rural isolation more bearable and, more important, decreases the amount of travel that rural dwellers must undertake.[34]

In an age in which the postal service is normally relegated to the delivery of mail and strictly defined types of freight parcels, it is easy to overlook the fact that the post can also transport people. In Scotland the post office transports both goods and persons in rural areas.[35] Some years ago the Scottish post noted that its vans were underused—postal mail volume, as traditionally defined, was insufficient to use existing capacity fully. To correct this, the post office expanded the list of goods that could be carried by allowing vans to transport bread, meat, and other foodstuffs; newspapers; and the materials and final products of cottage industries. The incremental costs of such an action were small. The Scottish post office now has few rules about what can be carried; such decisions are left to the discretion of the local staff. Hence, some offices transport parcels that do not conform to postal regulations elsewhere. In the absence of reliable local transport in many of the remote areas of Scotland, the post office also uses small eleven-seat buses rather than vans on certain rural routes to carry mail, some cargo, and passengers.[36]

The Postal Service and Rural Banking

Many rural areas in developing countries are isolated from banking services of any kind. Urban banks sometimes hesitate to enter rural areas and smaller towns because of the high cost of maintaining offices, the small per capita savings potential, and the small size of credit requirements (relative to urban areas). The lack of small town or rural banks does not, however, imply that rural inhabitants do not require banking services. Many farmers and businesspersons periodically have surplus cash, which can be profitably placed in savings accounts; require facilities to transfer funds to creditors; and require regular sources of credit. Farmers in developing countries, in particular, are often at the mercy of private lenders who offer capital at usurious rates, when such capital is available at all.[37] As a result, postal banking systems can, acting in concert with other government agencies or private banks, provide credit for rural development needs at competitive market interest rates.

The post office has been important in establishing banking facilities in several countries. The British, French, German, several other Western European, and Japanese post offices traditionally have offered savings account facilities for small savers, and this practice has been followed in many former colonies. The post can provide such additional rural services largely because investment and operating costs are smaller than those of other postal functions; the labor-intensive nature of the postal service can be exploited to provide a variety of services requiring face-to-face contact. When the postal service maintains counter and postal box services in a community, the addition of various banking services does not involve substantial incremental costs. Hence, postal savings facilities have been established in smaller towns and rural areas at a much lower cost than would be incurred for similar services at conventional commercial banks. The Universal Postal Union has contended that such facilities, because of their closeness with the people, actually encourage habits that lead to savings and thrift.[38] Rural post offices are also sometimes used as a means to transfer funds, either through giro (checking) accounts or through the sale of money orders; as a credit facility; and as a means for rural inhabitants to pay their taxes.[39]

The Universal Postal Union also notes that postal giro accounts provide governments with funds for capital or operating expenses, since they tend to be stable over time.[40] The proceeds from such accounts are sometimes used to finance national development programs, as in India where approximately 10 percent of the national development plan was at one time financed from postal savings.[41] The funds from these accounts are also sometimes used to finance further postal expansion.[42]

Typical examples of postal savings banks are found in Egypt and India. In Egypt, where three out of five citizens live in rural villages, the Egyptian Postal Savings Bank has encouraged small savings accounts through the use of "thrift accounts" (accounts specifically designed for small savers).[43] In India in 1970, the Indian Post Office Savings Bank had Rs19,160 million[44] in deposits, with 24 million depositors. This bank has no limit on the number of withdrawals—a policy that attracts small deposits.

In the past twenty years, however, the growth of postal financial services in many countries has stopped. In 1979, only about one-third to half of developing countries offered giro or postal savings services, and the number of services had not grown since 1970.[45] Although the average deposits in postal giro and savings accounts have increased since 1970, the growth in the assets of postal savings banks has been slower than the growth in the gross domestic products of countries offering the service, indicating a relative decrease in real terms in these assets. Overall, savings banks and giro accounts together represent only about 13 percent of the national savings in developing countries where they are offered, compared with 40 percent in industrial countries where they are offered.[46]

The Universal Postal Union suggests that one of the reasons for the slow development of postal banking is the restrictive policy of postal and finance ministry authorities concerning tariffs, interest rates, and the use of the funds collected.[47] In developing countries, some governments have borrowed money from the postal savings bank at nominal or no interest without recording the interest savings as revenues to be credited to the post office—an "accounting error" that vastly underestimates the post office's contribution to government revenues. These governments frequently make no provisions to repay the "borrowed" funds.[48] Although such practices generate savings in interest paid by the national government, they tend to limit the amount of funds available for rural credit, to limit the extent to which competitive interest rates can be paid to small depositors, and to cast doubt on the overall integrity of the postal system.

Competition and Monopoly in Posts and Telecommunications

In the United States in the early years of rural free delivery service, when alternative forms of freight carriage were not widely available, the postal system was an important source of rural freight transport.[49] Today, however, almost all freight delivery services in the United States and many other industrial countries are privately owned, and in recent years numerous private message and parcel handling services have appeared, often in direct competition with government postal services.

Such change indicates that as economic development takes place and as the volume of message, parcel, and freight traffic increases, private carriers, which may have stronger incentives to control costs and manage their operations efficiently, sometimes provide cheaper and better service than government monopolies. In New Zealand, for example, a network of commercial delivery services that carries farm supplies, commodities (including food), and the mail has evolved to serve rural areas.[50] One observer noted that "these private transport operators are able to carry out the delivery of mail in conjunction with the delivery of other goods at a much lower cost than would be the case if the Post Office carried out the work."[51] The post office in several developing countries has also by necessity adopted such techniques. For example, in Ethiopia, Kenya, and Tanzania, among others, the post office contracts private transporters to carry mail and postal parcels to remote places along with cargo and passengers.

Part of the reason private carriers can sometimes effectively undercut the government postal service is that in many instances the objectives of private carriers are more limited and better focused. Private carriers can concentrate on achieving a low-cost and reliable movement of information, parcels, goods, freight, and possibly people. In providing such service they have a greater flexibility to choose the most efficient mix of cargo and inputs in terms of both labor and technology. Government monopolies, however, tend to be affected by a wide variety of policies made by officials who know little about the practical aspects of the business and who generally try to achieve multiple objectives. Partly because of this, they have less incentive to provide good service, minimize costs, pay competitive wages, and generate increasing revenues; many postal authorities can rely on revenue subsidies from government whenever they have financial shortfalls for whatever reason. Also, government postal monopolies rarely keep their accounts on a commercial basis. Hence, no one knows how much the service or portions of the service actually cost.

Also, governments themselves sometimes use the fact that the postal service is labor intensive and generates employment as a reason to prevent the introduction of new cost-reducing technology. They sometimes contend that automatic letter sorting and routing equipment, mobile post offices, modern computers, and electronic mail, data, and funds transfer would reduce employment.[52] Although this argument may have some short-run merit in countries with surplus labor, the objective of generating employment has been frequently carried to an extreme to the detriment of efficient and innovative postal service.[53] The possibility that an efficient, well-managed postal system could generate much more employment indirectly in business, commerce, and agriculture than it could possibly generate directly tends to be overlooked. For example, recently in the Côte d'Ivoire, the reliability of the postal service was improved by reorganizing the mail transport system within the post office and by making minor improvements in sorting. As a result, the volume of letters handled in Abidjan increased from 100,000 to more than 200,000 a day in only two years.

Technological innovation is, of course, another factor contributing to the likelihood that increased competition will emerge in the postal sector in developing countries. As outlined earlier in this appendix, as technology and consumer demand have evolved, so have the means for providing numerous transfer functions. The development of new telecommunications technology has not only provided opportunities for more efficient transmission of messages, it has also fostered the emergence of new business firms, institutions, and organizations to provide various types of postal-related services, often in direct competition with the government post office.

If there is a lesson in all of this, it might be that it is becoming less appropriate to condone the postal system as a static government monopoly. Rather, it might better be thought of in a functional context as a service that can be supplied with varying degrees of coverage and efficiency by different institutional and market structures. Evidence reviewed in the main text shows that for increased productivity and labor specialization to occur, rural as well as urban areas increasingly require efficient and rapid movement of information and goods. In some developing countries the postal monopoly has served a purpose and can continue to improve; in other situations mixed public or private systems that transfer data and other forms of messages electronically and carry the post, other goods, and perhaps even people might be more effective. The introduction of public and private electronic mail and data systems and the established private parcel and freight delivery in North America, Western Europe, Japan, Australia, and many developing countries show that the actual mix of competition and monopoly can be determined mostly by market demand, the mix of technological and service opportunities, and the potential for creating additional incentives for efficient management.

Other things being equal, since competition stimulates management and service efficiency, competition might be allowed to develop for the supply of postal and other communication functions wherever feasible—particularly in the provision of transport (including transport of postal and freight materials and of people) and some telecommunications services (electronic message, data, and funds transfer). Of course, some government regulation would still be necessary to assure widespread access to postal, telecommunications, and other services. However, the possibility that increased competition might foster the greater overall development of such functions, might further the innovative integration of communication and transport alternatives for specific needs, and could serve as a benchmark for judging efficiency in the remaining government-administered operations deserves increased attention.

If, however, public ownership of and monopoly in the primary postal and telecommunications functions are maintained, and if the two services are separated organizationally and operated commercially for the management and staffing reasons outlined in chapter 3, the two entities might be allowed to compete in some of the more capital-intensive communication areas (information transfer). Interaction and cooperation would, however, probably best serve them both by providing more labor-intensive rural services such as letter and parcel delivery, rural banking and funds transfer, the operation of PCOs, and so forth. A key for the efficient management of each entity would be to keep separate commercially oriented accounts for each individual service so that its financial position could be monitored, and informed management decisions could be made.

Notes

Note: The author acknowledges helpful comments from Jeremy J. Warford.

1. One major exception to this was the private development of telegraphy and telephony in the United States.

2. However, writing a letter was not an important alternative to telephone communication in the Chilean and the 1981 Indian surveys.

3. In rural Egypt, however, the long-distance telephone links were extremely congested, and inhabitants of the nine villages had very limited access to both the postal system and telephones; only five of the villages had post offices, three had small telephone exchanges, five had a few scattered private telephone lines, and one had no access to a telephone.

4. Between 1970 and 1976 the number of telephones in Great Britain increased from 14 to 21 million. Telephone density increased from 25 to 39 telephones per 100 persons.

5. Analysis of a sample of about 500 respondents from rural areas in Lincolnshire, England, showed that for social contacts, "contacts by post form the basis for a network of distant friends and relatives who are infrequently visited, whereas visits and phone calls appear to be interdependent modes of social contact, especially over short and middle distances." See Clark and Unwin (1981), p. 55. Results consistent with these are described in Thorngren (1977), pp. 374–85.

6. For an overview of international private electronic mail systems and other communications networks, see Elbert (1990).

7. Rapid technological advances have greatly extended the spread of facsimile message communications systems. Commercially available analog facsimile machines, using standard telephone lines, are available for less than $1,000 and can transmit one page every 20 seconds. Even faster machines, using specially conditioned telephone lines, are able to transmit digitally much higher volumes of information. Although these machines vary in cost, they offer the opportunity to expand electronic mail systems within an enormous array of commercial and government enterprises.

8. National Research Council (1981).

9. Lee and Meyburg (1981).

10. It has been contended, however, that the U.S. Postal Service will over the longer term be unable to compete in the electronic message transfer market in the United States partly because of its inexperience in competing with private business firms and partly because (a) it is behind private industry in technology development, (b) it has performed too little market research, and (c) it would be required to provide a standard set of services in competition with a set of diverse and unstandardized offerings by other firms. Sorokin (1980), p. 122.

11. Pool (1979), pp. 187–88.

12. Langley and Pisharodi (1981).

13. Lee and Meyburg (1981).

14. In 1982, the government of Singapore, in a move somewhat counter to that discussed in chapter 3, merged the Telecommunications Authority of Singapore with the Post

Office. One of the principal reasons for this merger was that within the highly urbanized Singapore area most postal mail is used to transfer messages and thus could be eligible for electronic mail service; packages or parcels represent a relatively small portion of total mail handled. Another reason unique to the urban Singapore environment was that telecommunications staff outnumber postal staff.

15. Much has been written on the general need for information for rural development. For case studies on rural uses of information, see McAnany (1980), pp. 3–18; *Development Communication Report* (1988); Parker and others (1989); and Hudson (1984).

16. Fuller (1964), p. 301.

17. Postal and telecommunications services are also complementary in the administration of health care programs; see appendix A.

18. U.S. Government (1990), p. 5.

19. Fuller (1964), p. 312.

20. Fuller (1964), p. 312.

21. Such objectives are still partly reflected in the preferential postal rates given to magazines, newspapers, and brochures in many countries.

22. U.S. Government (1902), p. 16, and U.S. Government (1901), p. 25.

23. Fuller (1964), p. 294.

24. Bates (1982).

25. Telephone teaching is also an increasingly viable option for some forms of distance-related education in developing countries. The major advantage is that it provides immediate two-way interaction between teachers and students. Such teaching, of course, also relies extensively on the postal service. See Bates (1982) and app. I of that book. For a description of the history and practice of telephone teaching in the United States, see Rao (1977), pp. 473–86.

26. For an informative discussion of distance learning programs in, among others, Brazil, England, Israel, Kenya, Korea, Malawi, Mauritius, Nicaragua, the former Soviet Union, and the United States, see Perraton (1982). See also Jenkins (1988); *Development Communication Report* (1988); Asian Development Bank (1987); and Lockheed, Middleton, and Nettleton (1991).

27. Personal interview by Douglas Goldschmidt with Tony Dodds, International Extension College, London, Eng., 1979.

28. Dodds (1982).

29. An issue that has been raised to defend rural telecommunications investment, and indirectly to attack rural postal spending, concerns the incidence of rural literacy in developing countries. Where literacy is low, the importance of the post for carrying information may be questioned. Unfortunately, few developing countries have reliable data on this issue. Although the levels of household (as opposed to individual) literacy may in many instances be higher than is generally documented, few developing countries have reliable data on either household literacy or village literacy (for example, the existence of people in the village who will read and write for others).

30. The post has also frequently been used to subsidize rural transport facilities through postal carriage contracts. Even today, air service to rural villages in Alaska is highly dependent on U.S. government postal contracts, which historically were used as much to support continued rural transport as to ensure mail carriage.

31. P. V. Degraw, Fourth Assistant Postmaster General. Quoted in U.S. House of Representatives (1912).

32. See U.S. Congress (1971), pp. 29–31.

33. One small example of the effect of the rural free delivery system on rural transportation was noted by the U.S. Congress in 1914. The Post Office found that in the northern regions of the country the inhabitants didn't like to go out after major snow storms until the road had been traveled. The mail carrier was generally the first person out and, hence,

provided the opening of the road, which then elicited other travel, a minor, but interesting, externality. See U.S. Congress (1913)

34. *Union Postale* (1974).

35. Carpenter (1974).

36. Similar delivery systems run by private transporters are discussed in the last section of this appendix.

37. Donald (1976), chaps. 8–12.

38. Universal Postal Union (1974), p. 13.

39. Universal Postal Union (1974), p. 13.

40. Universal Postal Union (1974), p. 13.

41. A. M. Narula (1972).

42. Simonelli (1967), p. 31.

43. Chaffar (1974).

44. Approximately $2.3 million at early 1982 exchange rates.

45. Universal Postal Union (1980), pp. 11–12.

46. Universal Postal Union (1980), p. 13.

47. Universal Postal Union (1980).

48. World Bank data.

49. Fuller (1964).

50. These privately owned New Zealand delivery services correspond closely in function to the publicly owned Scottish post office and rural Swedish post office services discussed earlier.

51. Williams (1972).

52. A similar argument is sometimes used to oppose telecommunications investment; telecommunications systems are felt to be too capital intensive.

53. The government's objective of generating employment usually is accompanied by the necessity of holding down wages of postal employees. This makes it difficult for the postal system to hire and retain well-qualified staff, which, in turn, directly contributes to poor efficiency and management of the system.

Appendix C

A Survey of Rural Public Call Offices in Costa Rica

IN 1970 A PROGRAM was initiated by Instituto Costarricense de Electricidad (ICE), Costa Rica's power and telecommunications utility, to install public call office (PCO) telephones on a concessionaire basis in small towns and villages without telephone access. By early 1974, eighty-two small towns had been provided with at least one PCO. However, an unexpectedly rapid increase in costs and a decrease in financial operating surpluses from urban and long-distance telecommunications services reduced the resources available to continue the program. In the face of these financial constraints, ICE, in conjunction with the University of Costa Rica and with input from World Bank staff, initiated a study of rural PCO telephone demand, use, and benefits. The initial objective of the study was to provide input needed to clarify priorities for expanding this rural service further.

When the study was being designed, unresolved conceptual problems about how to identify and measure telephone benefits as well as data and resource constraints arose. Therefore, the analysis of benefits was confined primarily to estimating the short-run price elasticity of demand for PCO telephone use and to identifying some of the characteristics of PCO users, the purposes for which PCO telephones were used, the patterns of use, and the substitutability of PCO telephones for other modes of communication. For the study, a rural PCO telephone possessed the following characteristics:

a. It was administered by an ICE concessionaire, who was usually the owner of a business (a shop, bar, cafe, small hotel, restaurant, and so forth). In a few instances the concessionaire was an employee of the local municipality, a local telegraph operator, or a member of the rural police

b. It was a public access telephone, that is, the telephone did not belong to any particular subscriber; it belonged to ICE and could be used by anyone who could pay the call charge

c. No private telephone subscribers were located in the village where the PCO telephone was located or within approximately 10 kilometers of the village.

Most of the rural telephones were connected to the national network through single- or multichannel ultra or very high-frequency radio equipment.

Two types of data were gathered for the study. First, secondary data on the economic, demographic, and telephone use characteristics of the villages were obtained from three sources: a housing and population census and an agricultural census, both carried out in 1973 by the Costa Rican Department of Statistics and Census, and PCO call traffic information collected and tabulated by ICE. The data covered ninety-two rural PCO telephones in eighty-two villages. Relatively detailed census information was available for sixty-four of the villages; somewhat incomplete data were available for the rest. Call traffic data were compiled by ICE both before and after a tariff increase, which raised the call charge for each metered pulse from 12 to 15 centimos on January 1, 1975.

To supplement the secondary data, primary data were collected by (a) telephone concessionaires who interviewed telephone users immediately following the completion of calls in eleven villages for seventeen weeks during about ten months in 1974 and early 1975 and (b) two University of Costa Rica students who interviewed callers following calls made at PCO telephones in three villages (Puerto Cortes, Santa Clara, and Llano Grande) for one week in October 1974.

Three aspects of the study are reported here:

a. Estimation of the short-run price elasticity of demand for PCO telephone calls based on the observed volume of call traffic before and after the January 1975 tariff increase

b. Statistical analysis of the variations in telephone use among sample villages to identify some of the economic and demographic determinants of use at the village level

c. Examination of the survey data to determine who used rural PCOs (what are the characteristics of users) and for what purposes.

Estimation of Price Elasticity of Demand for Calls

To derive a simple estimate of the price elasticity of the demand for PCO calls, call traffic before the January 1975 tariff increase (Q_1) was measured as the number of metered pulses generated by a village PCO telephone from September through December 1974. Call traffic after the January 1, 1975, tariff increase (Q_2) was measured as the number of metered pulses from January through April 1975. Given unit-pulse prices (P_1 and P_2) before and after the price increase (12 centimos and 15 centimos per pulse, respectively), an estimate of the price elasticity of demand for telephone traffic for each PCO was calculated using the standard definition for price elasticity:

$$\text{Price elasticity} = -\left[\frac{Q_2 - Q_1}{1/2(Q_1 + Q_2)}\right] \bigg/ \left[\frac{P_2 - P_1}{1/2(P_1 + P_2)}\right]$$

For the ninety-two PCO telephones located in the eighty-two rural villages, the estimated price elasticity per village averaged −0.29.

Price elasticity estimates calculated for this relatively short period—four months before and after the price change—have the advantage that few other factors changed significantly in the villages in that short period of time. A disadvantage, however, is that the elasticity estimates could have been biased in several ways. On the one hand, PCO users might not have had sufficient time to adapt their behavior to the higher price. For example, they could have continued some of their old calling and travel habits without fully perceiving or adjusting to the new price. In this case the price elasticity would have been underestimated. On the other hand, although call traffic did on average initially fall following the price increase, this could have been a relatively short-term reaction, which might have been followed after several months by a return to previous levels. In this case the estimated price elasticity would have been an overestimate of the price elasticity of demand that would be relevant over a longer period.

Bias could also have resulted from the presence of a secular trend in telephone use. For example, in the villages where the telephone had been installed less than six months before the price change, call traffic increased relatively rapidly as villagers became accustomed to using the new facility. This upward trend partly offset the decrease in call traffic resulting from the price change; the average estimated price elasticity of demand for the sixty-five villages that had a PCO in service for six months or more was −0.51.

Seasonal variations in local economic activity also affected the estimates. After removing villages that had acquired a PCO within six months of the January 1, 1975, price change, the remaining villages that had positive price elasticity estimates were in tourist areas where the period from January to March was the busiest season or were relatively far from the capital city of San José and were partly involved in tourism. The average estimate of the price elasticity of demand for villages in the tourist area was +0.29, whereas faraway villages also partly involved in tourism had an average price elasticity of +0.44.[1]

Excluding from the sample the villages involved in seasonal tourism and those far from San José and partly influenced by seasonal tourism (twenty-six of the eighty-two villages in the sample), the average estimated price elasticity of demand for the remaining fifty-six villages was −0.57. Table C-1 groups the remaining villages by the primary demographic characteristics or type of economic activity in their area. As can be seen, however, the estimated price elasticity among these villages still varied widely.

The concessionaire survey of eleven villages, all of which had access to a PCO telephone for six months or more, showed that the average duration of calls decreased from 3.90 minutes between September and December 1974 before the price changed to 3.42 minutes between January and April 1975 after the price changed. This 12.3 percent reduction of the average duration, when compared with the 25 percent increase in call charge, gives roughly an average price elasticity of de-

Table C-1. *Average Estimated Price Elasticity of Demand for Rural* PCO *Telephone Traffic (Metered Pulses) for Selected Groups of Villages in Costa Rica, 1974–75*

Type of area	Average estimated price elasticity of demand
Frontier or border	−1.23
Island	−1.19
Coffee and sugarcane	−1.03
Banana and cocoa	−0.45
Cereal and cattle	−0.43
Overall average	−0.57

Note: Excluding villages with less than six months of telephone access and those located in tourist and distant tourist-related areas, all of which were experiencing major growth in telephone traffic or were subject to predictable seasonal increases in traffic.

Source: Traffic data supplied by ICE.

mand for call duration of −0.49.[2] This estimate is not far from both the average call traffic price elasticity for PCO telephones in service more than six months (−0.51) and the average for older and nontourism-related PCO telephones (−0.57). This implies that PCO users on average may have reacted to the increase in call charges by reducing the duration of their calls rather than the number of calls actually placed.

Table C-2 shows that the effect of the price change on call duration varied considerably with the primary purpose of the call. Calls related to agriculture and industrial activities remained virtually unchanged, whereas those related to professional services, emergency services, and inquiries about or discussions of health decreased about 20 percent in average duration. The negative price elasticity of demand for the duration of calls varied by purpose of call between −0.02 and −0.84. The price elasticity of demand for call duration was positive for calls pertaining to work opportunities.

Table C-2. *Average Duration of Outgoing Calls and Price Elasticities, by Primary Purpose of Call, for Eleven Concessionaire-Survey Villages in Costa Rica, 1974–75*

	Duration (minutes)		Price elasticity of demand for call duration
Primary purpose of call	September to December 1974	January to April 1975	
Family, kin, and friends	3.97	3.56	−0.41
Inquiry about health[a]	3.35	2.75	−0.72
Commerce	3.88	3.40	−0.45
Agriculture and industry	3.83	3.81	−0.02
Government administration	4.24	3.95	−0.27
Professional services	3.97	3.16	−0.82
Work opportunities[b]	3.12	3.34	0.28
Sports, religion, study	3.67	3.13	−0.59
Emergency services	3.61	2.85	−0.84
Unknown	3.61	3.14	−0.77
Average	3.90	3.42	−0.49

Note: On average, the concessionaire survey was taken once every two to three weeks during roughly seventeen survey weeks. All of the village surveys overlapped at least during the period between September 1974 and April 1975.

a. Encompasses calls to family or friends inquiring about health matters or discussing someone's health.

b. Calls to inform about, or obtain information on, work opportunities.

Source: World Bank data.

Determinants of Rural PCO Telephone Use

In an initial cross-sectional analysis of the determinants of PCO telephone use in rural Costa Rica, an attempt was made to specify and test an explanatory model of the form:

$$Q = f(P, Y, X_1 \ldots X_n)$$

where Q = quantity of telephone use per capita, P = price for using the telephone, Y = average yearly income of the caller, the caller's household, or the village of the caller, and X_i = selected other economic and demographic variables that could be associated with variations in use.

As can be seen, two of the most important explanatory variables were assumed to be price and income. Given the data available, however, it was not possible to identify completely acceptable proxies for these two variables nor for the quantity of use.

The problem of data existed partly because telephone calls are not a homogeneous good. To examine empirically the relation between the price paid for a good and the quantity of that good consumed at different prices, price must be varied for a specific uniform good. However, PCO telephone calls were of different durations and made over different distances, at different times of the day and week, and in some cases with varying technical quality. Each telephone call was essentially a different good for which consumers would presumably be willing to pay a different price.

The information on call charges was also inadequate to construct a totally acceptable proxy for price. In rural Costa Rica, as in many countries, calls are charged on the basis of duration, distance, and time of day. The price of a daytime call was 12 centavos per pulse before January 1, 1975, and 15 centavos per pulse afterward. Call distances were divided into three ranges: short distance, with a pulse frequency of one per twenty seconds; medium distance, with a pulse frequency of one per ten seconds; and longer distance, with a pulse frequency of one per five seconds. When the census data were collected (before January 1975), calls of short, medium, and longer distance cost 36 centimos, 72 centimos, and 144 centimos per minute, respectively.

Hence, although the posted price paid by users was uniform (the price per pulse was constant), the price paid varied only with the nonhomogeneity of the calls (with time of day, call duration, and distance), and the price variable was the average call charge paid in the village, not the more relevant price paid by each caller. As a result, an adequate proxy for the price variable could not be satisfactorily derived from the available data.

Deriving a suitable income proxy was also problematic. For many villages the closest proxy was a piece of census information for each village called "average wage per month." This variable only reflected wages of wage earners, however, and in some villages the proportion of recorded wage earners to total population was relatively small. It was not clear whether or not their earnings varied among villages in a roughly similar manner as average village incomes.

Given these and other problems in deriving adequate proxies, it was decided that the best feasible course of action was simply to search empirically for variables that were associated with differences in traffic among villages. The two search techniques used were stepwise regression analysis and factor analysis.

Statistical Regression Analysis

For the regression exercise, three variables were specified to reflect telephone use (Q):

a. The average monthly number of pulses per capita (or, equivalently, the average call revenue per capita) generated by a village
b. The average weekly number of calls per capita initiated by a village
c. The average weekly minutes of call time per capita generated by a village.

The pulse was tabulated by ICE for the sixty-four rural villages for which relatively detailed census information was available. The number of calls and the duration of calls were obtained from the concessionaire's survey of users and were therefore available for only eleven villages.

None of these three variables was fully satisfactory as a measure of the quantity of telephone service that each village consumed during an average week or month. The pulses variable provided information affected by the number of calls, the duration of each call, and the distance of each call—the relative importance of the three factors being unknown. The number of calls variable and the minutes variable also provided mixed information on the duration, number, and distance of calls. Finally, none of the variables provided information on the quality of the transmission. Given these shortcomings, the three quantity-of-use variables were interpreted loosely as general proxies for village telephone use.

Twenty-four other economic and demographic variables were available or could be tabulated from census data to reflect differences in population characteristics and in economic activity among the sixty-four villages. These are listed in table C-3. All economic, demographic, and traffic data examined in this exercise refer to the period before the tariff increase.

In an initial analysis of these data, a stepwise multiple linear regression computer algorithm (least squares estimation) was used to select

Table C-3. *Economic and Demographic Variables Available for the Cross-Village Examination of Rural Public Telephone Use in Costa Rica, 1973–74*

Variable number	Description
1.	Locally owned farm area out of total farm area in the village zone[a]
2.	Cultivated area as a percent of total farm area in the village zone
3.	Quantity of fertilizer used per fertilized hectare in the village zone
4.	Number of locally owned farms as a percent of total farms in the village zone
5.	Percent of milk cattle out of total cattle in the village zone
6.	Index of expenditures on machinery in the village zone
7.	Average area of each farm in the village zone (hectares)
8.	Percent of houses in the village within a 2-kilometer radius of telephone
9.	Number of houses within a 2-kilometer radius of telephone
10.	Percent of population 0 to 6 years old
11.	Percent of population 7 to 14 years old
12.	Percent of population 15 to 19 years old
13.	Percent of population 20 to 24 years old
14.	Percent of population 25 years and older
15.	Illiterate population as a percent of population 10 years and older
16.	Percent of employers and self-employed out of the economically active population in the village zone
17.	Average wage per month of wage earner in the village zone (colones)
18.	Average number of rooms per house in the village zone
19.	Percent of houses with septic tanks in the village zone
20.	Percent of houses with radios
21.	Number of shops, offices, health clinics, bars, and so forth in the village where the rural telephone is located
22.	Students as a percent of the population 7 to 24 years old
23.	Distance from the village to San José (kilometers)
24.	Total population in the village zone

a. The village zone includes the village and surrounding area in which the population is assumed to have convenient access to the village. In Costa Rica this was generally an area within a 10-kilometer radius of the telephone.

Source: Department of Statistics and Census, Costa Rica.

a small set of statistically significant regressors (independent variables), which together explained the largest proportion of the variation among the sixty-four villages in the values of the dependent variables (telephone use). This involved regressing, in turn, the three variables reflecting telephone use on the twenty-four economic and demographic variables. Three equations that emerged from these stepwise regressions are shown in table C-4. In these equations,

Q_1 = average monthly pulses per capita
Q_2 = average weekly number of calls per capita
Q_3 = average weekly minutes of calls per capita
X_{23} = distance between the village and San José (kilometers)
X_{17} = average monthly wage for wage earners
X_{20} = percent of houses with radios
X_9 = percent of houses in the village within 2 kilometers of the telephone
X_{22} = percent of the population between 7 and 24 years old who are students.

These three regression equations revealed a consistent direction of association between the independent variables (regressors) and the telephone use variables.[3] All of the regressors were statistically significant at the 0.05 level except for the wage variable (X_{17}) in equation 1, the distance variable (X_{23}) in equation 2, and the radios variable (X_{20}) in equation 3. The direction of association between the regressors and the telephone use variables was roughly consistent with what was expected in the following ways. First, if differences in the wage variable (X_{17}) roughly reflected differences in relative income among the villages, the positive relation between the three telephone use variables and the wage variable showed that more traffic per capita tended to be generated by higher-income villages than by lower-income villages.

Second, on the one hand, the positive relation between the telephone pulses per capita variable (Q_1) and the distance to San José (X_{23}) was in line with the fact that in general most calls were made to San José (see table C-17) and that, for a given call, the frequency of pulses was by definition a function of the distance of the call. On the other hand, a positive relation was also found between distance to San José and both the number of calls (Q_2) and the duration of calls (Q_3). This is consistent with the likelihood that the greater the distance over which communication took place, the more the telephone was the least-cost way to communicate; the more distant villages had fewer feasible alternatives for communicating rapidly with the major

Table C-4. *Regression Equations in Which Selected Economic and Demographic Variables Are Regressed on Proxies for Village Telephone Traffic*

Equation number	Sample size	Dependent variable	Constant term	X_{23}
1	64	Q_1	5.652	0.039 (2.86)
2	11	Q_2	0.729	0.005 (1.89)
3	11	Q_3	−4.130	0.024 (2.29)

— Not available.

Note: Values of t-statistics based on students' distribution of t are presented in parentheses below the regression coefficients.

Source: World Bank data.

economic, financial, and government center in the country, or the alternatives were more expensive.

Third, the negative relation between telephone use and the proportion of houses with radios (X_{20}) was more difficult to explain, although telephones and radios were, to some extent, substitutes. The prevalence of radios as a means of receiving certain types of information (weather forecasts, agricultural reports, market prices, and political events) may have decreased the need for telephone use. The interviews of individual users in the three-village sample partly supported this possibility (see table C-13).

Equation 1 is the only equation that examined differences among the sixty-four villages. The inclusion in this equation of independent variables representing the percentage of village households within 2 kilometers of the telephone (X_9) and the percentage of village population between 7 and 24 years old who were students (X_{22}) significantly improved the explanatory power of the model. The positive association between the population concentration variable (X_9) and the telephone use variable (Q_1) suggested that the more accessible the telephone, the more frequently it was used.[4] To the extent that the student variable (X_{22}) reflected the general educational level of the village, its positive association with telephone use (Q_1) perhaps indicated that use increased with the general educational level of the population. A related explanation for the significance of this variable is that students had a wider geographical community of interest or a greater familiarity with the telephone than did most villagers.[5]

X_{17}	X_{20}	X_9	X_{22}	Adjusted coefficient of determination
0.008	−0.168	0.068	0.002	0.451
(1.82)	(−2.96)	(2.36)	(2.47)	
0.004	−0.026	—	—	0.648
(2.41)	(−2.24)			
0.022	−0.060	—	—	0.655
(3.27)	(−1.20)			

Factor Analysis

Among the reservations concerning this stepwise multiple linear regression exercise is that the final selection of three or five regressors was influenced by the size and pattern of the collinearity among the original data. As a result a further exercise was undertaken to identify regressors that explicitly represent the underlying patterns of association among the original twenty-four economic and demographic variables. Factor analysis was used to help identify those underlying patterns.

Factor analysis is a statistical technique that aids in summarizing the general patterns of association among variables on the basis of their intercorrelations. Factor analysis can help disentangle complex relations in a set of data and present the data summarized in major and distinct regularities. Although the algebra of factor analysis is somewhat complicated, the technique is generally to extend into n dimensions an examination of the extent to which vectors representing measurable characteristics tend to cluster when examined for several observations. Highly related characteristics tend to cluster together, whereas unrelated characteristics can be set at right angles to one another. Algebraically, these clusters are set forth as vectors or linear combinations of the original variables that explain the maximum possible amount of variation present, given that the variables are combined linearly. After redundancies in the original sets of data are identified and eliminated, proxies for the remaining variables or characteristics can be introduced into least squares regression analysis.[6]

Summary results of a factor analysis of the twenty-four economic and demographic variables are presented in table C-5. The columns of figures are the factor loadings, that is, the coefficients or weights of

Table C-5. The First Five Factors from a Factor Analysis of Twenty-four Economic and Demographic Variables Compiled for Sixty-four Villages in Costa Rica, 1973–74

Variable number	Description	Factor 1	Factor 2	Factor 3	Factor 4	Factor 5
1.	Percent locally owned farm area out of total farm area in the village zone[a]	-0.006	-0.017	-0.150	0.381	0.420
2.	Cultivated area as a percent of total farm area in the village zone	0.299	-0.112	-0.348	0.055	0.041
3.	Quantity of fertilizer used per fertilized hectare in the village zone	0.534	-0.031	0.122	0.062	0.156
4.	Number of locally owned farms as a percent of total farms in the village zone	-0.221	-0.243	0.244	0.187	0.509
5.	Percent of milk cattle out of total cattle in the village zone	0.228	-0.399	-0.191	0.215	-0.148
6.	Index of expenditures on machinery in the village zone	0.513	-0.194	-0.041	0.079	0.169
7.	Average area of each farm in the village zone (hectares)	-0.265	0.643	-0.108	-0.100	-0.063
8.	Percent of houses in the village within a 2-kilometer radius of a telephone	-0.106	0.455	-0.048	-0.158	0.360
9.	Number of houses within a 2-kilometer radius of a telephone	0.814	0.141	0.127	-0.079	0.091
10.	Percent of population 0 to 6 years old	-0.287	0.307	0.429	0.594	-0.333
11.	Percent of population 7 to 14 years old	0.025	-0.216	0.761	-0.326	0.002
12.	Percent of population 15 to 19 years old	0.035	-0.186	-0.312	-0.074	0.141

13.	Percent of population 20 to 24 years old	0.111	0.381	−0.631	0.428	0.239
14.	Percent of population 25 years and older	0.216	−0.285	−0.483	−0.618	0.123
15.	Illiterate population as a percent of population 10 years and older	−0.611	0.278	0.031	0.132	0.164
16.	Percent of employers and self-employed out of the economically active population in the village zone	−0.423	−0.150	0.197	−0.173	0.261
17.	Average wage per month of wage earner in the village zone (colones)	0.403	0.519	−0.070	−0.336	−0.285
18.	Average number of rooms per house in the village zone	0.345	−0.255	−0.028	0.230	−0.433
19.	Percent of houses with septic tanks in the village zone	0.388	0.447	−0.044	0.028	−0.468
20.	Percent of houses with radios	0.245	−0.374	−0.002	−0.040	0.080
21.	Number of shops, offices, health clinics, bars, and so forth in the village where the rural telephone is located	0.676	0.198	0.350	0.016	0.311
22.	Students as a percent of the population 7 to 24 years old	−0.004	0.024	0.207	−0.343	−0.142
23.	Distance from the village to San José (kilometers)	−0.131	0.522	0.089	−0.293	0.323
24.	Total population in the village zone	0.705	0.098	0.340	0.315	0.261
	Percent of total variance explained	22.4	16.0	12.6	10.9	10.4
	Cumulative percent of total variance explained	22.4	38.3	50.9	61.8	72.3

a. The village zone includes the village and surrounding area in which the population is assumed to have convenient access to the village. In Costa Rica this was generally an area within a 10-kilometer radius of the telephone.

Source: World Bank data.

each variable in a linear combination with all variables in the analysis. The factor loadings represent the degree of association between the individual variables and the vector made up of a weighted combination of all the variables, which is sometimes referred to as a factor, characteristic vector, or eigenvector. Each factor represents an independent dimension of the total variation of all variables in the analysis. The absolute value of a factor loading may be interpreted as an ordinal measure of the degree to which each variable is involved in each factor or cluster of variables.

As shown in table C-5, the first factor explains approximately 22 percent of the total variation among all twenty-four variables. The size of the factor loadings on the first factor shows that population (variable 24), number of houses within 2 kilometers of a telephone (9), number of shops (21), illiterate population (15), fertilizer use (3), and machinery expenditures (6) were highly interrelated and together reflected a dimension in which highly ranked villages tended to be larger, to have a better educated population, and to make relatively greater expenditures on machinery and fertilizer. Factor 1 might then be designated as reflecting a general dimension of village size, education, and technology use.

The second factor explains an additional 16 percent of the variation among the twenty-four variables, bringing the cumulative variation explained to 38 percent. Given the relatively high loadings of variables 7, 8, 17, 19, and 23, this vector seems to represent a general dimension in which highly scored villages tended to be high-income, densely clustered, large-farm villages located far from San José.

The third and fourth factors are dominated by variables describing the age distribution of the population. Villages ranked highly on the third vector had a relatively high proportion of population less than 15 years old, whereas villages ranked highly on the fourth vector had a relatively higher proportion of their population in the age group less than 6 years old or between 20 and 24 years old.

The fifth and last factor reviewed here adds an additional 10 percent to the explained variation, bringing the total variation explained by the five factors to approximately 72 percent. Villages ranked highly according to this vector tended to have a relatively large proportion of locally owned farms and farm areas that had small houses and no septic tanks.

Given that these five factors represent five underlying patterns of association among the twenty-four variables examined, highly loaded variables from these factors can be selected as proxies for each factor.

The following five variables have the largest absolute loading on each of the five vectors and, as a result, were chosen to represent the general dimension reflected by each vector.

X_9 = Number of houses within a 2-kilometer radius of a telephone. This variable represents the general dimension of village size, education, technology use (factor 1)

X_7 = Average area in hectares of each farm in the village zone. Villages in which farm areas were larger tended to be higher-income, densely clustered, large-farm villages relatively farther from San José (factor 2)

X_{11} = Percent of population between 7 and 14 years old (factor 3)

X_{14} = Percent of population 25 years and older (factor 4)

X_4 = Locally owned farms as a percent of the total number of farms in the village zone. Villages in which the percentage of locally owned farms and locally owned farm areas was relatively large tended to have small houses with no septic tanks (factor 5).

These five variables were regressed, in turn, on the three telephone use variables previously examined, with Q_1 expressed as average monthly revenue (pulses times 0.12 colones) per capita rather than as pulses. The results of the three regressions (linear with least squares estimation) are presented in table C-6.

Although the direction of the association between X_9, X_7, X_{11}, and X_4 and the PCO call traffic variables is consistent among the three regressions, only X_9 and X_7 are statistically significant at the 0.05 level in any of the regressions. Given this, it can be tentatively concluded that villages that exhibited greater amounts of PCO telephone use per capita tended to be larger, higher-income farming villages relatively far from San José, in which the farms were larger and the population was relatively better educated and engaged in more technically progressive agriculture.

Results of the Determinants Analysis

In summary, the cross-sectional analysis of the determinants of PCO telephone use in Costa Rican villages yielded interesting but not necessarily surprising results. If it is assumed that the per capita telephone use variables that were examined correspond roughly to crude indexes of benefits that villages derive from telephones, it might then be concluded that the rural villages in Costa Rica that tended to re-

Table C-6. *Regression Equations in Which Economic and Demographic Variables Selected through Factor Analysis Are Regressed on Proxies for Village Telephone Traffic*

Equation number	Sample size	Dependent variable	Constant term	X_9
4	64	Q_1	2220.40	7.11 (1.95)
5	11	Q_2	−3.72	0.002 (0.001)
6	11	Q_3	−10.37	0.007 (0.005)

Note: Standard errors are presented in parentheses below the regression coefficients.
Source: World Bank data.

ceive the greatest benefits from rural PCO telephone service had on average some of the following characteristics:[7]

a. They were relatively large. Several variables relating to village size were highly weighted on the factor represented by X_9

b. The village population was relatively better educated and engaged in more progressive agricultural techniques. Several variables relating to both of these characteristics were highly loaded on the factor represented by X_9, and the student variable (X_{22}) was significant in equation 1

c. The village was better off economically. The wage variable was statistically significant in equations 2 and 3 and was highly loaded on the factor represented by X_7. In addition, X_7 itself was the farm size variable, which, other things being equal, could partly reflect the stock of wealth per family

d. The telephone was relatively convenient to use for a greater proportion of the population. The percentage of households within 2 kilometers of the telephone was statistically significant in equation 1 and highly loaded on the second factor represented by X_7

e. They were relatively far from San José. This suggests that as distance as well as the cost of some other forms of rapid communication increased, telephones increasingly became the least-cost solution for communicating rapidly in rural Costa Rica. The distance from San José variable was significant in equation 1 and was highly loaded on the vector represented by X_7.

X_7	X_{11}	X_{14}	X_4	Adjusted coefficient of determination
13.20	−136.98	−77.37	37.59	0.210
(8.82)	(135.50)	(109.51)	(27.08)	
0.019	−0.001	0.055	0.017	0.880
(0.003)	(0.001)	(0.038)	(0.010)	
0.073	−0.107	0.151	0.067	0.561
(0.026)	(0.330)	(0.320)	(0.088)	

Characteristics of PCO Users

The determinants exercise provided some insights into the types of villages that might receive the greatest immediate benefits from rural telephone investment in Costa Rica.[8] This and the following sections review some of the characteristics of the persons using PCO telephones in rural areas. The basic questions examined are who uses the telephones and for what purposes?

To address these questions two sources of data were examined. One was the responses obtained in the eleven-village concessionaire survey from which the number and duration of calls were tabulated. The second source was the October 1974 survey of telephone users in three villages: Puerto Cortes, Santa Clara, and Llano Grande. Selected economic and demographic characteristics of the three villages are shown in table C-7. Survey information on the characteristics of rural PCO telephone users is presented below.

Employment Characteristics

In the three user-survey villages approximately 97 percent of the callers who answered the question on employment status claimed to be employed. As can be seen from table C-8, more than 60 percent of those persons were wage earners, rather than self-employed or employers of others. With regard to their occupations, the largest percentage overall (32.7 percent) consisted of agricultural workers followed closely by businesspersons, professionals, and technicians.

Table C-7. *Selected Characteristics of Three Villages in Costa Rica in Which Surveys of Public Telephone Use Were Made, 1974*

Characteristic	Puerto Cortes[a]	Santa Clara	Llano Grande
Location	Southern Costa Rica	Northern Costa Rica	Central valley
Main economic activity	River port; banana plantation	Cattle breeding; cereal growing; technical school	General agriculture; vegetables; flowers; dairy products
Percent of houses within 2 kilometers of the telephone	91	39	35

a. Puerto Cortes is the largest of the three villages.
Source: World Bank data.

Table C-8. *Employment Characteristics of Callers from a User Survey of Three Rural Villages in Costa Rica, 1974*

Employment characteristics of callers	Puerto Cortes	Santa Clara	Llano Grande	Total
Employment status of the caller				
Wage earner	199 (52.0)	73 (65.8)	131 (74.9)	403 (60.5)
Self-employed	157 (41.0)	30 (27.0)	36 (20.6)	223 (33.5)
Employer	17 (4.5)	0 (0.0)	0 (0.0)	17 (2.5)
Principal occupation of the caller				
Professionals, technicians, and similar workers	98 (25.9)	37 (42.0)	1 (0.8)	136 (23.2)
Higher public officials	14 (3.7)	0 (0.0)	0 (0.0)	14 (2.4)
Administrative personnel and similar workers	45 (11.9)	7 (7.9)	0 (0.0)	52 (8.9)
Businesspersons and similar occupations	120 (31.7)	5 (5.7)	14 (11.6)	139 (23.7)
Workers in the service sector	3 (0.8)	6 (6.8)	5 (4.1)	14 (2.4)
Agricultural workers	65 (17.2)	29 (32.9)	98 (81.0)	192 (32.7)
Nonagricultural laborers	33 (8.7)	4 (4.5)	3 (2.5)	40 (6.3)

Note: For each question the number of affirmative answers is presented first and the percentage of answers falling in that category is presented below in parentheses. There are 674 total observations: 383 in Puerto Cortes, 114 in Santa Clara, and 177 in Llano Grande.
Source: World Bank data.

The occupations of the callers predictably varied according to the economic characteristics of the village in which they lived. In Puerto Cortes, a relatively large river port, businesspersons constituted the greatest proportion of callers (31.7 percent). In Santa Clara, professionals and technicians, to some extent from the local technical school, constituted the most frequent callers, whereas in Llano Grande, an agricultural area, agricultural workers made 81 percent of all calls.

Income Level

The average income of income earners in the three user-survey villages was approximately ₡372.23 ($43.79) a month in 1973.[9] The mean income of persons making PCO telephone calls who answered the income question was ₡377.87 ($44.45) a month. This suggests that at least in these three villages rural telephone access was not dominated by either high- or low-income users. Table C-9 shows the distribution of calls according to the income of the caller for the three villages. The median income of the callers who answered the income question was somewhat less than the mean income. Of course, the median income for the villages as a whole might also have been less than the mean.

Information on income and calls broken down for each of the three villages is presented in table C-10. Again, as shown in the more aggregate data, relatively lower-income villagers participated heavily in PCO telephone use.[10]

Table C-9. *Summary Distribution of Calls, by Income Class of Caller, for Three Rural Villages in Costa Rica, 1974*

Income group (colones per month)	Number of calls	Percentage
No response	148	21.9
0–75	23	3.4
75–150	95	14.1
150–225	87	12.9
225–300	102	15.1
300–375	37	5.5
375–450	23	3.4
450–525	58	8.6
525–600	30	4.5
600–800	26	3.9
800+	45	6.7
Total	674	100.0

Source: World Bank data.

Table C-10. *Distribution of Calls and Income of Caller, by Village, for Three Rural Villages in Costa Rica, 1974*
(colones per month)

Village	Income group	Average income of income earners	Percentage of total calls
Puerto Cortes	0–300	180.69	35.1
	300–600	431.31	32.2
	600+	1,063.33	12.9
	No response[a]	—	19.8
Santa Clara	0–300	188.00	17.8
	300–600	425.88	11.0
	600+	774.79	19.1
	No response	—	52.1
Llano Grande	0–300	182.39	79.2
	300–600	463.17	4.8
	600+	1,154.00	0.8
	No response	—	15.2

— Not available.

a. No response includes no answers and those who claimed no income because they did not work. The latter group comprised about 3 percent of the total.

Source: World Bank data.

Distance Traveled to Make Calls

Both the eleven-village concessionaire survey and the three-village user survey provided information on the distances PCO telephone users traveled to make calls. The user survey revealed that overall 78.6 percent of all calls were made by individuals who traveled at most 500 meters to make the call. The number within 500 meters ranged from 70.8 percent of the callers in Puerto Cortes (the largest community of the three) to 99.1 percent in Santa Clara. The concessionaire survey showed that on average 94.8 percent of the calls were made by individuals who lived within approximately 2 kilometers of the telephone. These results reaffirm the telephone convenience factor observed in the statistical examination of the determinants of telephone use.

Frequency of Calls

Table C-11, which summarizes the frequency of calls made during the sample week, suggests that PCO telephone use in the three user-survey villages was not dominated by any one small group of users.

Table C-11. *Percentage of Calls Made by the Same Caller for Three Rural Villages in Costa Rica, 1974*

Number of calls made during the sample week by the same person	Puerto Cortes	Santa Clara	Llano Grande	Average
1–2	78.8	82.1	76.1	78.8
3–4	8.4	13.4	12.7	10.3
5–10	11.3	4.5	7.0	9.1
11–15	1.0	0.0	1.4	0.9
15+	0.5	0.0	2.8	0.9
Total	100.0	100.0	100.0	100.0

Source: World Bank data.

Almost 79 percent of the callers made only one or two calls during the week, and more than 89 percent made four calls or less. Table C-12, however, which compares the mean monthly income of callers with the frequency of calls, indicates that users who made more than two calls a week tended to have at least average or above-average incomes.[11] Hence, although the benefits of the rural telephones were not generally restricted to any particular income group or small group of PCO users, the 21 percent who did use the telephone three or more times in the week tended, on average, not to be among the lowest-income callers.

Table C-12. *Average Monthly Income of Callers According to the Number of Calls Made during the Sample Week for Three Rural Villages in Costa Rica, 1974*
(colones)

Number of calls made during the sample week by the same person	Puerto Cortes	Santa Clara	Llano Grande	Average
1–2	350.90	393.13	159.12	321.45
3–4	429.79	520.80	170.75	369.89
5–10	511.13	404.00	172.50	439.82
11–15	a	a	a	303.67a
15+	a	a	a	300.00a

a. Sample size is too small for the estimate to be reliable.
Source: World Bank data.

Table C-13. *Alternative Communication Preferences for Callers in Three Rural Villages in Costa Rica, 1974*

Alternative method of communication if PCO telephone were not available	Puerto Cortes	Santa Clara	Llano Grande	Total
Would not have communicated	64 (16.8)	53 (46.5)	3 (1.7)	120 (17.9)
Telegram	107 (28.1)	15 (13.2)	0 (0.0)	122 (18.2)
Radio	132 (34.6)	0 (0.0)	16 (9.1)	148 (22.1)
Post office	57 (15.0)	45 (39.5)	131 (74.9)	233 (34.8)
Other telephone	5 (1.3)	1 (0.9)	14 (8.0)	20 (3.0)
Other	16 (4.2)	0 (0.0)	11 (6.3)	27 (4.0)

Note: For each question the number of affirmative answers is presented first and the percentage of answers falling in that category is presented below in parentheses. There are 674 total observations: 383 in Puerto Cortes, 114 in Santa Clara, and 177 in Llano Grande.
Source: World Bank data.

Preferences for Alternative Forms of Communication

Callers in the three user-survey villages were also asked the means of communication they would have used if the PCO telephone had not been available in the village. The results of the responses are presented in table C-13. Although responses varied widely among villages, the first preference of almost 35 percent of the respondents was the postal service. The next most frequently listed alternative was the radio, primarily because it was the dominant alternative in Puerto Cortes. Third overall was a telegram. On average almost 18 percent of the callers would not have communicated at all if the telephone had not been available; the range was a large 2 to 46 percent. Overall, the wide variation in choice of alternative communication suggests that preferences, or more likely feasible alternatives, tended to be location specific.

Purpose and Characteristics of Calls

Table C-14 summarizes information from the three user-survey villages on the primary purposes for which telephone calls were made.

Table C-14. *Primary Purpose of* PCO *Calls Made in Three Rural Villages in Costa Rica, 1974*

Primary purpose of call	Puerto Cortes	Santa Clara	Llano Grande	Total
Family, kin, and friends	180	84	82	346
	(47.4)	(73.7)	(46.3)	(51.5)
Inquiry about health, health- related topics, or someone's health	30	0	6	36
	(7.9)	(0.0)	(3.4)	(5.4)
Commerce	72	2	26	100
	(18.9)	(1.7)	(14.7)	(14.9)
Agriculture and industry	16	0	28	44
	(4.2)	(0.0)	(15.8)	(6.6)
Government administration	33	16	10	59
	(8.7)	(14.0)	(5.6)	(8.8)
Professional services	18	3	3	24
	(4.7)	(2.6)	(1.7)	(3.6)
Informing of, or obtaining information on, work opportunities	10	4	3	17
	(2.6)	(3.5)	(1.7)	(2.5)
Sports, religion, study, and other	20	4	16	40
	(5.3)	(3.5)	(9.0)	(6.0)
Emergency services	1	1	3	5
	(0.3)	(0.9)	(1.7)	(0.7)

Note: For each question the number of affirmative answers is presented first and the percentage of answers falling in that category is presented below in parentheses. There are 674 total observations: 383 in Puerto Cortes, 114 in Santa Clara, and 177 in Llano Grande.
Source: World Bank data.

Table C-15 presents similar information from the eleven concessionaire-survey villages, which are grouped according to prominent economic or demographic characteristics. These two tables show that although the relative importance of the reasons for calling did vary a few percentage points from village to village, overall the reasons given were roughly consistent.

Both surveys showed that the most frequent purpose for making a rural PCO telephone call was communicating with family or friends, for whatever reason. Calls related to trade and commerce ranked second in frequency, and government administration calls were third in the three-village survey, while sports, religion, and study were third in the eleven-village survey.

Table C-15. *Percentage of Calls Generated in Eleven Concessionaire-Survey Villages in Costa Rica, Grouped by Characteristics of the Area, by Purpose of Call, 1974–75*

Primary purpose of call	Distant	Frontier; border	Coffee; sugarcane	Banana; cocoa	Cereal; cattle	Primarily tourist	Total
Family, kin, and friends[a]	67.8	58.3	45.6	54.8	58.1	74.7	59.7
Inquiry about health, health-related topics, or someone's health	4.3	3.2	12.1	5.7	6.0	5.8	5.9
Commerce	9.3	20.5	10.6	10.6	14.1	11.5	12.6
Agriculture and industry	5.2	1.4	5.9	3.3	3.0	1.2	3.1
Government administration	1.8	3.0	3.9	2.7	6.1	1.0	3.7
Professional services	0.6	0.1	2.6	1.1	1.6	0.6	1.1
Sports, religion, study, and other	1.0	0.5	11.1	4.0	6.8	0.8	4.7
Emergency services	0.1	0.3	0.1	0.2	0.5	0.2	0.3
No reply	7.9	12.4	8.2	13.4	4.0	4.3	8.6

a. Includes calls to discuss work opportunities.
Source: World Bank data.

The fact that economic and demographic factors affect the types of calls made from a village can be seen from the results of the concessionaire survey, which are presented in table C-15. Calls to family and friends were highest in the tourist and the more distant areas that were also associated partly with tourism—almost 75 and 68 percent, respectively, of all calls made from these areas. A partial explanation for this could be that individuals in distant areas felt relatively isolated and thus relied more heavily on the telephone as a means of personal contact. Also, since the business of distant tourist areas is people, a higher percentage of tourist-generated personal calls might be expected from such areas. A relatively large proportion of commercial calls were made from border areas, perhaps because these areas are often associated with trade. Agricultural or industrial calls were made with approximately the same frequency from the three crop-producing areas and the distant areas; the share of such calls was relatively smaller in border and tourist areas.

These results essentially reinforce what was observed earlier in the examination of the employment characteristics of callers in the three-village user sample. The share of agricultural or industrial calls tended to be average or higher in crop-producing areas, commercial calls were highest in trade-oriented border areas, and calls to family or friends tended to dominate tourist areas. Similarly, calls about religion, sports, government administration, and professional services accounted for higher shares in the villages surrounding San José. Hence, overall, the reasons for rural PCO telephone use predictably tended to reflect the economic and demographic characteristics of the area.

Survey information on the distances that callers traveled to the PCO telephone by purpose of the call showed that although the distances traveled were on average about the same, individuals tended to travel longer distances to make certain types of calls. Findings from the three-village user survey showed that on average individuals traveled relatively longer distances to make calls about agriculture or industry, health, and emergency services. Similar results for calls about agriculture or industry and emergency services were found in the eleven-village concessionaire survey (see table C-16); callers making agricultural or industrial and emergency services calls traveled 81 and 58 percent farther, respectively, than the average distance traveled. Both surveys indicated that individuals calling family or friends traveled about average distances. Agricultural or industrial callers probably travel relatively farther because farmers or farm workers are likely to be located farther from the center of the villages, where PCO telephones

Table C-16. *Mean Distance That Callers Traveled to Make Calls, by Purpose of Call, for Eleven Concessionaire-Survey Villages in Costa Rica, 1974–75*
(kilometers)

Primary purpose of call	Callers residing more than 2 kilometers from the telephone	All callers
Emergency services	9.00	0.41
Inquiry about health	8.16	0.36
Family, kin, and friends[a]	8.73	0.27
Commerce	7.56	0.26
Agriculture, industry	9.39	0.47
Government administration	6.99	0.12
Sports, religion, study, and other	7.14	0.30
Professional services	7.29	0.41
No answer	9.28	0.12
Total	8.46	0.26

a. Includes calls to discuss work opportunities.
Source: World Bank data.

Table C-17. *Percentage of Calls to San José, by Primary Purpose of Call, for Eleven Concessionaire-Survey Villages in Costa Rica, 1974–75*

Primary purpose of call	Florencia T.C.	Guapiles	Llano Grande	Miramar	Orosi	Penas Blancas
Family, kin, and friends[a]	49	70	43	45	53	67
Inquiry about health	40	71	37	42	49	64
Commerce	8	71	53	44	48	73
Agriculture, industry	—	73	35	39	45	59
Government administration	100	69	28	39	50	62
Professional services	50	72	38	36	25	80
Sports, religion, study, and other	37	65	45	47	55	69
Emergency services	—	40	—	49	—	38
No answer	37	67	40	41	53	66
Total percentage of calls going to San José	38	69	44	44	50	68

— Not available.
a. Includes calls to discuss work opportunities.
Source: World Bank data.

tend to be located. This is, of course, opposite the situation for calls about government administration. Finally, callers seeking emergency services may travel relatively longer distances because such calls are viewed as being more important, and thus individuals are willing to travel farther to make them.

Information on the most frequent destination of rural calls, by the primary purpose of the call, is provided in table C-17. On average, 60 percent of all calls made in the eleven concessionaire-sample villages went to San José. The two most frequent types of calls to San José were related to commerce and to professional services. This, of course, is consistent with the fact that San José is the business and government center of Costa Rica. The least-frequent reason for making calls to San José was for emergency services. This might indicate that many emergencies must be dealt with quickly, and therefore a neighboring town or provincial center is a more appropriate place to seek assistance.

Finally, information was also collected on how the primary purpose of calls varied with the duration of calls as well as with the day of the week and the time of day the calls were made. Table C-18 shows the average duration of calls made in each of the eleven concessionaire villages. The shortest duration was on average for calls seeking emergency services, when time is of the essence. At the other extreme were

Poasitocitos Playas Jaco	Puerto Cortes	Puerto Jiménez	Santa Clara	San Vito	Total percentage of calls to San José
60	66	50	47	67	60
48	78	45	45	83	58
69	76	50	57	74	65
55	80	49	54	83	59
55	57	23	45	68	55
54	73	36	62	74	55
80	100	100	50	50	50
46	79	60	53	77	64
49	63	51	44	67	61
59	67	50	48	69	60

Table C-18. *Mean Duration of Calls, by Primary Purpose of Call, for Eleven Concessionaire-Survey Villages in Costa Rica, 1974–75*
(minutes)

Primary purpose of call	Call duration
Family, kin, and friends[a]	3.70
Inquiry about health	3.09
Commerce	3.62
Agriculture and industry	3.79
Government administration	4.00
Professional services	3.39
Sports, religion, study, and other	3.19
Emergency services	2.88
No answer	3.48

a. Includes calls to discuss work opportunities.
Source: World Bank data.

calls relating to government administration. This would be consistent with situations in which public sector employees are reimbursed for their official calls and therefore lose some of the incentive to economize on telephone call charges.

Table C-19 shows how the purposes for which calls were made varied by day of the week. The major changes in daily calling patterns seem to be that calls related to commerce, agriculture, government administration, and professional services tended to fall off on the weekend, whereas the proportion of calls to family and friends made on the weekends tended to increase. This is as would be expected when a five- or six-day workweek is followed by a one- or two-day weekend holiday.

Table C-20 shows types of calls made, by the time of day. Once again, calls about commerce, agriculture, government administration, and professional services varied together and tended to be made during the working day (6 a.m.–6 p.m.), with the largest number occurring during the morning (6 a.m.–12 noon). For no category were more calls made at night (6 p.m.–6 a.m.), when reduced call charges were supposed to be in effect.[12] Calls to family and friends and those relating to inquiries about health and religion, study, and sports did not fall off as much as did business-related calls after 6 p.m., but they did decline. Table C-21, which summarizes calling time information from the three user-survey villages, shows roughly the same general patterns, although differences among the time slots were not as pronounced as they were in the eleven concessionaire-survey villages. The

Table C-19. Percentage Distribution of Calls per Day, by Primary Purpose of Call, for Eleven Concessionaire-Survey Villages in Costa Rica, 1974–75

Primary purpose of call	Monday	Tuesday	Wednesday	Thursday	Friday	Saturday	Sunday
Family, kin, and friends[a]	55.8	56.9	58.0	56.4	57.8	63.6	69.2
Inquiry about health	5.6	5.7	5.8	6.1	5.4	5.7	7.2
Commerce	15.0	14.5	13.6	13.5	13.1	11.6	7.0
Agriculture and industry	3.8	3.8	2.9	3.5	3.3	2.5	2.0
Government administration	4.8	5.0	4.3	4.6	4.4	2.2	0.8
Professional services	1.4	1.6	1.4	1.3	1.3	0.7	0.1
Sports, religion, study, and other	4.9	4.1	4.7	4.0	5.3	4.9	5.0
Emergency services	0.3	0.3	0.8	0.1	0.1	0.2	0.3
No answer	8.4	8.2	8.6	9.9	9.0	8.1	8.0
Total	100.0	100.0	100.0	100.0	100.0	100.0	100.0

a. Includes calls to discuss work opportunities.
Source: World Bank data.

Table C-20. *Percentage Distribution of the Times of Day during Which Calls Are Made, by Primary Purpose of Call, for Eleven Concessionaire-Survey Villages in Costa Rica, 1974–75*

Primary purpose of call	Morning (6 a.m.–12 noon)	Afternoon (12 noon–6 p.m.)	Night (6 p.m.–6 a.m.)
Family, kin, and friends[a]	35.6	36.8	27.6
Inquiry about health	39.4	34.9	25.7
Commerce	49.4	39.7	10.9
Agriculture and industry	47.3	35.9	16.8
Government administration	52.3	40.4	7.4
Professional services	50.8	40.5	8.7
Sports, religion, study, and other	36.6	38.2	25.3
Emergency services	46.7	30.4	22.8
No answer	36.2	40.0	23.8

a. Includes calls to discuss work opportunities.
Source: World Bank data.

Table C-21. *Summary Characteristics of Calls for Three Rural Villages in Costa Rica, 1974*

Call characteristics	Puerto Cortes	Santa Clara	Llano Grande	Total
Call placed between				
6 a.m.–12 noon	143	28	68	239
	(37.5)	(24.6)	(38.4)	(35.6)
12 noon–6 p.m.	140	32	54	226
	(36.7)	(28.1)	(30.5)	(33.6)
6 p.m.–6 a.m.	98	54	55	207
	(25.7)	(47.4)	(31.1)	(30.8)
Outgoing call	352	73	114	539
	(91.9)	(64.0)	(64.4)	(80.0)
Outgoing collect call	61	21	3	85
	(17.3)	(28.8)	(2.6)	(15.8)
Call made on own behalf	359	113	160	632
	(93.7)	(99.1)	(90.4)	(93.8)
Trip to town or PCO telephone location made only to place a call	346	113	172	631
	(91.0)	(99.1)	(97.7)	(94.0)

Note: Numbers in parentheses are percentage of total calls.
Source: World Bank data.

one exception to the general pattern was in Santa Clara, perhaps because staff and students of the technical school were in session during much of the daytime hours.

Table C-21 also shows that outgoing calls comprised approximately 80 percent of all calls made in the three user-survey villages. For the eleven concessionaire-survey villages outgoing calls comprised 76 percent of all calls, perhaps because it is inconvenient to receive incoming calls at a village PCO. In the three-village sample, almost 16 percent of the outgoing calls were collect calls, and approximately 94 percent of the trips made by callers to the establishment in which the telephone was located were made specifically to use the telephone. In these villages at least, outgoing calls tended to be planned in advance.

Factor Analysis of User Characteristics and Purposes of Use

In the above sections, information on characteristics of persons using rural PCO telephones and the purposes of those calls was compiled and reviewed on a two-dimensional basis. An attempt was also made to examine the relations within the user/usage survey data more thoroughly to understand the size and direction of the collinearity among the individual variables. The exercise involved a factor analysis of thirty-eight economic, demographic, and telephone use variables collected in the user survey of Puerto Cortes, Santa Clara, and Llano Grande.

The results of this factor analysis were for the most part disappointing in that they generally revealed only obvious or trivial relations. For example, if more calls were made by self-employed persons and businesspersons, fewer calls were made by wage earners. One of the few interesting results that emerged was that on the fourth vector in the three-village factor analysis and the third vector in the individual Puerto Cortes analysis the highly loaded variables suggested that callers who lived farther than the average distance from the telephone tended to take more time to reach the telephone and complete the call and therefore tended to make the trip not only for calling, but for other purposes as well. Furthermore, since they made these trips for multiple purposes, they tended to allocate a relatively small portion of the cost of the trip to the call.[13]

Also the second vector in the three-village analysis and vectors one or two in each of the three individual village analyses indicated that call charges tended to be lower for outgoing calls. The meaning of this result is unclear, but the relation was consistent for all three villages. It could indicate that call charges on incoming collect calls tended to be higher than average since it takes additional time to find and bring the person being called to the PCO, or that when someone takes the unusual step of calling a village PCO collect, the reasons for the call are sufficiently complex that the relevant information takes longer to convey.

Summary of the Costa Rican Rural PCO Telephone Study

In evaluating the priorities for investment in a rural telephone program, it is useful to understand the type and amount of benefits associated with alternative configurations of investment and the segments of the population and of the economy that are likely to benefit the most. This examination of the rural PCO telephone program in Costa Rica as it existed in 1974 and 1975 provided information relating to both of these questions.

Evidence suggested that benefits to individuals using PCO telephones tended to be greatest in rural villages that possessed one or more of the following (sometimes collinear) characteristics:[14]

a. Village income per capita was higher than average
b. The village had a relatively large population
c. The village was located relatively far from the major economic, social, and government center of San José
d. The educational level of the population was above average
e. The population tended to be clustered more closely around the site where the telephone was located.

Within villages telephone use appeared to be relatively widespread with no one small group dominating the others. There also appeared to be no major income bias among the callers; overall both higher- and lower-income villagers used the PCO.

The purposes for which calls were made tended to be correlated with the economic base and demographic characteristics of the village. By far the most prominent purpose for which calls were made, however, was to maintain contact with family or friends, for whatever reasons. Hence, in addition to facilitating economic and government administrative activity, the rural public telephone program also improved the general quality of life of rural residents by allowing them to stay in touch and exchange information with family and friends.

A smaller but significant proportion of calls were to discuss health matters with family and friends and for emergency purposes, and, on average, people tended to travel farther to make such calls. This suggests that the telephone's spatial zone of influence varied somewhat by the type of call.

Finally, the most common substitute for the PCO telephone was the postal service. PCO telephones did, however, evidently generate new activity and communication. Overall, approximately 18 percent of the callers in the three user-survey villages stated that if the telephone had not been available, they would not have communicated with the person they called.

Notes

Note: The analysis and write-up presented in this appendix were done by Robert J. Saunders. Antonio Canas of ICE directed the collection of telephone use data provided by ICE. Marco Tristán of the University of Costa Rica directed the tabulation of the census data and the collection of the survey data. Inputs to the initial study design and analysis were made by R. J. Saunders, S. C. Littlechild, M. Tristán, and J. J. Warford. Research assistance or comments from M. Gellerson, K. Challa, J. Lu, and B. Wellenius are also acknowledged.

1. Telephone traffic records for the previous twenty-four months for villages involved at least partly in tourism provided evidence that the positive price elasticities observed for several of these villages were likely the result of predictable seasonal variations resulting from tourism.

2. Such a calculation assumes that the distribution of calls according to destination and time of day remained unchanged.

3. The signs of the regressor or independent variable coefficients are the same for all three equations.

4. This conclusion is consistent with survey results from other countries reviewed in chapter 12.

5. Both of these explanations are also supported by the survey results outlined in chapter 12.

6. For a more detailed discussion of factor analysis, see Harmon (1967) and Anderson (1958).

7. The factors examined were interrelated so that it is not totally valid to speak of the effects of any one of them in isolation.

8. This statement assumes that the volume of PCO use per capita is directly proportional to the size of benefits derived, which may not always be a valid assumption.

9. National Census of 1973 and Céspedes (1971).

10. The survey data on income for Santa Clara could be biased downward. The interviewers reported that the no-response rate was high partly because some of the technical school employees were reluctant to reveal their incomes.

11. It is possible that concessionaires making multiple telephone calls themselves partly brought about this result.

12. Although reduced call charges were supposed to be in effect between 6 p.m. and 6 a.m., several concessionaires charged the same tariff throughout the twenty-four-hour per-

iod and therefore made larger profits on night calls. The fact that most concessionaires closed their business at night also affected the number of night calls, since the PCOs could not be used.

13. Example 21 in chapter 9 outlines an attempt to estimate some travel and time costs associated with PCO telephone calls in Puerto Cortes.

14. A summary of these results was first published in Saunders and Warford (1977).

Appendix D

Shadow Pricing
for Telecommunications

THE BORDER PRICING APPROACH to shadow pricing, outlined by Little and Mirrlees and elaborated by Squire and van der Tak, is used here as a basis for explaining the relevance of shadow pricing for estimating telecommunications costs.[1] With the border pricing approach, the numeraire or unit of account is defined to be uncommitted public income at border prices, that is, basically a foreign exchange numeraire expressed in units of the local currency converted at the official rate of exchange. Therefore, directly imported inputs, such as cables or exchange equipment, whose foreign exchange costs are known, are already in border prices (converted at the official exchange rate). Locally purchased inputs, however, which are expressed in domestic market prices, need to be converted to border prices by an appropriate conversion factor.[2] The ways in which commodities normally traded internationally (tradables) and those typically produced locally (nontradables) may be analyzed are summarized below.

For tradables, which are those items for which there is a world supply of imports and a global demand for exports, the cost, insurance, and freight border price for imports and the border price for exports may be used, with a suitable adjustment made for the marketing margin. The free trade assumption is not required to justify the use of border prices, since domestic price distortions can in effect be adjusted for by netting out the effect of all taxes, duties, subsidies, and so forth. In the telecommunications sector the most important tradables are capital goods, such as switching and transmission equipment, and subscriber apparatus and cables.

413

For nontradable commodities, primarily labor and land, the associated border-priced marginal social cost (MSC) is the relevant resource cost, unless the input is supplied through the decreased consumption of other users, in which case the border-priced marginal social benefit (MSB) of this forgone consumption would be a more appropriate measure of social cost. More generally, if both effects are present, a weighted average of MSC and MSB should be used, with the MSB tending to dominate in the short-run supply-constrained situation and the MSC being more important in the longer run when output may be expanded. The MSC may be determined by successively decomposing the nontradable into its constituent inputs, through several rounds, until all that remains are the ultimate tradable inputs, which can be valued at border prices, and nontraded primary factors such as labor and land, which may also be shadow priced as discussed below. The MSB for intermediate and final consumption should be evaluated in terms of the forgone social profit and consumer surplus, respectively, plus the income transfer effects of any changes in price. In practice, the MSC is often used because of data constraints.

In the telecommunications sector the most important nontradable input is labor. The procedure for estimating the shadow wage rate (SWR) can be illustrated by considering the use of unskilled labor in a labor-surplus country, for example, workers employed in constructing buildings or laying cable. The forgone output of workers used in the telecommunications project is the dominant component of the SWR. Complications arise because the original income earned may not reflect the marginal product of labor, and, further, for every new job created, more than one worker may give up his old employment. Seasonal activities, such as harvesting, must also be allowed for. In theory, if the laborer has to work harder in his new job than in his old one, then the disutility of forgone leisure must also be included in the SWR, but in practice this component is ignored. Overhead costs incurred (for example, transport) should also be considered. The foregoing can be summarized by the following basic equation:

$$\text{SWR} = a\,m + c\,u$$

where m and u are the forgone marginal output and overhead costs of labor in domestic prices, and a and c are corresponding factors for converting these values into border prices.

These effects on consumption patterns also need to be considered. Suppose a worker receives a wage W_n in his new job and that the income forgone is W_o, both in domestic prices. (W_o may not necessarily be equal to the marginal output forgone, m.) Assuming, as is likely for low-income workers (and also in the interests of simplicity), that all the increase in income $(W_n - W_o)$ is consumed, this increase in consumption will result in a resource cost to the economy of b $(W_n - W_o)$, where b is the border-priced MSC of increasing consumption (in domestic prices) by one unit. The increased consumption also provides a benefit given by w $(W_n - W_o)$, where w represents the MSB (in border prices) of increasing domestic-priced, private sector consumption by one unit. Therefore,[3]

$$\text{SWR} = a\ m + c\ u + (b - w)(W_n - W_o).$$

In the case of land inputs, the appropriate shadow value placed on this primary factor depends on location. In most cases, it is assumed that the market price of urban land is a good indicator of its economic value in domestic prices, and the application of an appropriate conversion factor, for example, the standard conversion factor, to this domestic price will yield the border-priced cost of urban land inputs. Rural land that has an alternative use in agriculture may be valued at its opportunity cost, that is, the net benefit of forgone agricultural output. In the case of telecommunications, this can usually be assumed to be negligible.

The shadow price of capital may be reflected in the discount rate or accounting rate of interest (ARI), which is defined as the rate of decline in the value of the numeraire over time. An appropriate discount rate would be the opportunity cost of capital (OCC), which refers to the rate of return to the economy as a whole resulting from an additional unit of public investment and may be used as a proxy for the ARI in the pure efficiency price model. A simple formula for ARI, which includes consumption effects, is

$$\text{ARI} = \text{OCC}\ [s + (1 - s)w\ /\ b]$$

where s is the fraction of the yield from the original investment that will be saved and reinvested.

Use of such an explicit shadow pricing framework would allow corrections for major distortions in the market mechanism. The shadow prices of inputs may then be incorporated into a given formula to yield the basic marginal costs in border prices. However, these border-priced marginal costs have to be divided by an appropriate conversion factor to arrive at the domestic-priced marginal costs, which may be used as a basis for determining the domestic price of telecommunications services, as in several of the examples cited in chapters 8 and 9.

Ideally, this conversion factor should depend on the alternative use of income by each type of consumer, but in practice some "average" proxy like b or the standard conversion factor is often used.

Notes

1. Little and Mirrlees (1976) and Squire and van der Tak (1975).

2. For inputs that are not important enough to merit individual attention or that lack sufficient data, a standard conversion factor may be used. It is equal to the official exchange rate divided by the more familiar shadow exchange rate. For a more detailed discussion of alternative interpretations of the shadow exchange rate, see Squire and van der Tak (1975).

3. The consumption term $(b - w)$ disappears if, at the margin, (a) society is indifferent to the distribution of income (or consumption), so that everyone's consumption has equivalent value; and (b) private consumption is considered to be as socially valuable as uncommitted public savings, that is, the numeraire.

Appendix E

World Bank Loans and Credits

Table E-1. *Loans and Credits by the World Bank and the International Development Association for Telecommunications Projects, 1962–89*
(millions of U.S. dollars)

Fiscal year and country	Source of funding	Loan or credit amount	Total project cost
1962			
Ethiopia	Bank	2.9	6.2
1963			
India	IDA	42.0	122.0
1964			
Costa Rica	Bank	9.9	12.6
El Salvador	Bank	9.5	13.6
Subtotal		19.4	26.2
1965			
India	IDA	33.0	228.0
1966			
Venezuela	Bank	37.0	100.0
Ethiopia	Bank	4.8	10.8
Subtotal		41.8	110.8
1967			
Jamaica	Bank	11.2	18.3
East African Community	Bank	13.0	26.7
Colombia	Bank	16.0	27.6
Subtotal		40.2	72.6
1968			
China, Rep. of	Bank	17.0	50.0
Singapore	Bank	3.0	9.5
Papua New Guinea	Bank	7.0	15.4
Subtotal		27.0	74.9

417

1969

Malaysia	Bank	4.4	49.0
Upper Volta	IDA	0.8	1.2
Pakistan	IDA	16.0	42.0
Ethiopia	Bank	4.5	25.4
India	Bank/IDA	55.0	361.0
Subtotal		80.7	478.6

1970

Costa Rica	Bank	6.5	9.5
Nepal	IDA	1.7	4.2
Singapore	Bank	11.0	37.1
Yugoslavia	Bank	40.0	95.0
Pakistan	IDA	15.0	35.3
East African Community	Bank	10.4	28.3
Subtotal		84.6	209.4

1971

Indonesia	IDA	12.8	22.1
Iran	Bank	36.0	149.2
India	IDA	78.0	290.7
Colombia	Bank	15.0	39.1
Malaysia	Bank	18.7	94.0
Venezuela	Bank	35.0	294.5
Subtotal		195.5	889.6

1972

Iraq	Bank	27.5	39.7
Guatemala	Bank	16.0	21.1
Costa Rica	Bank	17.5	32.2
El Salvador	Bank	9.5	12.7
Fiji	Bank	2.2	5.7
Mali	IDA	3.6	4.3
Subtotal		76.3	115.7

1973

Papua New Guinea	Bank	10.0	17.2
Bangladesh	IDA	7.3	12.1
Thailand	Bank	37.0	102.8
Senegal	Bank	6.3	8.9
Nepal	IDA	5.5	7.9
East African Community	Bank	32.5	53.3
India	IDA	80.0	534.1
Subtotal		178.6	736.3

1974

Iran	Bank	82.0	194.0
Upper Volta	IDA	4.5	5.6
Ethiopia	IDA	21.4	37.1
Côte d'Ivoire	Bank	25.0	53.6
Costa Rica	Bank	23.5	54.2
Trinidad and Tobago	Bank	18.0	30.5
Bangladesh	IDA	20.0	87.0
Subtotal		194.4	462.0

1975

Pakistan	IDA	36.0	67.5
Colombia	Bank	15.0	52.1
Guatemala	Bank	26.0	45.9
Egypt	IDA	30.0	173.4
Ethiopia	IDA	16.0	60.6
Ghana	Bank	23.0	29.5
Burma	IDA	21.0	30.9
Zambia	Bank	32.0	78.2
Subtotal		199.0	538.1

1976

Fiji	Bank	5.0	14.0
Syria	Bank	28.0	145.6
Thailand	Bank	26.0	146.1
Niger	IDA	5.2	6.5
Subtotal		64.2	312.2

1977

India	Bank	80.0	415.0
Colombia	Bank	60.0	167.7
Subtotal		140.0	582.7

1978

Lebanon	Bank	14.5	33.7
Egypt	IDA	53.0	210.0
Costa Rica	Bank	10.6	94.1
El Salvador	Bank	23.0	51.6
India	Bank	120.0	818.5
Subtotal		221.1	1,207.9

1979

Nepal	IDA	14.5	25.2
Thailand	Bank	90.0	307.2
Kenya	Bank	20.0	63.5
Subtotal		124.5	395.9

1980

Burma	IDA	35.0	93.0
Sri Lanka	IDA	30.0	36.3
Subtotal		65.0	129.3

1981

Burundi	IDA	7.7	9.1
Rwanda	IDA	7.5	17.5
Oman	Bank	22.0	97.2
Colombia	Bank	44.0	110.0
India	IDA	314.0	1,619.4
Subtotal		395.2	1,853.2

1982

Uruguay	Bank	40.0	204.8
Tanzania	IDA	27.0	47.0
Cameroon	Bank	7.5	11.1
Pakistan	IDA	40.0	266.1
Mali	IDA	13.5	25.3
Egypt	Bank	64.0	141.3
Upper Volta	IDA	17.0	40.6
Thailand	Bank	142.1	492.2
Kenya	Bank	44.7	117.9
Subtotal		395.8	1,346.3

1983

Bangladesh	IDA	35.0	61.8
Uganda	IDA	22.0	26.0
Subtotal		57.0	87.8

1984

Algeria	Bank	128.0	312.4
Guatemala	Bank	30.0	208.0
Thailand	Bank	8.5	..
Subtotal		166.5	520.4

1985

Ethiopia	IDA	40.0	151.8
Kenya	Bank	32.6	86.1
Nepal	IDA	22.0	65.3
Oman	Bank	23.0	227.8
Philippines	Bank	4.0	4.8
Subtotal		121.6	535.8

1986

Côte d'Ivoire	Bank	24.5	116.7
Laos	IDA	3.9	7.8
Senegal	IDA	22.0	156.9
Subtotal		50.4	281.4

1987

Burundi	IDA	4.8	25.4
Hungary	Bank	70.0	833.1
India	Bank	345.0	2,050.0
Indonesia	Bank	14.5	17.9
Morocco	Bank	·125.0	674.5
Pakistan	Bank	100.0	817.2
Tanzania	IDA	23.0	60.0
Subtotal		682.3	4,478.1

1988

Jordan	Bank	36.0	338.1

1989

Benin	IDA	16.0	65.3
Ecuador	Bank	45.0	330.0
Fiji	Bank	8.1	47.9
Ghana	IDA	19.0	173.0
Togo	IDA	16.0	44.6
Uganda	IDA	52.3	58.8
Western Samoa	IDA	4.6	16.3
Subtotal		161.0	735.9
Total[a]		3,933.0	16,875.4

a. Does not include four small loans totaling $24.25 million made before 1962.
Source: World Bank data.

The Canadian Radio-Television and Telecommunications Commission Costing Methodology: Cost Allocation Categories

THE EIGHT BROAD CATEGORIES established by the Canadian federal telecommunications regulatory agency, the CRTC, in 1989 as part of the Cost Inquiry Phase III are presented here in detail. Although the requirements for an appropriate approach to the situation in each country are necessarily different, this case illustrates how costs can be allocated in order to analyze a telecommunications tariff (see chapter 15 for a complete discussion of costing methodologies).

Access (A)

The facilities associated with Access and their associated costs are composed of three separate components: the subscriber premises equipment, the loop, and the serving central office equipment.

Subscriber Premises Equipment

Costs related to the equipment and connections provided on a monopoly basis pursuant to "Attachment of Subscriber-Provided Terminal Equipment," TELECOM Decision CRTC 82-14, November 23, 1982, are included in the Access category. In some cases, costs associated with terminal equipment giving the subscriber access to a service that can only be provided by the carrier are included in this category. A specific example is the private branch exchange line card required by a centralized system for reporting emergencies.

All other subscriber premises equipment and connection costs are included under the categories of Competitive Network or Competitive Terminal. Unless otherwise specified and affirmed by the carriers, the boundary between the Access and Competitive Terminal categories is the demarcation point on the customer's premises.

Loop

The loop includes all costs associated with (a) the provision by the company, for any purpose, of a facility connecting the customer's premises with the nearest serving central office and (b) the termination and protection of that facility at both the customer's premises and the serving central office.

These interconnecting facilities normally consist of outside plant but also include related equipment required to derive the interconnecting capability, such as subscriber line carrier systems and line concentrator equipment. The interconnection may also be provided by a radio system, in which case the loop includes all costs associated with the radio base station equipment and facilities connecting the serving central office.

Serving Central Office Equipment

Costs associated with all central office switching equipment not sensitive to traffic but needed to give customers access to the public switched telephone network are included under Access.

Costs associated with equipment located in a central office required specifically to derive the loop facility, such as subscriber line carriers and line concentrators, are included under Access, while those associated with equipment for specific services, such as circuit conditioning and signaling equipment, are included as appropriate under the categories of Monopoly Local, Monopoly Tool, or Competitive Network.

Monopoly Local (ML)

The Monopoly Local service category includes costs related to the provision, operation, and maintenance of the local switching equipment and interoffice transmission facilities that are required to establish and maintain communication services within the local calling area and that generally are not provided by another supplier.

Monopoly Toll (MT)

The Monopoly Toll service category includes costs related to the provision, operation, and maintenance of the switching equipment and transmission facilities that are required to establish and maintain communication services between local calling areas and that generally are not provided by another supplier.

Competitive Network (CN)

The Competitive Network category includes costs related to the provision, operation, and maintenance of the facilities that are required to establish and maintain communication services and that are, or can be, provided by another supplier, with the exception of the facilities included under Access.

Costs associated with terminal equipment (equipment that is specific to competitive network service and that is provided by the company as an integral part of the service) are included in this category.

Competitive Terminal—Multiline and Data (CT-MD)

As specified in Decision CRTC 86-5, costs and revenues associated with the Competitive Terminal category are subdivided into two distinct service categories: Competitive Terminal—Multiline and Data; and Competitive Terminal—Other.

The CT(MD) category includes costs associated with key telephone systems, private branch exchange systems, and all telephone sets behind key and private branch exchange systems, including those behind Centrex systems subject to the final classification of proprietary sets and consoles. Also included are costs associated with all data terminal equipment at the customer's premises that is not integral to the operation of the channel provided by the carrier. This category also includes costs associated with inside wiring on the customer's side of the service demarcation point that is associated with multiline and data terminals. Costs associated with terminal equipment that may be located on a customer's premises but that is provided by the company as an integral part of a network service shall be included as appropriate under the categories of Competitive Network, Monopoly Toll, or Monopoly Local.

Competitive Terminal—Other (CT-O)

Costs associated with terminal equipment other than those identified above are included in the Competitive Terminal—Other category. This category consists primarily of costs associated with single-line telephone sets.

Other (O)

The Other service category includes the causal costs associated with activities and services that, in general, do not relate directly to the provision of telecommunications services and are not included in other service categories. Examples include the following:
- Building space rented to others
- Subscription charges for tariffs
- Arrangements for cable television lessees
- Billing service arrangements for CNCP
- Communications seminars
- Data processing for BCE

Common (C)

The Common category includes costs that cannot be causally related to a particular category of service and that, with the support of empirical evidence, are shown to be fixed.

References

The word "processed" describes works that are reproduced from typescript by mimeograph, xerography, or similar means; such works might not be cataloged or commonly available through libraries or might be subject to restricted circulation.

African Medical and Research Foundation. "AMREF Reports: Progress in Development of Various AMREF Projects for the Period January–June 1979." Nairobi, Kenya, 1979. Processed.

Alleman, J. H. *The Pricing of Local Telephone Services.* Special Publication no. 77-14. Washington, D.C.: U.S. Department of Commerce, Office of Telecommunications, April 1977.

Alleman, James H., and Richard D. Emmerson, eds. *Perspectives on the Telephone Industry: The Challenge for the Future.* New York: Harper and Row Publishers, Ballinger Division, 1989.

Ambrose, W., P. R. Hennemeyer, and J. P. Chapon. "Privatizing Telecommunications Systems: Business Opportunities in Developing Countries." Discussion Paper no. 10. International Finance Corporation, Washington, D.C., 1990.

American Telephone and Telegraph Corporation. *The World's Telephones.* Morris Plains, N.J.: AT&T Long Lines Overseas Department, August 1980, June 1981, and July 1982.

―――. *The World's Telephones: January 1987–1988.* Whippany, N.J., 1989.

―――. *The World's Telephones: January 1989.* Whippany, N.J., 1990.

Analysys, Ltd. "Final Report of Phase 2." Prepared for CEC DGXIIIF ORA Program. Cambridge, Eng., 1990. Processed.

―――. *A Study of the Economic Implications of Stimulating Applications of IT&T in Rural Areas,* 2 vols. CEC DGXIIIF ORA Program. Cambridge, Eng., 1989. Processed.

Anderson, T. W. *An Introduction to Multivariate Statistical Analysis.* New York: John Wiley and Sons, 1958.

Anonymous. "Restructuration du secteur des télécommunications dans Niger." Paper presented at a seminar on telecommunications restructuring sponsored by the World Bank, the ITU, and ACCT, Tunisia, May 1992.

Arrow, Kenneth J. "The Economics of Information." In Michael L. Dertouzos and Joel Moses, eds., *The Computer Age: A Twenty-Year View.* Cambridge, Mass.: M.I.T. Press, 1980.

———. "Political and Economic Evaluation of Social Effects and Externalities." In Julius Margolis, ed., *The Analysis of Public Output*. New York: National Bureau of Economic Research, Columbia University Press, 1970.

Ashby, Jacqueline, Steve Klees, Douglas Pachico, and Stuart Wells. "Alternative Strategies in the Information/Education Projects." In Emile G. McAnany, ed., *Communications in the Rural Third World*. New York: Praeger Publishers, 1980.

Asian Development Bank. *Distance Education in Asia and the Pacific*, 2 vols. Manila, 1987.

Baer, Walter S. "Defining Productivity Gains: Telecommunications and Productivity." *Telecommunications Policy* (December 1981), pp. 329–30.

Baeza, S., J. Bunster, and O. Schenone. "Evaluación económica de la inversión en teléfonos públicos de larga distancia en zonas rurales." Documento de Trabajo no. 37. Instituto de Economía, Universidad Católica de Chile, November 1975. Processed.

Balson, David, Robert Drysdale, and Bob Stanley. *Computer Based Conferencing Systems for Developing Countries*. Report of a workshop held in Ottawa, Canada, October 26–30, 1981. Ottawa: International Development Research Centre, 1982.

Barry, Mamadou Pathe, and Mohamed Sylla. "Le cas de la Guinée." Paper presented at a seminar on telecommunications restructuring sponsored by the World Bank, the ITU, and ACCT, Tunisia, May 1992.

Bates, Tony. "Options for Delivery Media." In Hilary Perraton, ed., *Alternative Routes to Formal Education: Distance Teaching for School Equivalency*. Baltimore, Md.: Johns Hopkins University Press, 1982.

Baumol, William J. "Reasonable Rules for Rate Regulation: Plausible Policies for an Imperfect World." In A. Phillips and O. Williamson, eds., *Prices: Issues in Theory, Practice, and Public Policy*. Philadelphia: University of Pennsylvania Press, 1968.

Baumol, William J., and David F. Bradford. "Optimal Departures from Marginal Cost Pricing." *American Economic Review*, vol. 60 (June 1970), pp. 265–83.

Bebee, E. L., and E. T. W. Gilling. "Telecommunications and Economic Development: A Model for Planning and Policy Making." *Telecommunications Journal*, vol. 43, no. 8 (August 1976), pp. 537–43.

Beesley, M. E. "The Value of Time Spent Travelling: Some New Evidence." *Economica*, vol. 32 (May 1965), pp. 174–85.

Beesley, M. E., and S. C. Littlechild. "The Regulation of Privatized Monopolies in the United Kingdom." *Rand Journal of Economics*, vol. 20, no. 3 (Autumn 1989).

Beier, G., A. Churchill, M. Cohen, and B. Renaud. "The Task for the Cities of the Developed Countries." World Bank Staff Working Paper no. 209. Washington, D.C., July 1975.

Bell Canada. *Cost Inquiry Phase III Manual*. Hull, Quebec, 1989.

Bethesda Research Institute. "Regulatory Systems for the 1990s: A Comparative Analysis of Price Caps and Rate of Return Regulation." Bethesda, Md., 1988. Processed.

428

Bigham, Fred G. "Integrating Telecommunications Costing by Class of Service with Incentive Regulation." Paper presented at NARUC Advanced Regulatory Studies Program, Williamsburg, Va., February 1990. Processed.

Bigham, Fred G., and G. W. Wall. "A Canadian Approach to the Determination of Broad Categories of Revenues and Costs." Paper presented at the conference on telecommunications costing in a dynamic environment, San Diego, Calif., 1989. Processed.

Blair, Michael L. "VSAT Systems in Developing Countries." Proceedings of the Pacific Telecommunications Conference. Honolulu, Hawaii, February 1988.

Block, Clifford. "Satellite Linkages and Rural Development." In Heather E. Hudson, ed., New Directions in Satellite Communications: Challenges for North and South. Norwood, Mass.: Artech House, 1985.

Block, Clifford, Douglas Goldschmidt, Anwar Hafid, Gerald C. Lalor, and Angel Velásquez. "Satellite Telecommunications for Rural Development: The AID Rural Satellite Program and Its Projects in Indonesia, Peru, and the Caribbean." Proceedings of the Pacific telecommunications conference, Honolulu, Hawaii, January 1984.

Booz Allen and Hamilton. "A Microeconomic Study of the Benefits of Improved Telephone Service in Selected Areas of the Philippines, 1984." London, Eng. Processed.

————. "Study of the Economic Benefits of New Telecommunications Services in Costa Rica, 1986." London, Eng. Processed.

Boiteux, M. "La tarification des demandes en pointe." Revue Générale d'Électricité, vol. 58 (1949), pp. 321–40. Translated as "Peak Load Pricing." Journal of Business, vol. 33 (April 1960), pp. 157–79.

Bower, L. B. "Telecommunications Market Demand and Investment Requirements." Telecommunications Journal, vol. 39 (November 3, 1972), pp. 172–81.

Bowers, D. A., and W. F. Lovejoy. "Disequilibrium and Increasing Costs: A Study of Local Telephone Service." Land Economics (February 1965), pp. 36–38.

Bradshaw, Ted K. "Rural Development and Telecommunications Potential and Policy." Working Paper no. 524. Berkeley: Institute of Urban and Regional Development, University of California, October 1990.

Bruce, Robert R. "Franchising and Subcontracting for Services and Facilities: New Options for Attracting New Sources of Investment." Roundtable on Eastern European telecommunications, Badacsonytomaj, Hungary, May 1991.

Bruce, Robert R., Jeffrey P. Cunard, and Mark Director. The Telecom Mosaic: Assembling the New International Structure. London: International Institute for Communications, 1988.

Bryan, Elizabeth B., and David A. Evans. "Access to Calls." Background papers for a seminar on social research and telecommunications planning, TELECOM, Australia, August 1979. Processed.

Burkhead, Jesse, and Jerry Miner. Public Expenditure. Chicago: Aldine Publishing Company, 1971.

Carpenter, Trevor C. "The Rural Postal Services of Scotland." Union Postale, vol. 5 (1974), pp. 66–68.

Carsberg, Sir Bryan. "The Liberalization of Telecommunications: Experience in the United Kingdom." Roundtable on Eastern European telecommunications, Badacsonytomaj, Hungary, May 1991.

_____. "Regulating Private Monopolies and Promoting Competition." *Long Range Planning,* vol. 19, no. 6 (1986).

Casasus, Carlos. "Privatization of Telecommunications: The Case of Mexico." In Björn Wellenius and Peter Stern, eds., *Implementing Reforms in the Telecommunications Sector: Lessons from Recent Experience.* Washington, D.C.: World Bank, 1991.

Castilla, Adolfo, Maria Cruz Alonso, and Carlos Tirado. *Telecomunicaciones y desarrollo en España e Iberoamérica.* Madrid: FUNDESCO, 1989.

CCITT. *Blue Book,* vol. 2. Geneva: ITU, 1965.

_____. *GAS-5 Handbook: Economic Studies at the National Level in the Field of Telecommunications.* Geneva: ITU, 1968.

_____. *GAS-5 Handbook: Economic Studies at the National Level in the Field of Telecommunications.* Geneva: ITU, 1976.

_____. *National Telephone Networks for the Automatic Service, Third Plenary Assembly, Geneva, 1964.* Geneva: ITU, 1965.

_____. "Optimum Allocation and Use of Scarce Resources in Order to Meet Telecommunications Needs in Urban or Rural Areas of a Country." GAS-5 *Economic Studies (1981–84),* no. 8. Geneva: ITU, 1984a.

_____. "Socio-Economic Aspects of Telecommunications Development in Isolated and/or Underprivileged Areas of Countries." GAS-5 *Economic Studies (1981–1984),* no. 4. Geneva: ITU, 1984b.

Center for Science and Technology for Development. ATAS *Bulletin: New Information Technologies and Development,* issue 3. New York: United Nations, June 1986.

Céspedes, V. H. *Costa Rica: Income Distribution and Consumption of Some Foods.* San José: University of Costa Rica, 1971.

Chaffar, Y. A. "Some Aspects of Postal Administrations' Role in Financing Development Plans." In "Proceedings of the 1973 Postal Executives' Seminar, Tokyo, March 1–20, 1973." Universal Postal Union, Tokyo, 1974. Processed.

Chandler, A. D. *The Visible Hand: The Managerial Revolution in American Business.* Cambridge, Mass.: Harvard University Press, 1977.

Chan-Kil, Chung. "Evaluation Study on Rural Telephone Expansion Project under IBRD Loan." Korea Rural Economics Institute, Seoul, December 1979. Processed.

Chapuis, Robert J. *100 Years of Telephone Switching.* Part I, *Manual and Electromechanical Switching,* Studies in Telecommunications vol. 1. Amsterdam: North-Holland, 1982.

_____. "Telephony Is a Heavy Industry." *Telecommunications Journal,* vol. 42-X (1975).

Chapuis, Robert J., and A. Joel. *Electronic Computers and Telephone Switching: A Book of Technological History.* Studies in Telecommunication no. 13. New York: Elsevier Science Publishers/North-Holland, 1990.

430

Charles, Jeff. "Approaches to Teleconferencing Justification: Towards a General Model." *Telecommunications Policy* (December 1981), p. 297.

Christaller, Walter. *Central Places in Southern Germany.* Translated by Carlisle W. Baskin. Englewood Cliffs, N.J.: Prentice-Hall, 1966.

Chu, Godwin C., Alfian Chote Srivisal, and Boonlert Supadhiloke. "Rural Telephone in Indonesia and Thailand." *Telecommunications Policy* (June 1985), pp. 159–69.

Clark, David, and Kathryn Unwin. "Telecommunications and Travel: Potential Impact in Rural Areas." *Regional Studies,* vol. 15 (1981), pp. 47–56.

Cleevely, D. D. "Modelling the Role of Telecommunications in Regions of a Developing Country." Working Paper. Department of Engineering, Division of Control and Management Systems, University of Cambridge, Cambridge, Eng., February 1979. Processed.

Cleevely, D. D., and G. Walsham. "Interim Report on Modelling the Role of Telecommunications within Regions of Kenya." Department of Electrical Engineering, University of Cambridge, Eng., May 1980. Processed.

Clippinger, J. H. "Can Communication Development Benefit the Third World?" *Telecommunications Policy* (September 1977).

Cole, Barry, ed. *After the Break-Up: Assessing the New Post-AT&T Divestiture Era.* New York: Columbia University Press, 1991.

Communications Studies and Planning International. "The Impact of Telecommunications on the Performance of a Sample of Business Enterprises in Kenya." Research report commissioned by the ITU. New York, August 1981. Processed.

——. "A Study of the Economic Benefits of Improved Telephone Service in the Philippines." *Information, Telecommunications, and Development.* Geneva: ITU, 1986.

Communications Studies Group. "The Effectiveness of Person-to-Person Telecommunications Systems." Long Range Research Report no. 003/ITF. Post Office Telecommunications, London, May 1975. Processed.

Contreras, Eduardo. "Brazil and Guatemala: Communications, Rural Modernity, and Structural Constraints." In Emile G. McAnany, ed., *Communications in the Rural Third World.* New York: Praeger Publishers, 1980.

Courville, Léon, Alain De Fontenay, and Rodney Dobell, eds. *Economic Analysis of Telecommunications: Theory and Applications.* New York: North-Holland, 1983.

Cowhey, P. "The Political Economy of Telecommunications Reform." In Björn Wellenius and Peter Stern, eds., *Implementing Reforms in the Telecommunications Sector: Lessons from Recent Experience.* Washington, D.C.: World Bank, 1991.

Crandall, Robert W. *After the Breakup: U.S. Telecommunications in a More Competitive Era.* Washington, D.C.: Brookings Institution, 1991.

Crandall, Robert W., and Kenneth Flamm, eds. *Changing the Rules: Technological Change, International Competition, and Regulation in Communications.* Washington, D.C.: Brookings Institution, 1989.

Crockett, Bruce. "The Changing Role of Satellites in Telecommunications." *Tele-communications* (June 1989), pp. 37–45.

CRTC (Canadian Radio-Television and Telecommunications Commission). "Report of the Inquiry Officer with Respect to the Inquiry into Telecommunications Carriers' Costing and Accounting Procedures: Phase III—Costing of Existing Services." 1984. Processed.

Culham, P. G. "A Method for Determining the Optimal Balance of Prices for Telephone Services." OFTEL Working Paper no. 1. OFTEL, London, March 1987.

Cyert, R. M., and J. G. March. *A Behavioral Theory of the Firm*. Englewood Cliffs, N.J.: Prentice-Hall, 1963.

Demac, Donna, and Joseph Pelton, eds. *Telecommunications for Development: Exploring New Strategies—An International Forum*. Washington, D.C.: INTELSAT, 1986.

Department of Trade and Industry, United Kingdom. "Competition and Choice: Telecommunications Policy for the 1990s—A Consultative Document." HMSO, London, 1990.

Deutches Institut für Wirtschaftforschung. *Economic Evaluation of the Impact of Telecommunication Investment in the Communities*. Berlin, 1984.

Development Communication Report, vol. 63, no. 4 (1988).

Director General of Telecommunications, OFTEL. "The Regulation of British Telecom's Prices—A Consultative Document." London, 1988. Processed.

———. "Responses to OFTEL's Consultative Document on the Future Regulation of British Telecom's Prices." OFTEL Working Paper no. 3. London.

Dodds, Tony. "The Mauritius College of the Air." In Hilary Perraton, ed., *Alternative Routes to Formal Education: Distance Teaching for School Equivalency*. Baltimore, Md.: Johns Hopkins University Press, 1982.

Donald, Gordon. *Credit for Small Farmers in Developing Countries*. Boulder, Colo.: Westview, 1976.

Dorozynski, Alexander. *Doctors and Healers*. Ottawa: International Development Research Centre, 1975.

Drewer, T. M. "Econometric Models of International Communications." Report no. 28. British Post Office, Statistics and Business Research Department, London, 1973. Processed.

Economics Study Cell, Posts and Telegraphs Board, Ministry of Communications, India. "India's Rural Telephone Network." New Delhi, May 1981. Processed.

Elbert, Bruce R. *International Telecommunication Management*. Boston, Mass.: Artech House, 1990.

Engvall, Lars. "A Socio-Economic Study on Usage of Telephone Services." In ITU, *Information, Telecommunications, and Development*. Geneva, 1986.

Estabrooks, Maurice F., and Rodolphe H. Lamarche, eds. *Telecommunications: A Strategic Perspective on Regional, Economic, and Business Development*. Selected papers from a conference in Ottawa, Canada. Moncton, Canada: The Canadian Institute for Research on Regional Development, 1986.

Evans, D. A., and E. B. Bryan. "Access to Calls. A Perspective on the Economics of Telecommunications in Papua New Guinea." Study sponsored by the Papua New Guinea Department of Posts and Telegraphs and the World Bank. The Implementation and Management Group Party, Ltd., Sydney, Australia, December 1977. Processed.

Evans, John R., Karen Lashman Hall, and Jeremy Warford. "Shattuck Lecture—Health Care in the Developing World: Problems of Scarcity and Choice." *New England Journal of Medicine,* vol. 305 (November 1981), pp. 1117–27.

Fargo, Dan S. "International Telecom Spending Slows Down." *Telephony* (February 22, 1982), pp. 78–85.

Favout, R. "Improving Urban Highway Transportation through Electronic Route Guidance." In D. Brand, ed., *Urban Transportation Innovation.* New York: American Society of Civil Engineers, 1970.

Federal-Provincial-Territorial Task Force on Telecommunications, Ministry of Supply and Services, Canada. "The Effect of Changing Technology on the Structure of Costs for the Provision of Public Long-Distance Telephone Service." Consulting Report no. 3. (December) 1988a. Processed.

———. "The Effect of Toll Competition on Prices, Costs, and Productivity of the Telephone Industry in the United States." Prepared by David Chessler and Associates, 1988b. Processed.

Flamm, Kenneth. "Technological Advance and Costs: Computers versus Communications." In Robert W. Crandall and Kenneth Flamm, eds., *Changing the Rules: Technological Change, International Competition, and Regulation in Communications.* Washington, D.C.: Brookings Institution, 1989.

Fordyce, S. W. "NASA Experience in Telecommunications as a Substitute for Transportation." U.S. National Aeronautics and Space Administration Headquarters, Washington, D.C., April 1974. Processed.

Fox, R. W. B. "A Review of the Short-term Forecasting Model for Total Trunk Calls." Report no. 27. British Post Office, Statistics and Business Research Department, London, 1973. Processed.

Frantz, R. X-Efficiency: *Theory, Evidence, and Applications.* Boston, Mass.: Kluwer Academic Publishers, 1988.

Fuller, Wayne. RFD: *The Changing Face of Rural America.* Bloomington: Indiana University Press, 1964.

Gellerman, Robert F. "Subscriber Financing of Telecommunications Investment." *Telecommunications Policy,* vol. 10, no. 1 (March 1986).

Gellerman, Robert F., and S. Ling. "Linking Electricity with Telephone Demand Forecast: A Technical Note." IEEE *Transactions on Communications,* vol. COM-24, no. 3 (March 1976), pp. 322–24.

General Motors Research Laboratories and Delco Radio Division. "A Design for an Experimental Route Guidance System." Prepared for the Bureau of Public Roads, Federal Highway Administration, Contract no. FM-11-6626. 1962. Processed.

Giaoutzi, M., and P. Nijkamp, eds. *Informatics and Regional Development.* Aldershot, Eng.: Avebury, 1988.

Gille, Laurent. "Growth and Telecommunications." In ITU, *Information, Telecommunications, and Development*. Geneva, 1986.

Gillespie, Andrew E., John B. Goddard, Mark E. Hepworth, and Howard Williams. "Information, Communications Technology, and Regional Development: An Information Economy Perspective." *Science, Technology, Industry Review*, OECD, no. 5 (April 1989), pp. 85–111.

Gillick, David. "Telecommunications Policy in the U.K.: Myths and Realities." *Telecommunications Policy*, vol. 15, no. 1 (February 1991), pp. 3–9.

Globerman, Steven. "Economic Factors in Telecommunications Policy and Regulation." Paper presented at IRPP conference on competition and technological change: the impact on telecommunications policy and regulation in Canada, Toronto, 1984. Processed.

Goddard, J., and R. Pye. "Telecommunication and Office Location." In R. C. Smith, ed., "Impacts of Telecommunications on Planning and Transport." Research Report no. 24. Department of Transport and Department of the Environment, London, July 1978. Processed.

Goldschmidt, Douglas. "Financing Telecommunications for Rural Development." *Telecommunications Policy* (September 1984), pp. 181–203.

————. "Telephone Communications, Collective Supply, and Public Goods: A Case Study of the Alaskan Telephone System." Ph.D. diss., University of Pennsylvania, Philadelphia, 1978.

Goldschmidt, Douglas, Victor Forsythe, and Heather Hudson. "An Evaluation of the Medex/Guyana Two-Way Radio Pilot Project." Academy for Educational Development, Washington, D.C., December 1980. Processed.

Goldschmidt, Douglas, Heather E. Hudson, and Wilma Lynn. "Two-Way Radio for Rural Health Care: An Overview." U.S. Agency for International Development, Office of Education, Development Support Bureau, Washington, D.C., May 1980. Processed.

Goldschmidt, Douglas, Karen Tietjen, and Willard D. Shaw. *Design and Installation of Rural Telecommunications Networks: Lessons from Three Projects— Technology*. Washington, D.C.: Academy for Educational Development, AID Rural Satellite Program, 1987.

Golladay, Frederick. "Health Problems and Policies in the Developing Countries." World Bank Staff Working Paper no. 412. Washington, D.C., August 1980.

Gordon, D. A., and H. C. Wood. "How Drivers Locate Unfamiliar Addresses—An Experiment on Route Finding." *Public Roads* (June 1970), pp. 44–47.

Gorelik, M. A., and I. B. Efimova. "The Economic Efficiency of Development of Long Distance Telephone Communication." *Vestnik Svyazi*, Moscow, no. 5 (1977), pp. 28–30.

Gorelik, M. A., I. B. Efimova, and E. Kareseva. "How to Determine the Economic Efficiency of Long Distance Telephone Communication." *Vestnik Svyazi*, Moscow, no. 7 (1975), pp. 26–28.

Gorelik, M. A., and E. Kareseva. "Standards and Assessment of the Economic

434

Efficiency of Long Distance Telephone Communications." *Vestnik Svyazi*, Moscow, no. 8 (1975), pp. 30–31.

Green, James H. *The Dow Jones–Irwin Handbook of Telecommunications.* Homewood, Ill.: Dow Jones–Irwin, 1986.

Greenhut, M. L. *Microeconomics and the Space Economy.* Chicago: Scott Foresman Company, 1963.

――――. *A Theory of the Firm in Economic Space.* New York: Appleton-Century-Crofts, 1970.

Guiscard Ferraz, J. L. "The Infrastructural Role of the Telecommunications Sector and Its Impact on Economic Development." In ITU, *World Telecommunications Forum 1987, Part IV.* Geneva, 1987.

Hagget, P. *Locational Analysis in Human Geography.* London: Edward Arnold, 1965.

Halina, J. W. "Communications and the Economy: A North American Perspective." *International Social Science Journal,* vol. 32, no. 2 (1980), pp. 266–67.

Hall, Peter. "Moving Information: A Tale of Four Technologies." In J. Brotchie, M. Batty, and Peter Hall, eds., *New Technologies and Spatial Systems.* London: Unwin Hyman, 1990.

――――. "Transportation." *Urban Studies,* vol. 6, no. 3 (1969), pp. 408–35.

Hansen, Suella, David Cleevely, Simon Wadsworth, Hilary Bailey, and Oliver Bakewell. "Telecommunications in Rural Europe: Economic Implications." *Telecommunications Policy* (June 1990), pp. 207–22.

Hardy, Andrew Peter. *The Role of the Telephone in Economic Development.* Stanford, Calif.: Institute for Communications Research, Stanford University, January 1980.

Harkness, R. C. "Communications Substitutes for Intra-Urban Travel." Joint ASCE-ASME Transportation Engineering Meeting, Reprint no. 1453. Seattle, Wash., 1971. Processed.

Harmon, Harry H. *Modern Factor Analysis,* 2d ed. Chicago: University of Chicago Press, 1967.

Hartley, Nicholas M., and Peter Culham. "Telecommunications Prices under Monopoly and Competition." *Oxford Review of Economic Policy,* vol. 4 (summer), no. 2 (1988), pp. 1–19.

Hazelwood, A. "Optimum Pricing as Applied to Telephone Service." *Review of Economic Studies,* vol. 18 (1950–51), pp. 67–78. Reprinted with amendments as "Telephone Service." In Ralph Turvey, ed., *Public Enterprise: Selected Readings.* Harmondsworth, Eng.: Penguin Books, 1968.

Helling, A. G. Henrik. "Tariffs for National and International Telecommunications Services." Asia Pacific Telecommunity, Bangkok, March 10, 1981. Processed.

Heymann, H. J. "The Relationship between Telecommunications and the National Economy: A Pragmatic Approach to Its Quantitative Determination Using a Macro-Economic Cross-Sectional Analysis." In ITU, *World Telecommunications Forum 1987, Part IV.* Geneva, 1987.

Hills, Alex. "Feasibility of Mobile Satellite Service in Alaska." *Telecommunications Policy* (December 1988), pp. 369–78.

Hills, Jill. "The Telecommunications Rich and Poor." *Third World Quarterly*, vol. 12, no. 2 (April 1990), pp. 71–90.

Hobday, Mike. "Telecommunications—A 'Leading Edge' in the Accumulation of Digital Technology? Evidence from the Case of Brazil." *Information Technology for Development*, Oxford University Press, vol. 1, no. 1 (1986).

Hoffman, Kurt, and Michael G. Hobday. "The Third World and U.S. Telecommunications Policy." In Paula R. Newberg, ed., *New Directions in Telecommunications Policy*, vol. 2. Durham, N.C.: Duke University Press, 1989.

Hollas, D. F., and R. S. Hereen. "An Estimation of the Deadweight and X-Efficiency Losses in the Municipal Electric Industry." *Journal of Economics and Business*, vol. 34 (1984), pp. 269–81.

Hudson, Heather E. *Communication Satellites: Their Development and Impact.* New York: Free Press, 1990.

_____. "Telemedicine: Some Findings from the North American Experience." Academy for Educational Development, Washington, D.C., June 1980. Processed.

_____. *When Telephones Reach the Village: The Role of Telecommunications in Rural Development.* Norwood, N.J.: Ablex Publishers, 1984.

Hudson, Heather E., Douglas Goldschmidt, Edwin B. Parker, and Andrew Hardy. "The Role of Telecommunications in Socioeconomic Development." Report prepared for the ITU by Keewatin Communications. Washington, D.C., May 1979. Processed.

Hudson, Heather E., Andrew P. Hardy, and Edwin B. Parker. "Impact of Telephone and Satellite Earth Station Installations on GDP." *Telecommunications Policy*, vol. 6, no. 4 (December 1982), pp. 300–07.

Hudson, Heather E., and Edwin B. Parker. "Medical Communication in Alaska by Satellite." *New England Journal of Medicine*, vol. 289, nos. 1351–56 (December 20, 1973).

Hughson, Terri L., and Paul S. Goodman. "Telecommuting: Corporate Practices and Benefits." *National Productivity Review*, vol. 5 (Autumn 1986), pp. 315–24.

Huntly, L. R. "Some Ideas Regarding Economics of Telecommunications Engineering." *Electrical Communication*, vol. 42, no. 1 (1967), pp. 6–21.

Implementation and Management Group Party, Ltd. "A Study of Remote Area Telecommunications in the Northern Territory," vol. 2. Report prepared for TELECOM Australia. Sydney, May 1980. Processed.

Integrated Development Systems. "Economics of Telecommunications in Nepal." Prepared for Nepal Communications Corporation. Kathmandu, August 1980. Processed.

INTELSAT. *Annual Report.* Washington, D.C., 1988.

_____. *Project SHARE: A Final Report and Evaluation.* Washington, D.C., 1988.

International Development Research Centre. *Sharing Knowledge for Development: IRDC's Information Strategy for Africa.* Ottawa, 1989.

International Energy Agency. *Energy Balances of Paris:* OECD *Countries, 1987– 1988.* Paris: OECD.

ILO. "Employment Income and Equality: A Strategy for Increasing Productive Employment in Kenya." Geneva, 1972. Processed.

————. "Studies in Development: A Program of Development, Equity, and Grants for the Philippines," vol. 1, March Report. Geneva, 1972. Processed.

Isard, Walter. *General Theory: Social, Political, Economic, and Regional.* Cambridge, Mass.: M.I.T. Press, 1969.

ITU. *Benefits of Telecommunications to the Transportation Sector of Developing Countries: A Case Study in the People's Democratic Republic of Yemen.* Geneva, March 1988a.

————. *The Changing Telecommunications Environment: Policy Considerations for the Members of the* ITU. Report of the Advisory Group on Telecommunications Policy. Geneva, February 1989.

————. *Contribution of Telecommunications to the Earnings/Savings of Foreign Exchange in Developing Countries: Case Studies of 20 Kenyan Firms.* Geneva, 1988b.

————. "Development of Manpower Resources for Telecommunications Administrations in Africa." Paper prepared for the third African telecommunications conference, Monrovia, Liberia, December 8–19, 1980a. Processed.

————. *Information, Telecommunications, and Development.* Geneva, 1986.

————. *The Missing Link: Report of the Independent Commission for World Wide Telecommunications Development.* Geneva, 1984.

————. "Restructuring of Telecommunications in Developing Countries: An Empirical Investigation with ITU's Role in Perspective." Geneva, May 1991.

————. *Socio-Economic Benefits of Improved Telecommunications in Developing Countries: Results of a Research Study in Vanuatu.* Geneva, 1988c.

————. "Some Aspects of Multi-National Telecommunication Training in Developing Areas." Paper prepared for the third African telecommunications conference, Monrovia, Liberia, December 8–19, 1980b.

————. *World Telecommunications Forum 1987, Part IV.* Symposium on economic and financial issues: the role of telecommunications in the infrastructure and its impact on economic growth, Geneva, 1987.

————. *Yearbook of Public Telecommunication Statistics.* 17th ed. Geneva, 1990.

Ivanek, Ferdo, Timothy Nulty, and Nikola Holcer. *The Impact of Technological Progress on Telecommunications Equipment Manufacturing in* NICS. World Bank Technical Paper no. 145. Washington, D.C.: World Bank, 1991.

Jabif, R. "Análisis y modelación de la demanda telefónica comercial." Departamento de Electricidad, Universidad de Chile, Santiago, 1971. Processed.

Jenkins, Janet, ed. *Commonwealth Co-operation in Open Learning: Background Papers.* London: Commonwealth Secretariat, 1988.

Jipp, A. "Wealth of Nations and Telephone Density." *Telecommunications Journal* (July 1963), pp. 199–201.

Johansen, Robert, Kathleen Hansell, and David Green. "Growth in Teleconferencing: Looking beyond the Rhetoric of Readiness." *Telecommunications Policy* (December 1981), p. 289.

Johansen, Robert, J. Vallee, and Kathleen Spangler. *Electronic Meetings.* Reading, Mass.: Addison-Wesley, 1979.

Johnson, Leland L. "Price Caps in Telecommunications Regulatory Reform." Rand Note no. N-2894-MF/RC. Rand Corporation, Santa Monica, Calif., 1989.

Jonscher, Charles. "Benefits, Costs, and Optimal Prices in Telecommunications Services." Draft working paper. Harvard University, Cambridge, Mass., undated. Processed.

_____. "Economic Benefits of Telecommunications Investments: A Synthesis of Quantitative Evidence." In ITU, *World Telecommunications Forum 1987, Part IV.* Geneva, 1987.

_____. "Productivity and Growth of the Information Economy." In Meheroo Jussawalla and Helene Ebenfield, eds., *Communication and Information Economics: New Perspectives.* New York: Elsevier Science Publishers, 1984.

_____. "A Theory of Economic Organization." Ph.D. diss., Harvard University, Cambridge, Mass., 1980.

Jussawalla, Meherro. "The Information Economy and Its Importance for the Development of Pacific Region Countries." In ITU, *Information, Telecommunications, and Development.* Geneva, 1986.

Jussawalla, Meheroo, and Helene Ebenfield, eds. *Communication and Information Economics: New Perspectives.* New York: Elsevier Science Publishers, 1984.

Jussawalla, Meheroo, Donald Lamberton, and Neil Karunaratne, eds. *The Cost of Thinking: Information Economies of Ten Pacific Countries.* Norwood N.J.: Ablex Publishing, 1988.

Kahn, Alfred E. *The Economics of Regulation: Principles and Institutions,* vol. 1. New York: John Wiley and Sons, 1970.

Kahn, Alfred E., and W. Shew. "Current Issues in Telecommunications Regulation: Pricing." *Yale Journal on Regulation* (Spring 1987).

Kamal, Ahmed, Ali E. Hillal Dessouki, and Ithiel de Sola Pool. "Communication System in Rural Egypt." Communication Needs for Rural Development Research Report no. 11. Cairo University, 1980. Processed.

Karunaratne, Neil D. "An Input-Output Approach to the Measurement of the Information Economy." In Meheroo Jussawalla, Donald M. Lamberton, and Neil D. Karunaratne, eds., *The Cost of Thinking: Information Economies of Ten Pacific Countries.* Norwood, N.J.: Ablex Publishing, 1988.

Kasolojaona, Mamiharilala. "Colloque sur la restructuration du secteur de télécommunications Malagasy." Paper presented at a seminar on telecommunications restructuring sponsored by the World Bank, the ITU, and ACCT, Tunisia, May 1992.

Katsoulis, M. "Energy Impacts of Passenger Transportation." Bell Canada, Montreal, March 1974. Processed.

_____. "Travel/Telecommunications Substitution—Its Potential for Energy Conservation in Canada." Bell Canada, Montreal, February 1976. Processed.

Kaul, S. N. "Benefits of Rural Telecommunications in Developing Countries." Paper presented at the ITU/OECD expert meeting on telecommunications in developing countries, Paris, December 18–20, 1978. Conference Working Paper no. 8, Document no. CD/R (78) 39.8. OECD, Paris, 1978. Processed.

———. "Cost Return Analysis and Economies of Scale in Telephone Network Expansion, India." Economics Study Cell, P&T Board, New Delhi, March 1980. Processed.

———. "Planning of Telecommunications Services for Development." NCAER Margin, vol. 2, no. 3 (April 1979), pp. 71–80.

Kavanagh, N. J., and R. Smith. "Measurement of Benefits of Trout Fishing: Preliminary Results of a Study at Grafham Water." Journal of Leisure Research, vol. 1 (1969), pp. 316–32.

Khadem, Ramin. "An Economic Assessment of Mobile Satellite and Cellular Radio Systems." In ITU, World Telecommunications Forum 1987, Part IV. Geneva, 1987.

Kilgour, Mary Cameron. "The Telephone in the Organization of Space for Development." Ph.D. diss., Political Economy and Government, Harvard University, Cambridge, Mass., 1982.

King, John A. C. "The Privatization of Telecommunications in the United Kingdom." In Björn Wellenius, Peter A. Stern, Timothy Nulty, and Richard D. Stern, eds., Restructuring and Managing the Telecommunications Sector. Washington, D.C.: World Bank, 1989.

Kinsman, Francis. The Telecommuters. New York: John Wiley and Sons, 1987.

Kiss, Ferenc. "Productivity Gains in Bell Canada." In Léon Courville, Alain De Fontenay, and Rodney Dobell, eds., Economic Analysis of Telecommunications: Theory and Applications. Amsterdam: North-Holland, 1983.

Klein, A. "Tráfico telefónico interurbano y una tipología socioeconómica." Facultad de Ciencias Físicas y Matemáticas, Universidad de Chile, Santiago, 1971. Processed.

Kochen, Manfred. "Opportunity Costs in Computer Conferencing during and for Economic Development." In Meheroo Jussawalla and D. M. Lamberton, eds., Communication Economics and Development. Elmsford, N.Y.: Pergamon Press, 1982.

Kojina, Mitsuhiro, Junichiro Hoken, and Masaru Saito. "Report: The Use of Telephones in Sri Lanka." Telecommunications Policy (December 1984), pp. 335–38.

Kollen, J. H., and J. Garwood. "Travel/Communications Tradeoffs: The Potential for Substitution among Business Travellers." Headquarters Business Planning Group, Bell Canada, Montreal, April 1975. Processed.

Kottis, G. "The Service Sector in Greece: A Regional Perspective." FAST Paper no. 99. Commission of the European Communities, April 1986.

Krechmer, Ken. "The Hidden Costs of ISDN." Telecommunications (October 1989), pp. 25–34.

Laidlaw, Bruce. "The Evolution of Telecommunications Policy in the United Kingdom." In Björn Wellenius and Peter Stern, eds., Implementing Reforms in *the Telecommunications Sector: Lessons from Recent Experience.* Washington, D.C.: World Bank, 1991.

Lamberton, Donald. "Information Economics: 'Threatened Wreckage' or New Paradigm?" CIRCIT Working Paper no. 1990/91. Processed.

————. "The Theoretical Implications of Measuring the Communication Sector." In Meheroo Jussawalla and D. M. Lamberton, eds., *Communication Economics and Development.* Pergamon Policy Studies on International Development. Honolulu: East-West Center, 1982.

Lange, Siegfried, and Helmet Rempp. "Qualitative and Quantitative Aspects of the Information Sector." Karlsruhe Institut fur Systemtechnik und Innovationsforschung, Karlsruhe, Federal Republic of Germany, 1977. Processed.

Langley, C. John, Jr., and Rammohan Pisharodi. "Discussion of 'Resource Implications of Electronic Message Transfer in Letter-Post Industry,' by Alfred M. Lee and Arnim H. Meyburg." *Transportation Research Record,* vol. 812 (1981), pp. 63–64.

Lathey, C. E. "Telecommunications Substitutability for Travel: An Energy Conservation Potential." Report no. 75–58. Office of Telecommunications, U.S. Department of Commerce, Washington, D.C., 1975. Processed.

Lee, Alfred M., and Arnim H. Meyburg. "Resource Implications of Electronic Message Transfer in Letter-Post Industry." *Transportation Research Record,* vol. 812 (1981), pp. 59–63.

Leibenstein, Harvey. "Allocative Efficiency vs. 'X-Efficiency.'" *American Economic Review,* vol. 56 (June 1966), pp. 392–415.

————. "Competition and X-Efficiency." *Journal of Political Economy,* vol. 81 (May/June 1973).

Levy, Mildred B., and Walter Wadycki. "Lifetime versus One-Year Migration in Venezuela." *Journal of Regional Science,* vol. 12 (December 1972), pp. 407–14.

Lewis, W. A. "Fixed Costs." In Allen L. Unwin, ed., *Overhead Costs.* 1949. Reprinted in D. Munby, ed., *Transport.* Harmondsworth, Eng.: Penguin Modern Economics, 1968.

Linn, Johannes F. *Cities in the Developing World: Policies for Their Equitable and Efficient Growth.* New York: Oxford University Press, 1983.

Little, I. M. D., and J. A. Mirrlees. *Project Appraisal and Planning for Developing Countries.* New York: Basic Books, 1976.

Littlechild, S. C. "The Effects of Postal Responsibility and Private Ownership on the Structure of Telephone Tariffs: An International Comparison." University of Birmingham, Birmingham, Eng., June 1980. Processed.

————. *Elements of Telecommunications Economics.* Stevenage, Eng.: Peter Peregrinus, Ltd., 1979.

————. "Regulation of British Telecommunications Profitability." Department of Industry, London, 1983. Processed.

Lockheed, Marlaine E., John Middleton, and Greta Nettleton, eds. "Education and Technology: Sustainable and Effective Use". PHREE Background paper 91/32. World Bank, Washington, D.C., 1991.

Lönnström, Sven, Folke Marklund, and Ingemar Moo. *A Telephone Development Project*. 3d ed. Stockholm: L. M. Ericsson, 1975.

Luhan, P. "The Yearbook of Common Carrier Telecommunication Statistics, Published by the ITU: An Indispensable Tool for Gauging the Development of World Telecommunications." *Telecommunications Journal*, vol. 56 (November 1989), pp. 245-50.

Machlup, Fritz. *The Production and Distribution of Knowledge in the United States*. Princeton, N.J.: Princeton University Press, 1962.

Mack, M., and G. Lee. "TELECOM Australia's Experience with Photovoltaic Systems in the Australian Outback." *Telecommunications Journal*, vol. 56, no. 8 (1989), pp. 513-20.

Malgavkar, P. D., and V. K. Chebbi. "The Impact of Telecommunications Facilities on Rural Development in India." In Dan J. Wedemeyer and M. R. Ogden, eds., *Telecommunications and Pacific Development: Alternatives for the Next Decade*. New York: Elsevier Science Publishers and North-Holland and PTC, 1988.

Malien, Habyalimana. "Colloque sur la restructuration du secteur des télécommunications de Rwanda." Paper presented at a seminar on telecommunications restructuring sponsored by the World Bank, the ITU, and ACCT, Tunisia, May 1992.

Mansell, Robin. "The Role of Information and Telecommunication Technologies in Regional Development." *Science, Technology, Industry Review*, OECD, no. 3 (April 1988), pp. 135-73.

―――. *Telecommunications Rate Restructuring: Issues, Patterns, and Policy Problems for OECD Countries*. Paris: OECD and ICCP Research Publications, 1990.

March, J. G., and H. A. Simon. *Organizations*. New York: John Wiley and Sons, 1958.

Martin, James. *Future Developments in Telecommunications*, 3d ed. Englewood Cliffs, N.J.: Prentice-Hall, 1991.

Martin, R. L., and R. Rowthorn, eds. *The Geography of De-Industrialization*. London: MacMillan, 1986.

Mayer, Martin. "The Telephone and the Uses of Time." In Ithiel de Sola Pool, ed., *The Social Impact of the Telephone*. Cambridge, Mass.: M.I.T. Press, 1977.

Mayo, John K., Gary R. Heald, Steven J. Klees, and Martha Cruz de Yáñes. *Peru Rural Communication Services Project: Final Evaluation Report*. Tallahassee: Center for International Studies, Learning Systems Institute, Florida State University, 1987.

Mazumdar, Dipak. "The Urban Informal Sector." World Bank Staff Working Paper no. 211. Washington, D.C., July 1975, and *World Development*, vol. 4, no. 8 (1976).

McAnany, Emile G. "The Role of Information in Communicating with the Rural Poor: Some Reflections." In Emile G. McAnany, ed., *Communications in the Rural Third World*. New York: Praeger Publishers, 1980.

―――, ed. *Communications in the Rural Third World*. New York: Praeger Publishers, 1980.

McKean, Roland N. "The Use of Shadow Prices." In Samuel B. Chase, Jr., ed., *Problems in Public Expenditure Analysis*. Washington, D.C.: Brookings Institution, 1968.

Medinikov, D. "What Is the Effect of Dispatcher Communication in Farming?" *Vestnik Svyazi*, Moscow, no. 4 (1975), p. 35.

Melo, José Ricardo. "Liberalization and Privatization in Chile." In Björn Wellenius and Peter Stern, eds., *Implementing Reforms in the Telecommunications Sector: Lessons from Recent Experience*. Washington, D.C.: World Bank, 1991.

Melody, William H. "Telecommunications Implications for the Structure of Development." In ITU, *World Telecommunication Forum 1987, Part IV*. Geneva, October 1987.

Mendis, Vernon L. B. "Phased Privatization with Proposed Foreign Participation: The Sri Lanka Experience." In Björn Wellenius, Peter Stern, Timothy Nulty, and Richard Stern, eds., *Restructuring and Managing the Telecommunications Sector*. Washington, D.C.: World Bank, 1989.

Meyer, John R., Robert W. Wilson, M. Alan Baughcum, Ellen Burton, and Louis Caouette. *The Economics of Competition in the Telecommunications Industry*. Cambridge, Mass.: Oelgeschlager, Gunn, and Hain, 1980.

Meyers, Robert A., ed. *Encyclopedia of Telecommunications*. San Diego, Calif.: Academic Press, 1989.

Miernyk, William H. *The Elements of Input-Output Analysis*. New York: Random House, 1965.

Ministry of Labour and Housing and Ministry of Physical Planning of Local Government, Sweden. "Planning Sweden—Regional Development Planning and Management of Land and Water Resources." Allmanna Forlaget, Stockholm, 1973. Processed.

Ministry of Supply and Services, Canada. "Elasticity of Demand for Long Distance Telephone Service." CRTC Consulting Report no. 5. Prepared by Steven Globerman Associates, Ltd., 1988. Processed.

Mitchell, Bridger M. "Incremental Capital Costs of Telephone Access and Local Use." Rand paper no. R-3764-ICTF. Rand Corporation, Santa Monica, Calif., 1989.

———. "Local Telephone Costs and the Design of Rate Structures." In Léon Courville, Alain De Fontenay, and Rodney Dobell, eds., *Economic Analysis of Telecommunications Theory and Application*. New York: Elsevier Science Publishers, 1983.

———. *Optimal Pricing of Local Telephone Service*. R-1962-MF. Santa Monica, Calif.: Rand Corporation, November 1976.

———. "Optimal Pricing of Local Telephone Service." *American Economic Review*, vol. 68, no. 4 (September 1978), pp. 517–37.

———. "Pricing Policies in Selected European Telephone Systems." In H. Dordick, ed., *Proceedings of the Sixth Annual Telecommunications Policy Research Conference*. Lexington, Mass.: Lexington Books, 1979.

————. "Pricing Subscriber Access to the Telephone Network." In A. Baughcum and G. R. Faulhaber, eds., *Telecommunications Access and Public Policy.* Paper no. P-6815. Santa Monica, Calif.: Rand Corporation, 1984.

Mitchell, Bridger M., W. G. Manning, and J. P. Acton. *Peak-Load Pricing: European Lessons for U.S. Energy Policy.* Cambridge, Mass.: Ballinger Publishing Company, 1978.

Mitchell, Bridger M., and Ingo Vogelsang. *Telecommunications Pricing: Theory and Practice.* Cambridge, Eng.: Cambridge University Press, 1991.

Mody, Ashoka. "Information Industries in the Newly Industrializing Countries." In Robert W. Crandall and Kenneth Flamm, eds., *Changing the Rules: Technological Change, International Competition, and Regulation in Communications.* Washington, D.C.: Brookings Institution, 1989.

Moon, A. E., and others. "Technology Assessment of Telecommunications/ Transportation Interactions," vol. 3, "Contributions of Telecommunications to Improved Transportation System Efficiency." Stanford Research Institute, Menlo Park, Calif., May 1977. Processed.

Moss, Mitchell L. "Urban Development in a Global Economy." In Maurice F. Estabrooks and Rodolphe H. Lamarche, eds., *Telecommunications: A Strategic Perspective on Regional, Economic, and Business Development.* Moncton, Canada: Canadian Institute for Research on Regional Development, 1986.

————, ed. *Telecommunications and Productivity.* Reading, Mass.: Addison-Wesley, 1981.

Muller, Jurgen, and Emilia Nyevrikel. "Closing the Capacity and Technology Gaps in Central and Eastern European Telecommunications." In Björn Wellenius and Peter Stern, eds., *Implementing Reforms in the Telecommunications Sector: Lessons from Recent Experience.* Washington, D.C.: World Bank, 1991.

Munasinghe, Mohan. *The Economics of Power System Reliability and Planning.* Baltimore, Md.: Johns Hopkins University Press, 1979.

Munasinghe, Mohan, and Mark Gellerson. "The Determination of Optimal Reliability Standards for Electricity Supply." *Bell Journal of Economics,* vol. 10, no. 1 (Spring 1979), pp. 353–64.

Munasinge, Mohan, and Jeremy J. Warford. "Shadow Pricing and Power Tariff Policy." World Bank Staff Working Paper no. 286. Washington, D.C., June 1978.

Narula, A. M. "Post Office Savings Bank in India: A Saga of Progress." *Union Postale,* vol. 4 (1972), p. 49.

National Council of Applied Economic Research, India. "Fertilizer Survey, 1975–76." New Delhi, India, 1976. Processed.

————. "Survey of Rural Public Call Offices." Study commissioned by India P&T May 1978. Processed.

National Research Council. *Review of Electronic Mail Service Systems Planning for the U.S. Postal Service.* Report to the U.S. Postal Service. Washington, D.C.: National Academy Press, 1981.

National Research Council, Board on Science and Technology for International Development. *Science and Technology Information Services and Systems in Africa.* Report of a workshop held in Nairobi, Kenya, April 1989. Washington, D.C.: National Academy Press, 1990.

National Research Council, Board on Telecommunications and Computer Applications, Commission on Engineering and Technical Systems. *Office Workstations in the Home.* Washington, D.C.: National Academy Press, 1985.

Nettleton, Greta. "Distance Training for Telecommunications Managers: A Case Study." Paper presented at the ICDE conference in Caracas, Venezuela, November 1990. Processed.

Nettleton, Greta, and Emile McAnany. "Brazil's Satellite System: The Politics of Applications Planning." *Telecommunications Policy,* vol. 13, no. 2 (June 1989), pp. 159–66.

Neumann, Karl-Heinz, and Thomas Schnoring. "Reform and Unification of Telecommunications in Germany." In Björn Wellenius and Peter Stern, eds., *Implementing Reforms in the Telecommunications Sector: Lessons from Recent Experience.* Washington, D.C.: World Bank, 1991.

Newberg, Paula R., ed. *New Directions in Telecommunications Policy,* 2 vols. Durham, N.C.: Duke University Press, 1989.

Nicol, Lionel Y. "Communications, Economic Development, and Spatial Structures: A Review of Research." Working Paper no. 404. Berkeley: Institute of Urban and Regional Development, University of California, 1983a.

————. "Communications, Economic Development, and Spatial Structures: A Theoretical Framework." Working Paper no. 405. Berkeley: Institute of Urban and Regional Development, University of California, 1983b.

Nicolai, C., and Björn Wellenius. "Estudio de inversiones en teléfonos rurales, V y VI región." Departamento de Electricidad, Universidad de Chile, Santiago, January 1979. Processed.

Nilles, J. M. "Development of Policy on the Telecommunications Tradeoff." Phase 11 Report. University of Southern California, Los Angeles, March 1974. Processed.

Nilles, J. M., F. R. Carlson, P. Gray, and G. J. Hanneman. *The Telecommunications-Transportation Tradeoff: Options for Tomorrow.* New York: John Wiley and Sons, 1976.

Noll, Roger G. *Introduction to Telephones and Telephone Systems,* 2d ed. Boston, Mass.: Artech House, 1991.

————. "Telecommunications Regulation in the 1990s." In Paula R. Newberg, ed., *New Directions in Telecommunications Policy,* vol. 1. Durham, N.C.: Duke University Press, 1989.

Nora, S., and A. Minc. *The Computerization of Society.* Cambridge, Mass.: M.I.T. Press, 1980.

Nordlinger, Christopher W. "Users of Public Telecommunications Facilities and Their Benefits in a Developing Country: A Case Study of Senegal." In ITU, *Information, Telecommunications, and Development.* Geneva, 1986.

Nourse, Hugh O. *Regional Economics.* New York: McGraw-Hill Book Company, 1968.

Nulty, Timothy. "Challenges and Issues in Central and Eastern European Telecommunications." In Björn Wellenius and Peter Stern, eds., *Implementing Reforms in the Telecommunications Sector: Lessons from Recent Experience.* Washington, D.C.: World Bank, 1991a.

————. "Emerging Issues in World Telecommunications." In Björn Wellenius, Peter Stern, Timothy Nulty, and Richard Stern, eds., *Restructuring and Managing the Telecommunications Sector.* A World Bank Symposium. Washington, D.C.: World Bank, 1991b.

O'Brien, Rita Cruise, E. Copper, B. Perkes, and H. Lucas. "Communications Indicators and Indicators of Socio-Economic Development, 1960–1970." Institute of Development Studies, University of Sussex, Eng., 1977. Processed.

Oeffinger, John C. "Merging Computers and Communications: A Case Study in Latin America." *Telematics and Informatics,* vol. 4, no. 3 (1987), pp. 195–210.

Okazaki, H. "Planning an Urban Area Digital Telephone Junction Network." NEC *Research and Development,* vol. 73 (April 1984), pp. 43–52.

Okundi, Philip O. "Pan-African Telecommunications Network: A Case for Telecommunications in the Development of Africa." *World Telecommunication Forum.* Geneva: ITU, October 1975.

Okundi, Philip O., and B. G. Evans. "Afrosat: Proposals for an African Domestic Satellite Communication System." Paper presented at the international conference on satellite communication systems technology, London, April 1975. Processed.

Okundi, Philip O., Arthur W. Ogwayo, and James P. Kibombo. "Rural Telecommunications Development in East Africa." *Proceedings of the International Telecommunications Exposition (INTELCOM 77),* vol. 1. Dedham, Mass.: Horizon House International, 1977.

Organization for African Unity. "Social Aspects of Telecommunications." Paper submitted to the third African telecommunications conference, Monrovia, Liberia, December 8–19, 1980. Processed.

OECD. *Information, Technology, and New Growth Opportunities.* Information, Computer, Communications Policy Series no. 19. Paris, 1989.

————. *Telecommunication Network Based Services: Policy Implications.* Information, Computer, Communications Policy Series no. 18. Paris, 1989.

————. *Trends of Change in Telecommunications Policy.* Information, Computer, Communications Policy Series no. 13. Paris, 1987.

OECD, Working Party on Telecommunications and Information Policies. *Performance Indicators for Public Telecommunications Operators.* Paris, 1990.

OFTEL. "Responses to OFTEL's Consultative Document on the Future Regulation of British TELECOM's Prices." OFTEL Working Paper no. 3. London.

Page, Michael. *The Flying Doctor Story, 1928–78.* Adelaide, Australia: Rigby, 1977.

Paine, Nigel. "The Likely Impact of New Forms of Communication and Information Technology on Distance Education." Document submitted to the Commonwealth Secretariat, August 1986. Processed.

Park, Rolla Edward, and Bridger M. Mitchell. *Optimal Peak Load Pricing for Local Telephone Calls*. Publication no. R-3404-1-RC. Santa Monica, Calif.: Rand Corporation, 1987.

Parker, Edwin B. "MicroEarth Station Satellite Networks and Economic Development." *Telematics and Informatics*, vol. 4, no. 2 (1987).

Parker, Edwin B., Heather E. Hudson, Don Dillman, and Andrew Roscoe. *Rural America in the Information Age: Telecommunications Policy for Rural Development*. Lanham, Md.: University Press of America and Aspen Institute, 1989.

Perraton, Hilary, ed. *Alternative Routes to Formal Education: Distance Teaching for School Equivalency*. Baltimore, Md.: Johns Hopkins University Press, 1982.

———. ed. "Distance Education: An Economic and Educational Assessment of Its Potential for Africa." World Bank Discussion Paper, Education and Training Series Report no. EDT 43. World Bank, Washington, D.C., December 1986.

Phillips, F., ed. *Thinkwork: Working, Learning, and Managing in a Computer-Interactive Society*. New York: Praeger, 1991.

Pierce, J. "Communication as an Alternative to Travel." *Proceedings of the IRE*, 50th anniversary issue (May 1962), pp. 225–31.

Pierce, William. "A Global-Domestic (GLODOM) Satellite System for Rural Development." *Telecommunications Journal*, vol. 46 (December 1979), pp. 745–47.

Pierce, William B., and Nicolas Jequier. *Telecommunications for Development*. Geneva: ITU, 1983.

Pierson, J. H. S. "Planning for Telecommunications Expansion: Making the Case for a Telecommunications Budget." Paper presented at INTELCOM '79, Dallas, Tex., March 1979, Pierson Consultants, Inc., New York. Processed.

Polishuk, P., R. Guenther, and J. Lawlor. "Cost Comparison of Microwave, Satellite, and Fibre Optic Systems." *Telecommunications Journal*, vol. 54, no. 2 (1987), pp. 114–22.

Pool, Ithiel de Sola. "The Communications/Transportation Tradeoff." In Alan Altschuler, ed., *Current Issues in Transportation Policy*. Lexington, Mass.: D. C. Heath, 1979.

———. "The Influence of International Communications on Development." Massachusetts Institute of Technology, Cambridge, Mass., February 1976. Processed.

———, ed. *The Social Impact of the Telephone*. Cambridge, Mass.: M.I.T. Press, 1977.

Pool, Ithiel de Sola, and Peter M. Steven. "Appropriate Telecommunications for Rural Development." In Indu B. Singh, ed., *Telecommunications in the Year 2000: National and International Perspectives*. Norwood, N.J.: Ablex Publishers, 1983.

Porat, Marc Uri. *The Information Economy: Definition and Measurement*. Special Publication no. 77–12(1). Washington, D.C.: U.S. Department of Commerce, Office of Telecommunications, May 1977.

446

Preece, Robert S. "The Role of Telecommunications in Economic Growth and Income Distribution." In ITU, *World Telecommunications Forum 1987, Part IV*. Geneva, 1987.

Prest, Alan, and R. Turvey. "Cost-Benefit Analysis: A Survey." *Economic Journal* (December 1965).

Pye, Roger. "Communications within the Scottish Office." Communications Studies Group Report no. W/74174/PY, University College, London, 1974. Processed.

Pye, Roger, and J. B. Goddard. "Telecommunications and Location." *Regional Studies*, vol. 11, no. 2 (1977).

Pye, Roger, and Gillian Lauder. "Regional Aid for Telecommunications in Europe: A Force for Economic Development." *Telecommunications Policy* (June 1987), pp. 99–113.

Pye, Roger, M. Tyler, and B. Cartwright. "Telecommunicate or Travel?" *New Scientist* (September 12, 1974).

Pye, Roger, and P. I. Weintraub. "Attitudes toward Business Travel and Teleconferencing." Impact Paper no. 12. In Richard Harkness, *Technology Assessment of Telecommunications/Transportation Interactions*, vol. 2. Menlo Park, Calif.: Stanford Research Institute, May 1977.

Pye Telecommunications, Ltd. "A Study of the Future Frequency Spectrum Requirements for Private Mobile Radio in the U.K." Cambridge, Eng., 1976. Processed.

Qvortup, Lars. "Community Teleservice Centres and the Future of Rural Society." Telematics Project, Odense University, Odense, Denmark, September 1989. Processed.

Race, Tim. "Going It Alone, Bit by Bit." *New York Times Magazine*, December 2, 1990, pp. 19–33.

Ramajo, Germán. "Presentation on Compañía de Teléfonos de Chile, S.A." In Björn Wellenius and Peter Stern, eds., *Implementing Reforms in the Telecommunications Sector*. Washington, D.C.: World Bank, 1991.

Rao, Paladugu V. "Telephone and Instructional Communications." In Ithiel de Sola Pool, ed., *The Social Impact of the Telephone*. Cambridge, Mass.: M.I.T. Press, 1977.

Reid, A. A. L. "Comparing Telephone with Face-to-Face Contact." In Ithiel de Sola Pool, ed., *The Social Impact of the Telephone*. Cambridge, Mass.: M.I.T. Press, 1977.

Renaud, Bertrand M. *National Urbanization Policies in Developing Countries*. New York: Oxford University Press, 1981.

Research Institute of Telecommunication and Economics. "Analysis of Economic and Social Effects of Telecommunications Centered on Users' Appraisal." Tokyo, 1974. Processed.

Rettig, Jack. "Converting to Measured Service the Low-Cost Way." *Bell Laboratories Record* (November 1981), pp. 281–83.

Richardson, Harry W. *Regional Economics*. New York: Praeger Publishers, 1969.

Rockoff, Maxine L. "The Social Implications of Health Care Communication Systems." IEEE *Transactions on Communications*, vol. COM-23, no. 10 (October 1975), pp. 1085–88.

Rohlfs, Jeffrey H., and Harry M. Shooshan. "Will Price Caps Correct Major Economic Flaws in the Current Regulatory Process?" Paper presented at the twentieth annual Williamsburg conference, Williamsburg, Va., 1988. Processed.

Roojsma, A. H. "Optimization of Digital Network Structures." In M. Akiyama, ed., *Teletraffic Issues in an Advanced Information Society*. Amsterdam: Elsevier Press, 1985.

Satellite Communication Services. *Teleconferencing Newsletter*, vol. 1, no. 1 (June 1981).

Saunders, Robert J. "Organizational Structures for Telecommunications." *Telecommunications Journal* (August 1982), pp. 481–87.

———. "Rural Telecommunications: Economic and Policy Implications." *Seminar on Rural Telecommunications, New Delhi, September 11–22, 1978*, vol. 1. Geneva: ITU, 1979.

Saunders, Robert J., and C. R. Dickenson. "Telecommunications: Priority Needs for Economic Development." *Telecommunications Journal*, vol. 46, no. 9 (September 1979), pp. 567–68.

Saunders, Robert J., and Jeremy J. Warford. "Evaluation of Telephone Projects in Less Developed Countries." *Telecommunications Journal*, vol. 46, no. 1 (January 1979), pp. 22–28.

———. "Telecommunications Pricing and Investment in Developing Countries." *Proceedings of the International Telecommunications Exposition (INTELCOM 77)*, vol. 1. Dedham, Mass.: Horizon House International, October 1977.

———. *Village Water Supply: Economics and Policy in the Developing World*. Baltimore, Md.: Johns Hopkins University Press, 1976.

Saunders, Robert J., Jeremy J. Warford, and Patrick C. Mann. "Alternative Concepts of Marginal Cost for Public Utility Pricing: Problems of Application in the Water Supply Sector." World Bank Staff Working Paper no. 259. Washington, D.C., May 1977.

Scherer, F. M. *Industrial Market Structure and Economic Performance*. Chicago, Ill.: Rand McNally College Printers, 1970.

Schoppert, D. W., and others. "Some Principles of Freeway Directional Signing Based on Motorists' Experiences." Highway Research Board Bulletin no. 244. National Academy of Sciences, National Research Council, Washington, D.C., 1960. Processed.

Schramm, Wilbur. *Big Media, Little Media*. Beverly Hills, Calif.: Sage Publications, 1976.

———. *Mass Media and National Development*. Stanford, Calif.: Stanford University Press, 1964.

Schultz, Richard J., and Peter Barnes, eds. "Local Telephone Pricing: Is There a Better Way?" Centre for the Study of Regulated Industries, Montreal, Canada, 1984.

448

Secretaría de Comunicaciones y Transportes, Dirección General de Telecomunicaciones, Coordinación General del Plan Nacional de Telefonía Rural, Mexico. "Plan Nacional de Telefonía Rural: Documento básico." Mexico City, June 1979. Processed.

Selwyn, Lee L. "A Perspective on Price Caps as a Substitute for Traditional Revenue Requirements Regulation." Paper presented at the twentieth annual Williamsburg conference, Williamsburg, Va., 1988. Processed.

Selwyn, Lee L., and Gregory F. Borton. "USP: Will It Really Decrease Local Service Costs?" *Telephony* (January 28, 1980), pp. 36–38.

Selwyn, Lee L., and Scott C. Lundquist. "Adapting Telecom Regulation to Industry Change: Promoting Development without Compromising Ratepayer Protection." IEEE *Communications Magazine* (January 1989).

Shapiro, P. D. "Telecommunications and Industrial Development." IEEE *Transactions on Communications*, vol. COM-24, no. 3 (March 1976).

Sherman, Roger. *The Regulation of Monopoly.* Cambridge, Eng.: Cambridge University Press, 1989.

Short, J. A. "Long Range Social Forecasts: Congestion and Quality of Service." Long Range Intelligence Bulletin no. 10. British Post Office Telecommunications, London, April 1976. Processed.

Short, John, E. Williams, and B. Christie. *Social Psychology of Telecommunications.* London: John Wiley and Sons, 1976.

Simon, Herbert A. "Theories of Decisionmaking in Economics." *American Economic Review*, vol. 49 (June 1959).

Simonelli, Albert Lozano. "El papel del servicio postal en la vida económica y cultural de los países en vías de desarrollo." In Unión Postal de las Américas y España, *Conferencias del I seminario* UPU-AUPAE, *Lima.* Montevideo: Oficina Internacional de la Unión Postal de las Américas y España, 1967.

Simpson, Alan. "The State of the Optical Art." *Telephony* (August 27, 1990), pp. 40–44.

Singh, Indu B., ed. *Telecommunications in the Year 2000: National and International Perspectives.* Norwood, N.J.: Ablex Publishers, 1983.

Skoog, R. A., ed. "The Design and Cost Characteristics of Telecommunications Networks." Bell Telephone Laboratories, 1980. Processed.

Smith, Robert. "Telecommunications and Technology: Current Trends." *Impacts of Telecommunications on Planning and Transport.* Research Report no. 24. London: Department of the Environment and Transport, 1978.

Sorokin, A. L. *The Economics of the Postal System.* Lexington, Mass.: Lexington Books, 1980.

Spence, A. M. "The Economics of Internal Organization: An Introduction." *Bell Journal of Economics*, vol. 6 (1975), p. 169.

Squire, Lyn. "Some Aspects of Optimal Pricing for Telecommunications." *The Bell Journal of Economics and Management Science* (Autumn 1973), pp. 515–25.

Squire, Lyn, and Herman G. van der Tak. *Economic Analysis of Projects*. Baltimore, Md.: Johns Hopkins University Press, 1975.

Srivisal, Chote, and Hiroyoshi Tamura. "Usage Patterns of Residential Telephones Based on Customer Interviews." Office of Economic Studies, Telephone Organization of Thailand, September 1980. Processed.

Stahl, H. J., and A. Burmeister. "Benefits of Telecommunication for Transportation Systems in Developing Countries." In ITU, *World Telecommunications Forum 1987, Part IV*. Geneva, 1987.

Steiner, Peter O. "Peak Loads and Efficient Pricing." *Quarterly Journal of Economics*, vol. 71 (November 1957), pp. 585–610.

Stephens, B. W., and others. "Third Generation Destination Signing: An Electronic Route Guidance System." Highway Research Record no. 265. National Academy of Sciences, National Academy of Engineers, Washington, D.C., 1968. Processed.

Stern, Eliott, and Richard Holti. *Distance Working in Urban and Rural Settings: A Study for the EEC FAST Program*. London: Tavistock Institute of Human Relations, 1986.

Stigler, George J. "The Economics of Information." *Journal of Political Economy*, vol. 69 (1961), pp. 213–25.

Stover, William J. *Information Technology in the Third World: Can I.T. Lead to Humane International Development?* Boulder, Colo.: Westview Press, 1984.

Sy, Jacques Hababib. "African Nations and Access to Telecommunications Services: Political Economy and Legal Issues." Paper presented at a seminar on telecommunications restructuring sponsored by the World Bank, the ITU, and ACCT, Tunisia, May 1992.

Taylor, Lester D. "Problems and Issues in Modeling Telecommunications Demand." In L. Courville, A. Fontenay, and R. Dobell, eds., *Economic Analysis of Telecommunications: Theory and Applications*. New York: North-Holland, 1983.

———. *Telecommunications Demand: A Survey and Critique*. Cambridge, Mass.: Ballinger, 1980.

Taylor, William E. "Local Exchange Pricing: Is There Any Hope?" In James H. Alleman and Richard D. Emmerson, eds., *Perspectives on the Telephone Industry: The Challenge for the Future*. New York: Harper and Row Publishers, Ballinger Division, 1989.

Telecom France, "How France Financed Its Telephone Program." no. 1 (May 1981), pp. 29–32.

Telephone Organization of Thailand. "Telephone Usage Survey." Economic Studies Group, Bangkok, October 1980. Processed.

Telephony "Using Satellite Teleconference to Spread Marketing Gospel Throughout a Big Company." (January 25, 1982), pp. 34–43.

Thery, Gerard. "On the Road to Making France's Telecommunications System the Best." *Telephony* (November 21, 1977), pp. 60–72.

Thomas, Brownlee. "The Role of Communications in Economic Development."

M.A. thesis, Department of Political Science, McGill University, Montreal, Canada, 1984.

Thorngren, Bertil. "How Do Contact Systems Affect Regional Development?" *Environment and Planning,* vol. 2 (1970).

―――. "Silent Actors: Communication Networks for Development." In Ithiel de Sola Pool, ed., *The Social Impact of the Telephone.* Cambridge, Mass.: M.I.T. Press, 1977.

Tiene, Drew, and Shingenari Futagami. "Educational Media in Retrospect." World Bank Discussion Paper, Education and Training Series Report no. EDT 58. Washington, D.C., February 1987.

Tietjen, Karen. AID *Rural Satellite Program: An Overview.* Washington, D.C.: Academy for Educational Development, 1987.

Tomey, J. F. "The Field Trial of Audio Conferencing with the Union Trust Company." Report of contract no. H2104R. New Rural Society Project, Fairfield University, Fairfield, Conn., 1974. Processed.

Tomlinson, M., and Björn Wellenius. "Towards Economic Pricing of Telecommunications in the Developing World." In ITU, *World Telecommunications Forum 1987, Part IV.* Geneva, 1987.

Tornato, P. "Substitution Transport Telecommunication." Ministère de l'Équipement, Paris, 1974. Processed.

Townsend, David, and Patricia D. Kravtin. "Industry Structure and Competition in Telecommunications Markets: An Empirical Analysis." Paper presented at the seventh international conference of the International Telecommunications Society, Boston, Mass., July 1988. Processed.

Trebbing, Harry M., ed. *Performance under Regulation.* East Lansing, Mich.: Institute of Public Utilities, 1968.

Trevains, S. J. "Audio Conferencing in the Civil Service: An Evaluation of the Use of the Remote Meeting Table." Report no. CSD/78IOO/TR. Communications Studies and Planning, Ltd., London, 1978. Processed.

Turvey, Ralph, and Dennis Anderson. *Electricity Economics: Essays and Case Studies.* Baltimore, Md.: Johns Hopkins University Press, 1977.

Tyler, Michael. "Implications for Transport." In R. C. Smith, ed., *Impacts of Telecommunications on Planning and Transport.* Research Report no. 24. London: Department of Transport and Department of the Environment, July 1978.

―――. "A Note on the Interpretation of the NASA Teleconference Usage and Impact Data." Communications Studies Group, University College, London, 1976. Processed.

Tyler, Michael, B. Cartwright, and D. Bookless. "The Economic Consequences of Energy Scarcity." *Intelligence Bulletin,* vol. 1. Long Range Studies Division, Post Office Telecommunications, Cambridge, Eng. (1974).

Tyler, Michael, B. Cartwright, and G. Bush. "Interaction between Telecommunications and Face-to-Face Communication, The Energy Factor." *Intelligence Bulletin,* vol. 3, Long Range Studies Division, Post Office Telecommunications, Cambridge, Eng. (1974).

Tyler, Michael, B. Cartwright, and H. A. Collings. "Prospects for Teleconference Services." *Intelligence Bulletin*, vol. 9, Long Range Studies Division, Post Office Telecommunications, Cambridge, Eng. (May 1977).

Tyler, Michael, Martin Elton, and Angela Cook. *The Contribution of Telecommunications to the Conservation of Energy Resources.* Special Publication no. 77–17. Washington, D.C.: U.S. Department of Commerce, Office of Telecommunications, July 1977.

Ungerer, Herbert. "The European Situation: An Overview." In Björn Wellenius and Peter Stern, eds., *Implementing Reforms in the Telecommunications Sector: Lessons from Recent Experience.* Washington, D.C.: World Bank, 1991.

Union Postale. "Swedish Rural Postmen Also Provide Social Service." Vol. 10 (1974), pp. 148–52.

United Nations. *International Comparisons of Inter-Industry Data.* Industrial Planning and Programming Series no. 2. New York, 1969.

———. "Radio-guidage automobile presque pas à pas." *Science et Vie*, Paris (February 1977).

Uno, Kimio. "The Communication Sector in Japan and Its Role in Economic Development." In Meheroo Jussawalla and D. M. Lamberton, eds., *Communication Economics and Development.* Elmsford, N.Y.: Pergamon Press, 1982.

U.S. Congress. *Development of the RFD System.* Congressional report. Washington, D.C.: GPO, 1971.

———. *RFD Roads.* Hearings before the Joint Committee on Federal Aid in the Construction of Postal Roads, December 20, 1913. Washington, D.C.: GPO, 1913.

U.S. Federal Communications Commission. "Further Notice of Proposed Rulemaking." CC Docket no. 87–313. Washington, D.C.

U.S. Government. *Report of the Postmaster-General.* Washington, D.C.: GPO, annual, 1900, 1901, 1902.

U.S. House of Representatives. *Rural Delivery Systems,* no. 54. Hearings before the Committee on Expenditures in the Post Office Department, February 14–17, 21–23, 1912 (Washington, D.C.: GPO, 1912).

Universal Postal Union. "Memorandum on the Role of the Post as a Factor in Economic, Social, and Cultural Development." Berne, Switzerland, 1974. Processed.

———. "Study on Postal Development: Report by the Executive Council of the 18th UPU Congress, Rio de Janeiro, 1979." Document no. 17(e). Berne, Switzerland, 1980. Processed.

Vignon, H. "La restructuration des télécommunications au Benin." Paper presented at a seminar on telecommunications restructuring sponsored by the World Bank, the ITU, and ACCT, Tunisia, May 1992.

Vogelsang, Ingo. "Price Cap Regulation of Telecommunications Services: A Long-Run Approach." Paper no. N-2704-MF. Rand Corporation, Santa Monica, Calif., 1988.

Wall, S. D. "Four Sector Time Series of the U.K. Labor Force, 1841–1971." Long

452

Range Studies Division, Post Office Telecommunications, London, 1977. Processed.

Wall Street Journal, European ed., vol. 1, no. 2 (February 1, 1983), p. 1.

Walsham, Geoffrey. "Models for Telecommunications Strategy in the LDCs." *Telecommunications Policy* (June 1979), pp. 109–10.

Walters, A. A. *An Introduction to Econometrics*. London: Macmillan, 1968.

Watson, F. V. V. "Restructuring of the Telecommunications Sector in Sri Lanka: Review of Progress, 1988–91." In Björn Wellenius and Peter Stern, eds., *Implementing Reforms in the Telecommunications Sector: Lessons from Recent Experience*. Washington, D.C.: World Bank, 1991.

Waverman, Leonard. "The Regulation of Intercity Telecommunications." In Almarin Phillips, ed., *Promoting Competition in Regulated Markets*. Washington, D.C.: Brookings Institution, 1975.

Webber, Melvin M. "A Telecommunications Strategy for New Cities of the 21st Century." Working Paper no. 330. Berkeley: Institute of Urban and Regional Development, University of California, October 1980.

Wedemeyer, Dan J., and M. R. Ogden, eds. *Telecommunication and Pacific Development: Alternatives for the Next Decade*. New York: Elsevier Science Publishers North-Holland and PTC, 1988.

Wein, Harold H. "Fair Rate of Return and Incentives ... Some General Conclusions." In Harry M. Trebbing, ed., *Performance under Regulation*. East Lansing, Mich.: Institute of Public Utilities, 1968.

Weisbrod, Burton A. "Concepts of Costs and Benefits." In Samuel B. Chase, Jr., ed., *Problems in Public Expenditure Analysis*. Washington, D.C.: Brookings Institution, 1968.

Wellenius, Björn. "Apuntes: Curso de planificación de sistemas de telecomunicaciones EL720." Departamento de Electricidad, Universidad de Chile, Santiago, 1971. Processed.

————. "Beginnings of Sector Reform in the Developing World." In Björn Wellenius, Peter Stern, Timothy Nulty, and Richard Stern, eds., *Restructuring and Managing the Telecommunications Sector*. A World Bank Symposium. Washington, D.C.: World Bank, 1989.

————. "The Changing World of Telecommunications: Policy Options in Developing Countries." In ITU, *World Telecommunications Forum 1987, Part IV*. Geneva, 1987.

————. "Communication, Systems, and Society: An Interdisciplinary Foundation for Telecommunication Studies." Ph.D. diss., Department of Electrical Engineering Science, School of Physical Science, University of Essex, Colchester, Eng., 1978.

————. "Concepts and Issues on Information Sector Measurement." In Meheroo Jussawalla, Donald Lamberton, and Neil D. Karunaratne, eds., *The Cost of Thinking: Information Economies of Ten Pacific Countries*. Norwood, N.J.: Ablex Publishing, 1988.

————. "Financing Telecommunications in the Developing World—Issues and Opportunities." *Proceedings of ITU TELECOM 86*. Nairobi, Kenya, 1986.

_____. "Hidden Residential Connections Demand in the Presence of Severe Supply Shortage." IEEE *Transactions on Communications Technology* (June) 1969a.

_____. "Income and Social Class in Residential Telephone Demand." *Telecommunications Journal*, vol. 36 (May) 1969b.

_____. "On the Role of Telecommunications in Development." *Telecommunications Policy* (March 1984), pp. 59–66.

_____. "The Role of Multilateral Development Financing Agencies." Paper presented at the eighth international conference of the International Telecommunications Society, Venice, Italy, March 1990a. Processed.

_____. "Telecommunications and Third World Development." In Maurice F. Estabrooks and Rodolphe H. Lamarche, eds., *Telecommunications: A Strategic Perspective on Regional, Economic, and Business Development*. Moncton, Canada: The Canadian Institute for Research on Regional Development, 1986.

_____. "Telecommunications in the Developing World: Current Trends and New Issues." Paper presented at a symposium to honor Professor Kenneth Cattermole, University of Essex, Colchester, Eng., September 1990b.

Wellenius, Björn, R. Budinich, and P. Moral. "Estimación preliminar de la demanda de servicio telex en Chile." Fundación Chile, January 1979. Processed.

Wellenius, Björn, and Peter Stern, eds. *Implementing Reforms in the Telecommunications Sector: Lessons from Recent Experience*. Washington, D.C.: World Bank, 1991.

Wellenius, Björn, Peter Stern, Timothy Nulty, and Richard Stern, eds. *Restructuring and Managing the Telecommunications Sector*. A World Bank Symposium. Washington, D.C.: World Bank, 1989.

Wenders, John T. *The Economics of Telecommunication: Theory and Policy*. Cambridge, Mass.: Ballinger Publishing Company, 1987.

Whitlock, Erik, and Emilia Nyevrikel. "The Evolution of Hungarian Telecommunications Policy." *Telecommunications Policy*, vol. 16, no. 3 (April 1992), pp. 249–58.

Wilkinson, G. F. "The Estimation of Usage Repression under Local Measured Service: Empirical Evidence from the GTE Experiment." In Léon Courville, Alain De Fontenay, and Rodney Dobell, eds., *The Economic Analysis of Telecommunications: Theory and Applications*. New York: Elsevier Science Publishers, 1983.

Williams, E., and I. Young. "The Choice of Teleconference or Travel amongst Loudspeaking Telephone Users." Report no. E/77077/WL. Communications Studies Group, University College, London, 1977. Processed.

Williams, F. C. W. "The Rural Mail Delivery System in New Zealand." *Union Postale*, vol. 4 (1972), pp. 51–54.

Williams, Frederick. *The New Telecommunications: Infrastructure for the Information Age*. Belmont, Calif.: Wadsworth Publishers, 1991.

Williams, Howard, and Andrew Gillespie. "Telecommunications and the Reconstruction of Regional Comparative Advantage." In ITU, *World Telecommunications Forum 1987, Part IV*. Geneva, 1987.

Wilson, Carol, and Czatdana Inan. "New Markets Developing." *Telephony* (January 22, 1989), pp. 32–48.

World Bank. *Annual Report*. Various years. Washington, D.C.

————. *World Bank Atlas 1988*. Washington, D.C., 1989a.

————. *World Development Report 1980*. New York: Oxford University Press, 1980.

————. *World Development Report 1990*. New York: Oxford University Press, 1990.

————. *World Development Report 1994—Infrastructure for Development*. New York: Oxford University Press, 1994.

————. "Telecommunications Development, Investment Financing, and the Role of the World Bank." Paper presented at the ITU plenipotentiary conference, Nice, France, 1989b. Processed.

————. "Telecommunications: World Bank Experience and Strategy." Washington, D.C., February 1992. Processed.

Wurtzel, Alan H., and Colin Turner. "Latent Functions of the Telephone: What Missing the Extension Means." In Ithiel de Sola Pool, ed., *The Social Impact of the Telephone*. Cambridge, Mass.: M.I.T. Press, 1977.

Yap, Lorene Y. L. "Internal Migration in Less Developed Countries: A Survey of the Literature." World Bank Staff Working Paper no. 215. Washington, D.C., September 1975.

Yatrakis, P. G. "Determinants of the Demand for International Telecommunications." *Telecommunications Journal*, vol. 39 (December 1972).

Yoshita, M., and H. Okazaki. "Digital Telephone Junction Network Planning System Considering Logical and Physical Network Aspects." In M. Akiyama, ed., *Teletraffic Issues in an Advanced Information Society*. Amsterdam: Elsevier Press, 1985.

Index